AF734684

Linear and Quasilinear Parabolic Systems

Sobolev Space Theory

Mathematical
Surveys
and
Monographs

Volume 251

Linear and Quasilinear Parabolic Systems

Sobolev Space Theory

David Hoff

AMERICAN MATHEMATICAL SOCIETY
Providence, Rhode Island

EDITORIAL COMMITTEE

Robert Guralnick, Chair
Natasa Sesum
Bryna Kra
Constantin Teleman
Melanie Matchett Wood

2010 *Mathematics Subject Classification.* Primary 35K51, 35K61.

For additional information and updates on this book, visit
www.ams.org/bookpages/surv-251

Library of Congress Cataloging-in-Publication Data

Names: Hoff, David Charles, 1948– author.
Title: Linear and quasilinear parabolic systems : sobolev space theory / David Hoff.
Description: Providence, Rhode Island : American Mathematical Society, [2021]
Series: Mathematical surveys and monographs, 0076-5376 ; volume 251 | Includes bibliographical references and index.
Identifiers: LCCN 2020036612 | ISBN 9781470461614 (paperback) | ISBN 9781470463205 (ebook)
Subjects: LCSH: Differential equations, Partial. | Differential equations, Parabolic. | AMS: Partial differential equations – Parabolic equations and systems [See also 35Bxx, 35Dxx, 35R30, 35R35, 58J35] – Initial-boundary value problems for second-order parabolic systems. | Partial differential equations – Parabolic equations and systems [See also 35Bxx, 35Dxx, 35R30, 35R35, 58J35] – Nonlinear initial-boundary value problems for nonlinear parabolic equations.
Classification: LCC QA377 .H62 2021 | DDC 515/.3534–dc23
LC record available at https://lccn.loc.gov/2020036612

Copying and reprinting. Individual readers of this publication, and nonprofit libraries acting for them, are permitted to make fair use of the material, such as to copy select pages for use in teaching or research. Permission is granted to quote brief passages from this publication in reviews, provided the customary acknowledgment of the source is given.

Republication, systematic copying, or multiple reproduction of any material in this publication is permitted only under license from the American Mathematical Society. Requests for permission to reuse portions of AMS publication content are handled by the Copyright Clearance Center. For more information, please visit www.ams.org/publications/pubpermissions.

Send requests for translation rights and licensed reprints to reprint-permission@ams.org.

© 2020 by the American Mathematical Society. All rights reserved.
The American Mathematical Society retains all rights
except those granted to the United States Government.
Printed in the United States of America.

∞ The paper used in this book is acid-free and falls within the guidelines
established to ensure permanence and durability.
Visit the AMS home page at https://www.ams.org/

10 9 8 7 6 5 4 3 2 1   25 24 23 22 21 20

To my family–may they all thrive!

Contents

Preface

In this monograph I present a systematic theory of weak solutions in Hilbert-Sobolev spaces of initial-boundary value problems for parabolic systems of partial differential equations. The goal is to provide a practical reference for fundamental results often considered well-known but often difficult to access in the desired generality, as well as to provide instructional material for students preparing for research in applied analysis.

The development begins in Chapter 2 with an abstract theory in which $\mathcal{H}_0$ and $\mathcal{H}_1$ are separable Hilbert spaces with $\mathcal{H}_1$ dense and continuously included in $\mathcal{H}_0$ and A is a mapping of the time axis into the set of bilinear forms on $\mathcal{H}_1 \times \mathcal{H}_1$. The problem is to construct a mapping u for which the equation

$$\langle u(t), v\rangle_{\mathcal{H}_0} = \langle \varsigma_0, v\rangle_{\mathcal{H}_0} + \int_0^t \big(- A(s)(u(s), v) + F(s)\cdot v\big) ds$$

holds for all test functions $v \in \mathcal{H}_1$, for all time t and for given $\varsigma_0 \in \mathcal{H}_0$ and F mapping time into the dual $\mathcal{H}_1^*$. This abstract theory is then applied in Chapters 3 and 5 to initial–boundary value problems for linear parabolic systems with $\mathcal{H}_0$ a closed subspace of $[L^2(\Omega)]^N$ for a given open set $\Omega \subset \mathbf{R}^n$ (not assumed to be bounded) and $\mathcal{H}_1$ a closed subspace of $[H^1(\Omega)]^N$ whose elements satisfy essential boundary conditions. The basic theory of Chapter 3 addresses questions of measurability, continuity and energy conservation in the absence of regularity, and includes special considerations for problems with symmetry and representations of solutions in terms of spectral measures. Higher-order regularity of solutions is discussed in Chapter 5 in three related but categorically different formulations: the dynamical systems (Bochner space) formulation in which solutions are mappings from the time axis into various Sobolev spaces, the weak derivative formulation in which solutions are locally integrable on the space-time cross product, and the continuous derivative formulation familiar from calculus. Chapter 5 also includes a discussion of instantaneous regularization of solutions whose data fail to satisfy compatibility conditions. (Chapter 4 is a brief excursion into higher order regularity theory for elliptic systems, required for the application in Chapter 5.)

This linear theory is then applied in chapter 6 to quasilinear systems in which A and F may depend on u. A basic theory of solutions with minimal regularity is given, corresponding to the basic theory for linear problems. This includes the theorems of Leray and Hopf for the Navier-Stokes equations as well as results for more general problems in which nonlinearities have specific, controlled growth as $|u| \to \infty$. Results for more highly regular solutions of quasilinear systems are then derived by iteration from the corresponding regularity theory for linear problems. This includes local-in-time existence with large compatible data, global existence for data near an attracting rest point, invariant regions for quasilinear systems, and a large data, global existence result in two and three space dimensions for certain systems with symmetry.

This book was written in a continuous, linear process in which the material itself dictated formulation, order of topics and level of detail. This choice was made to insure an internal consistency of notation, terminology and perspective that would not have been achieved by collating results from the literature (no claim is made to mathematical originality, however). Maximal generality was an important goal, and this has occasionally resulted in considerable but unavoidable technicality. I have therefore attempted to present proofs in such a way that the experienced reader can easily grasp the flow of the arguments without reading every detail, but at the same time the intrepid student will succeed in checking these details without undue difficulty.

The following are the major theorems of the book, listed here for convenient reference:

- Basic existence and uniqueness: Theorem 3.6 for general linear systems and Theorem 3.7 for systems which are symmetric to leading order.
- Spectral representations: Theorem 3.8.
- Regularity for linear elliptic system: Theorems 4.1 and 4.2.
- Higher order regularity for linear parabolic systems: Theorem 5.5 for compatible data, Theorem 5.8 for incompatible data.
- Theorems of Leray and Hopf for the Navier-Stokes system: Theorem 6.5.
- Local existence of minimally regular solutions of quasilinear systems with nonlinearities of polynomial growth: Theorem 6.6.
- Local existence of smooth solutions of quasilinear systems: Theorem 6.10.
- Global existence for dissipative quasilinear systems with small initial data, Theorem 6.12.
- Invariant regions for quasilinear systems, Theorem 6.13.
- Global existence for certain symmetric quasilinear systems with large data in two and three space dimensions, Theorem 6.14.

Few readers will read the book cover-to-cover. The following suggestions for reading plans may therefore be considered (the first three should include the basic material in Chapter 1 and sections 2.1., 2.2 and 3.1–3.3):

- Any of the following extensions of the basic theory:
  - systems with symmetry (sections 2.6, 2.7 and 3.4)
  - Navier-Stokes equations (sections 6.1 and 6.2)
  - quasilinear systems with polynomial growth (sections 6.1 and 6.3).
- Higher order regularity for linear parabolic systems (sections 2.2–2.4, 5.1 and 5.2) with the possible addition of initial layer regularization (sections 2.5 and 5.3).
- Higher order regularity for quasilinear parabolic systems (sections 2.2–2.4, 5.1 and 5.2 are prerequisite, 6.4 is essential, and any of 6.5, 6.6 or 6.7 may be included independently).
- Chapter 4 is a self-contained treatment of higher-order regularity for elliptic systems.

- Section A.4 is a self-contained treatment of calculus on Lipschitz hypersurfaces, including the construction of the Radon measure corresponding to surface area and the normal vector field, the trace theorem $W^{1,p}(\Omega) \to L^p(\partial\Omega)$ and the divergence theorem for $W^{1,p}$ integrands.
- Section A.5 is a self-contained treatment of certain spaces $L^{p,q}(\Omega \times I)$ where I is an interval in $\mathbf{R}$ and Ω is an open set in $\mathbf{R}^n$. The elements of these spaces are in canonical one-to one correspondence with those of the Bochner spaces $L^p(I; L^q(\Omega))$ but are measurable on the cross product. This correspondence finds application throughout the book in navigating different formulations of regularity.

In all cases students should be familiar with the basic facts of measure theory and functional analysis, including weak derivatives and Sobolev spaces. Spectral measures appear twice in the book but their use can be side-stepped with minimal loss by appeal to the spectral theorem for compact self-adjoint operators.

The Hilbert space framework chosen here has the virtues of relative simplicity and broad applicability. Other approaches, including semigroup theory, Green's functions, and the method of continuation (Schauder theory) have important uses as well, however, but could not be included. For these the reader may consult standard references such as Ladyzhenskaya et. al. [**26**], Friedman [**15**] and [**16**] and Evans [**12**].

Many of the perspectives and formulations in this book reflect the influences of collaborators, colleagues and students, too numerous to mention, with whom I have had the good fortune to interact over many years. My thanks to them all. Of course, responsibility for any shortcomings, deficiencies or errors lies solely with me.

David Hoff

CHAPTER 1

Introduction

> Mathematicians invent a setting with a handful of characters...[and] a few rules of interaction, and watch how the plot unfolds.
>
> Kelsey Houston-Edwards [**24**]

We begin with an example in which an elementary initial–boundary value problem is converted to its corresponding weak form, which is then generalized to an abstract Hilbert space framework. A brief review of basic concepts and notations from functional analysis follows and we conclude with a discussion of geometric properties of open sets in $\mathbf{R}^n$.

§1.1 Weak forms: An Example

Consider the following elementary problem: let Ω be a proper open subset of $\mathbf{R}^n$ and ς_0 a given function on Ω, and let $a_{ij}, 1 \leq i, j \leq n$, and g be given functions on $\Omega \times (0,T)$ where $T \in (0,\infty]$. A solution $u : \overline{\Omega} \times [0,T) \to \mathbf{R}$ to the initial–boundary value problem

$$(1.1)\qquad \begin{cases} u_t = (a_{ij}u_{x_i})_{x_j} + g, \ (x,t) \in \Omega \times (0,T), \\ u(x,t) = 0, \ (x,t) \in \partial\Omega \times [0,T), \\ u(x,0) = \varsigma_0(x), \ x \in \Omega \end{cases}$$

is then to be constructed (subscripts denote partial derivatives and summation over repeated indices is understood), apparently satisfying fairly strong differentiability conditions. These conditions may be overly restrictive, however, and may have little significance for the underlying application. It is therefore of interest that a weaker form is available, both for this problem and for the broader class to which it belongs, which is often more convenient and which may in fact be closer to the physical principles from which the system is derived. To obtain this weak form for the above example we multiply the differential equation by a function v of x which is zero on $\partial\Omega$ and formally integrate over $\Omega \times [0,t]$. Applying the divergence theorem, we find that

$$(1.2)\qquad \begin{aligned} \int_\Omega u(x,t)v(x)\,dx &= \int_\Omega \varsigma_0(x)v(x)\,dx \\ &\quad + \int_0^t \int_\Omega \big(- a_{ij}u_{x_i}v_{x_j} + gv \big) dx\,ds. \end{aligned}$$

An appropriate weak formulation is now apparent: assume that for every $t \in (0,T)$ a_{ij} is bounded and measurable on $\Omega \times (0,t)$, that $f \in L^2(\Omega \times (0,t))$ and that

$\varsigma_0 \in L^2(\Omega)$. We then say that u is a weak solution of (1.1) if u is weakly continuous from $[0,T)$ into $L^2(\Omega)$ (which means that the inner product on the left above is continuous in t for every v), that $u(0) = \varsigma_0$, that the Lebesgue equivalence class of u on $(0,T)$ is an element of $L^2((0,t); H_0^1(\Omega))$ (the precise meaning of which will be given below), and which satisfies (1.2) for every $v \in H_0^1(\Omega)$ and $t \in (0,T)$. It is clear that sufficiently smooth solutions of (1.1) are weak solutions in this sense, but a proof of the converse requires greater regularity. Little is lost, however, if weak solutions are shown to be unique, and the weak formulation has the clear advantage that broader classes of problems and solutions are allowed.

A compact Hilbert space formulation can be given and generalized as follows. If we define $A(t) : H_0^1(\Omega) \times H_0^1(\Omega) \to \mathbf{R}$ and $F(t) : H_0^1(\Omega) \to \mathbf{R}$ for this example by

$$A(t)(z,v) = \int_\Omega \left[a_{ij}(x,t) z_{x_i}(x) v_{x_j}(x)\right] dx \tag{1.3}$$

and

$$F(t) \cdot v = \int_\Omega g(x,t) v(x) dx$$

for specific representatives of a_{ij} and g, then the requirement (1.2) is that

$$\langle u(\cdot,t), v\rangle_{L^2(\Omega)} = \langle \varsigma_0, v\rangle_{L^2(\Omega)} + \int_0^t \left[-A(s)(u(\cdot,s),v) + F(s)\cdot v\right] ds \tag{1.4}$$

for every $v \in H_0^1(\Omega)$ and $t \in [0,T)$.

In Chapter 2 we study the abstract version of (1.4) in which the spaces $L^2(\Omega)$ and $H_0^1(\Omega)$ are replaced by general separable Hilbert spaces $\mathcal{H}_0$ and $\mathcal{H}_1$ with $\mathcal{H}_1$ continuously contained in $\mathcal{H}_0$, A is a fairly general mapping from $[0,T)$ into the set of bilinear forms on $\mathcal{H}_1 \times \mathcal{H}_1$ and F is a mapping from $[0,T)$ into the dual of $\mathcal{H}_1$. This abstract theory is then applied in Chapter 3 to generalizations of (1.1) in which *systems* of partial differential equations with quite general essential and natural boundary conditions are considered. Higher order regularity of solutions is examined in Chapter 5 and applications to quasilinear systems are discussed in Chapter 6.

§1.2 Basic Notations and Definitions

Here and throughout the book Ω will be an open set in $\mathbf{R}^n$ which need not be bounded unless boundedness is explicitly assumed. We denote by $C_{bdd}^k(\Omega)$ the Banach space of real-valued functions on Ω having continuous, bounded derivatives on Ω up to order $k \geq 0$ with norm

$$|\varphi|_{C_{bdd}^k(\Omega)} = \max_{|\alpha|\leq k} \sup_{x\in\Omega} |D^\alpha \varphi(x)|.$$

For $k \geq 0$ and $\lambda \in (0,1]$, $C_{bdd}^{k,\lambda}(\Omega)$ will denote the Banach space of elements φ in $C_{bdd}^k(\Omega)$ whose derivatives are Hölder continuous with exponent $\lambda \in (0,1]$; that is, whose $C_{bdd}^{k,\lambda}$ norm, given by

$$|\varphi|_{C_{bdd}^{k,\lambda}(\Omega)} \equiv |\varphi|_{C_{bdd}^k(\Omega)} + \sup_{|\alpha|\leq k} \sup_{x\neq y\in\Omega} \left|\frac{|D^\alpha\varphi(y) - D^\alpha\varphi(x)|}{|y-x|^\lambda}\right|,$$

is finite. Observe that if Ω is a proper subset of $\mathbf{R}^n$ and $\varphi \in C_{bdd}^{k,\lambda}(\Omega)$, then φ and its derivatives up to order k have continuous extensions to $\partial\Omega$ and that the suprema in the above two norms may then be taken over $\overline{\Omega}$. If $k = 0$ then no differentiability is required, so that the domain space need not be open; thus for any set $W \subseteq \mathbf{R}^n$, $C^{0,\lambda}_{bdd}(W)$ will denote the space of real–valued functions on W which are bounded and Hölder continuous with exponent λ. Finally, if $\mathcal{Q}$ is a subset of $\mathbf{R}^n \times \mathbf{R}$ and $\lambda, \lambda' \in (0,1]$, then $C^{0,(\lambda,\lambda')}_{bdd}(\mathcal{Q})$ will denote the set of functions ψ which are continuous and bounded on $\mathcal{Q}$ and for which

$$\sup_{(x_1,t_1)\neq(x_2,t_2)\in\mathcal{Q}} \frac{|\psi(x_2,t_2)-\psi(x_1,t_1)|}{|x_2-x_1|^{\lambda}+|t_2-t_1|^{\lambda'}}$$

is finite. Elements of $C^{0,(\lambda,\lambda')}_{bdd}(\mathcal{Q})$ have continuous, bounded extensions to the closure of $\mathcal{Q}$.

We denote by $C^k_c(\Omega)$ the set of elements in $C^k_{bdd}(\Omega)$ having compact support in Ω and by $C^\infty_c(\Omega)$ the space of infinitely differentiable functions having compact support in Ω. Next recall that if u is a locally integrable function on an open set $\Omega \subseteq \mathbf{R}^n$ and α is a multi-index, then the α–derivative of u in the sense of distributions is the linear map from $C^\infty_c(\Omega)$ to $\mathbf{R}$ given by $\varphi \mapsto (-1)^{|\alpha|}\int_\Omega u\, D^\alpha\varphi\, dx$. If there is a locally integrable function v such that $(-1)^{|\alpha|}\int_\Omega u\, D^\alpha\varphi\, dx = \int_\Omega v\,\varphi\, dx$ for all $\varphi \in C^\infty_c(\Omega)$ then v is the weak α–derivative of u on Ω and we write $D^\alpha u = v$. Thus the weak derivative, if it exists, gives a representation of the distribution derivative, which always exists.

The subspace of elements of $L^p(\Omega)$ whose weak derivatives in Ω up to order k exist and are also elements of $L^p(\Omega)$ is the Sobolev space $W^{k,p}(\Omega)$, in which the norm of an element u is given by $|u|_{W^{k,p}(\Omega)} = \left(\sum_{|\alpha|\le k} |D^\alpha u|_{L^p(\Omega)}\right)^{1/p}$ if $p<\infty$ and by $|u|_{W^{k,\infty}(\Omega)} = \max_{|\alpha|\le k} |D^\alpha u|_{L^\infty(\Omega)}$ if $p=\infty$; $H^k(\Omega)$ is the Hilbert space $W^{k,2}(\Omega)$; and $H^k_0(\Omega)$ is the closure with respect to the $H^k(\Omega)$ norm of the space $C^\infty_c(\Omega)$. Concise accounts of the relevant facts concerning Sobolev spaces are given in [**46**], [**12**] and [**19**], and a more complete exposition in [**1**].

Much of our work will concern systems of differential equations whose solutions will be elements of the cross product $[W^{k,p}(\Omega)]^N$ where $N \ge 1$. The norm of an element $u=(u^1,\dots,u^N)$ in the latter Banach space is given by $\left(\sum_j |u^j|^p_{W^{k,p}(\Omega)}\right)^{1/p}$ if $p<\infty$ and by $\max_j |u^j|_{W^{k,\infty}(\Omega)}$ if $p=\infty$. The space $[W^{k,2}(\Omega)]^N = [H^k(\Omega)]^N$ is a Hilbert space with inner product $\langle u,v\rangle_{[H^k(\Omega)]^N} = \sum_j \langle u^j, v^j\rangle_{H^k(\Omega)}$.

More generally, the norm of an element u in a normed vector space X is denoted by $|u|_X$ (the subscript is omitted when $X=\mathbf{R}^n$) and the inner product of two elements u and v in a Hilbert space $\mathcal{H}$ by $\langle u,v\rangle_{\mathcal{H}}$. The dual of a Banach space X is denoted by X^*, the action of an element $F\in X^*$ on an element $v\in X$ by $F\cdot v$, and the vector space of bounded linear transformations from one Banach space X_1 into another Banach space X_2 by $\mathcal{L}(X_1,X_2)$. The norm of such a transformation S is denoted by $\|S\|_{X_1,X_2}$, or in the case that $X_1=X_2=X$, simply by $\|S\|_X$.

A mapping A from $X_1\times X_2$ to $\mathbf{R}$ is *bilinear* if $A(u_1,\cdot)$ is linear on X_2 for every fixed $u_1\in X_1$ and $A(\cdot,u_2)$ is linear on X_1 for every fixed $u_2\in X_2$. The norm $\|A\|_{X_1\times X_2}$ of such a map is $\sup|A(u_1,u_2)|$ taken over unit vectors $u_j\in X_j$, and the vector space of bilinear forms with finite norm is denoted by $\mathcal{B}(X_1,X_2)$. If $\mathcal{H}$ and $\mathcal{H}'$ are Hilbert spaces and $A:\mathcal{H}\times\mathcal{H}'\to\mathbf{R}$ is bilinear, then $A\in\mathcal{B}(\mathcal{H},\mathcal{H}')$ if and only if there is an operator $S\in\mathcal{L}(\mathcal{H},\mathcal{H}')$ such that $A(u,u')=\langle Su,u'\rangle_{\mathcal{H}'}$ for all $u\in\mathcal{H}$ and $u'\in\mathcal{H}'$; if so, then S is unique and $\|S\|_{\mathcal{H},\mathcal{H}'}=\|A\|_{\mathcal{H}\times\mathcal{H}'}$.

If $I \subset \mathbf{R}$ is a nontrivial interval and X is a normed vector space, then $C(I;X)$, $C^\lambda(I;X)$ and $Lip(I;X)$ are the vector spaces of mappings u from I into X which are respectively continuous, Hölder continuous with exponent $\lambda \in (0,1)$, or Lipschitz continuous. Thus $u \in C^\lambda(I;X)$ if there is a constant C such that

$$|u(t_2) - u(t_1)|_X \le C|t_2 - t_1|^\lambda \tag{1.5}$$

for all $t_1, t_2 \in I$; and $u \in Lip(I;X)$ if (1.5) holds with $\lambda = 1$. The vector space $C^\lambda_{loc}(I;X)$ is the space of mappings which are in $C^\lambda(I';X)$ for every compact interval $I' \subseteq I$, and similarly for $Lip_{loc}(I;X)$. Finally, if $\mathcal{H}$ is a Hilbert space then $C(I;\mathcal{H}^{weak})$ (or $C(I;\mathcal{H}_{weak})$ if notationally convenient) is the vector space of weakly continuous mappings from I into $\mathcal{H}$; that is, the set of functions $u : I \to \mathcal{H}$ such that $t \to \langle u(t), v\rangle_\mathcal{H}$ is continuous on I for every fixed $v \in \mathcal{H}$.

§1.3 Weak Measurability

The following notions of weak measurability of mappings into the vector spaces of section 1.2 will be applied throughout:

Definitions 1.1. *Let $I \subseteq \mathbf{R}$ be a nontrivial interval. A mapping u from I into a Banach space X is weakly measurable if for every $F \in X^*$ the mapping $t \to F\cdot u(t)$ is (Lebesgue) measurable on I. A mapping F from I into the dual of a Banach space X is weak-$*$ measurable if for every $u \in X$ the mapping $t \to F(t)\cdot u$ is measurable on I. A mapping A from I into $\mathcal{B}(X_1 \times X_2)$, where X_1 and X_2 are Banach spaces, is weakly measurable if for every pair $(u_1, u_2) \in X_1 \times X_2$ the mapping $t \to A(t)(u_1, u_2)$ is measurable on I; and a mapping S from I into $\mathcal{L}(X_1, X_2)$ is weakly measurable if for all $u \in X_1$ the mapping $t \to S(t)u$ is weakly measurable into X_2.*

It is easy to check that if $\mathcal{H}$ and $\mathcal{H}'$ are Hilbert spaces, $A : I \to \mathcal{B}(\mathcal{H}, \mathcal{H}')$ and $A(t)(u,u') = \langle S(t)u, u'\rangle\rangle_{\mathcal{H}'}$ for all $u \in \mathcal{H}$ and $u' \in \mathcal{H}'$, then A is weakly measurable if and only if S is weakly measurable. In general, norms and inner products of weakly measurable mappings into separable spaces are measurable:

Theorem 1.2 *Let $I \subseteq \mathbf{R}$ be a nontrivial interval, X a separable Banach space and $\mathcal{H}$ and $\mathcal{H}'$ separable Hilbert spaces.*

(a) *If $F : I \to X^*$ is weak-$*$ measurable on I, then the mapping $t \to |F(t)|_{X^*}$ is measurable on I.*

(b) *If $F : I \to \mathcal{H}^*$ is weak-$*$ measurable on I and $u : I \to \mathcal{H}$ is weakly measurable on I, then the mapping $t \to F(t)\cdot u(t)$ is measurable on I.*

(c) *If $A \to \mathcal{B}(\mathcal{H} \times \mathcal{H}')$ is weakly measurable on I then the mapping $t \to \|A\|_{\mathcal{H}\times\mathcal{H}'}$ is measurable on I; and if u and v are weakly measurable mappings from I into $\mathcal{H}$ and $\mathcal{H}'$ respectively, then the mapping $t \to A(t)(u(t), v(t))$ is measurable on I.*

PROOF. If F is as in (a) and $\{u_j\}_1^\infty$ is countable and dense in X, then each $a_j(\cdot) \equiv |F(\cdot)\cdot u_j|/|u_j|_X$ is measurable on I, hence so is $|F(\cdot)|_{X^*} = \sup_j a_j(\cdot)$. The proof of the first assertion in (c) is similar, and for the second we let

$\{\varphi_j\}_1^\infty$ and $\{\psi_k\}_1^\infty$ be orthonormal bases for $\mathcal{H}$ and $\mathcal{H}'$ respectively and u and v as in (c); then

$$A(u,v) = \sum_{j,k} \langle u, \varphi_j\rangle_{\mathcal{H}} \langle v, \psi_k\rangle_{\mathcal{H}'} A(\varphi_j, \psi_k).$$

Each summand here is the product of measurable functions and so is measurable, hence so is the converging sum. The proof of (b) is similar. □

Notice that both (b) and (c) show in particular that $\langle u(\cdot), v(\cdot)\rangle_{\mathcal{H}}$ is measurable when u and v are weakly measurable. Also, it is easy to check that as a consequence of (c), if $S : I \to \mathcal{L}(\mathcal{H}, \mathcal{H}')$ is weakly measurable on I, then the mapping $t \to \|S\|_{\mathcal{H},\mathcal{H}'}$ is measurable on I, and if u is weakly measurable from I into $\mathcal{H}$, then the mapping $t \to S(t)u(t)$ is measurable from I into $\mathcal{H}'$. It is of interest that, in the presence of separability, weak measurability generally implies measurability in the usual sense; see Exercise 1.2.

§1.4 Vector-Valued L^p Spaces

Two mappings from an interval $I \subseteq \mathbf{R}$ into a nonempty set V are Lebesgue equivalent if they agree for almost all $t \in I$, and the equivalence class of such a map F is denoted by $[F]_{I,V}$, or if the context makes clear, simply by $[F]$. If V is the dual X^* of a separable Banach space X and if $F : I \to X^*$ is weak-$*$ measurable, then the mapping $t \to |F(t)|_{X^*}$ is measurable by the above theorem and we can therefore consider whether its equivalence class over I belongs to $L^p(I)$. Similar considerations apply to mappings from I into vector spaces $\mathcal{B}(\mathcal{H} \times \mathcal{H}')$ and $\mathcal{L}(\mathcal{H}, \mathcal{H}')$ of bilinear forms and linear operators:

Definitions 1.3. *Let $I \subseteq \mathbf{R}$ be a nontrivial interval and F a weak-$*$ measurable mapping from I into the dual X^* of a separable Banach space X. Then for $p \in [1, \infty]$ the vector space of equivalence classes $[F]_{I,X^*}$ of weak-$*$ measurable mappings $F : I \to X^*$ such that $|F|_{X^*} \in L^p(I)$ is denoted by $L^p(I; X^*)$ and the norm of an element F is defined by $|F|_{L^p(I;X^*)} = \big||F(\cdot)|_{X^*}\big|_{L^p(I)}$. The vector space of mappings which are in $L^p(I'; X^*)$ for every compact interval $I' \subseteq I$ is denoted by $L^p_{loc}(I; X^*)$.*

The spaces $L^p(I; \mathcal{B}(\mathcal{H} \times \mathcal{H}'))$, $L^p_{loc}(I; \mathcal{B}(\mathcal{H} \times \mathcal{H}'))$, $L^p(I; \mathcal{L}(\mathcal{H}, \mathcal{H}'))$ and $L^p_{loc}(I; \mathcal{L}(\mathcal{H}, \mathcal{H}'))$, where $\mathcal{H}$ and $\mathcal{H}'$ are separable Hilbert spaces, are defined in a similar way.

Observe the important distinction between, for example, $L^p_{loc}([0,T);X^*)$ and $L^p_{loc}((0,T);X^*)$.

In nearly every case occurring in this book the Banach space X will be either a separable Hilbert space, in which case weak and weak-$*$ measurability are equivalent, or X^* will be $L^q(\Omega)$ for $q \in (1, \infty]$ and Ω open in $\mathbf{R}^n$, in which case the predual X is the separable space $L^{q'}(\Omega)$ where $q' \in [1, \infty)$ is the Hölder conjugate. It is easy to see that $L^p(I; X^*)$ and $L^p(I; \mathcal{B}(\mathcal{H} \times \mathcal{H}'))$ are normed linear spaces but we do not consider here their completeness or characterizations of their duals; see [**9**] for a more complete discussion. One exception is the case that X is a separable Hilbert space and $p = 2$:

Lemma 1.4. *Let $I \subseteq \mathbf{R}$ be a nontrivial interval and $\mathcal{H}$ a separable Hilbert space. Then $L^2(I; \mathcal{H})$ is a Hilbert space with inner product*

$$\langle u, v \rangle_{L^2(I;\mathcal{H})} = \int_I \langle u(t), v(t) \rangle_{\mathcal{H}} dt.$$

PROOF. The proof of completeness is nearly identical to the proof of completeness of L^2-spaces of real-valued functions, *mutatis mutandi.* See [**13**], Theorem 6.5 or [**34**], Theorem 6.3.5 for example.

□

Integrals of Hilbert space–valued functions of t will appear frequently and are defined as follows:

Definition 1.5. *Let $\mathcal{H}$ be a separable Hilbert space, I a nontrivial interval in $\mathbf{R}$ and $u \in L^1(I; \mathcal{H})$. Then since*

$$\Big| \int_I \langle u(t), v \rangle_{\mathcal{H}} dt \Big| \leq |u|_{L^1(I;\mathcal{H})} |v|_{\mathcal{H}}$$

for all $v \in \mathcal{H}$, there is a unique element of $\mathcal{H}$ denoted by $\int_I u(t)dt$ such that

$$\int_I \langle u(t), v \rangle_{\mathcal{H}} dt = \langle \int_I u(t)\, dt, v \rangle_{\mathcal{H}}$$

for all $v \in \mathcal{H}$.

§1.5 Geometric Properties of Open Sets in $\mathbf{R}^n$

Standard notions of boundary regularity for open sets in $\mathbf{R}^n$ will be applied throughout the text. These follow closely those in Adams [**1**], section 4.4:

Definition 1.6. *Let Ω be an open set in $\mathbf{R}^n$, $n \geq 2$, not necessarily bounded.*

(a) *Ω satisfies the segment condition if either $\Omega = \mathbf{R}^n$ or if for every $x \in \partial\Omega$ there is a neighborhood $\mathcal{U}$ of x and a nonzero vector y such that $z + ty \in \Omega$ for every $z \in \overline{\Omega} \cap \mathcal{U}$ and $t \in (0, 1)$.*

(b) *Ω satisfies the cone condition if there is a cone $\mathcal{K}$, that is, a set of the form*

$$\mathcal{K} = \{x \in \mathbf{R}^n : |x| \leq r \text{ and } x \cdot \omega \geq \delta |x|\}$$

for some $r > 0$, $\delta \in (0, 1)$ and unit vector ω, such that every $x \in \Omega$ is the vertex of a cone contained in Ω and congruent to $\mathcal{K}$.

The segment condition is required for the proof that the set of restrictions to Ω of elements of $C_c^\infty(\mathbf{R}^n)$ is dense in $W^{k,p}(\Omega)$ for $p \in [1,\infty)$ (Theorem A.2 in the Appendix) and the cone condition is required for various results concerning continuous and compact imbeddings of Sobolev spaces into Lebesgue spaces (Theorems A.3–A.5). Neither of these conditions implies the other; see Exercise 1.6.

Further regularity properties of a given set are formulated in terms of the regularity properties of functions whose graphs comprise its boundary:

Definition 1.7. *Let Ω be a nonempty, proper open subset of $\mathbf{R}^n$, $n \geq 2$, not necessarily bounded. Assume that there is given an at most countable ensemble $\{(\mathcal{U}_i, R_i, r_i, s_i, \mathcal{B}_i, \psi_i)\}_i$ satisfying the following:*

- *$\mathcal{U}_i = R_i(\mathcal{B}_i \times (0, s_i))$ where $\mathcal{B}_i$ is a ball in $\mathbf{R}^{n-1}$ of positive radius r_i, $s_i > 0$, and R_i is a rigid motion (that is, $R_i(z) = a_i + \mathcal{Q}_i z$ where a_i is a constant vector and $\mathcal{Q}_i$ is an orthogonal matrix with positive determinant)*
- *$\{\mathcal{U}_i\}$ is a locally finite open cover of $\partial\Omega$*
- *ψ_i is a mapping from $\mathcal{B}_i$ to a compact subset of $(0, s_i)$*
- *$\Omega \cap \mathcal{U}_i$ is the image under R_i of the set*
$$\{(y,s) : y \in \mathcal{B}_i \text{ and } 0 < s < \psi_i(y)\}.$$

Then:

(a) *Ω is a Lipschitz domain if each ψ_i is Lipschitz continuous on $\mathcal{B}_i$.*

(b) *Ω is a uniformly Lipschitz domain if there are positive constants $\underline{r}$, ε_0 and L such that for all i, $r_i \geq \underline{r}$, $\psi_i \geq \varepsilon_0$ on $\mathcal{B}_i$ and the Lipschitz constant for ψ_i is bounded by L.*

(c) *Ω is a C^k_{bdd} (resp. $C^{k,\lambda}_{\text{bdd}}$) domain, where k is a nonnegative integer and $\lambda \in (0,1]$, if there are positive constants $\underline{r}$, $\underline{s} < \overline{s}$, $\varepsilon_0 < 2\underline{r} \wedge \overline{s}/2$ (the smaller of $2\underline{r}$ and $\overline{s}/2$), M and i_0 such that*
  - *for all i, $\underline{s} \leq s_i \leq \overline{s}$, $r_i \geq \underline{r}$ and $\psi_i(\mathcal{B}_i) \subset [\varepsilon_0, \overline{s} - \varepsilon_0]$*
  - *the C^k_{bdd} (resp. $C^{k,\lambda}_{\text{bdd}}$) norm of each ψ_i is bounded by M*
  - *the sets*
  $$R_i\big(\mathcal{B}_i' \times (\varepsilon_0/2, \overline{s} - \varepsilon_0/2)\big),$$
  where $\mathcal{B}_i'$ is the ball of radius $r_i - \varepsilon_0/2$ concentric with $\mathcal{B}_i$, also cover $\partial\Omega$
  - *every subcollection of $i_0 + 1$ of the sets $\mathcal{U}_i$ has empty intersection.*

In all cases we define $T_i : \mathcal{B}_i \to \partial\Omega \cap \mathcal{U}_i$ by

$$T_i(y) = R_i(y, \psi_i(y)) \tag{1.6}$$

and for C^k_{bdd} (resp. $C^{k,\lambda}_{\text{bdd}}$) domains Ω we denote by M^k_Ω (resp. $M^{k,\lambda}_\Omega$) an upper bound for $\underline{r}^{-1}$, $\underline{s}^{-1}$, $\overline{s}$, ε_0^{-1}, M and i_0.

The reader can easily check that C^k_{bdd} domains with $k \geq 1$ are $C^{0,1}$ domains, $C^{0,1}$ domains are uniformly Lipschitz, and uniformly Lipschitz domains satisfy the segment and cone conditions. We reiterate that C^k_{bdd} and $C^{k,\lambda}_{bdd}$ domains need not be bounded.

Boundary measures corresponding to surface area are constructed in Theorem 3.1 for Lipschitz domains and the trace and divergence theorems are stated in Theorems 3.2 and 3.3 for uniformly Lipschitz domains; proofs are given in section A.4 in the Appendix. These domains provide a convenient balance between generality, allowing for boundaries with corners and edges, and simplicity of analysis. In particular, the basic existence and uniqueness theory developed in Chapter 3 will apply to problems (1.4) posed on uniformly Lipschitz domains. Higher-order regularity of solutions will require more highly regular regions in which they are posed, however, and these are the C^k_{bdd} domains defined above.

We conclude with a construction of specialized partitions of unity subordinate to covers such as those in Definition 1.7. The following result applies in particular to covers of the boundaries of C^k_{bdd} domains but is much more general:

Theorem 1.8. *Let $\{\mathcal{U}_i\}_i$ be an at most countable, locally finite open cover of a nonempty proper subset X of $\mathbf{R}^n$. Assume that there is a positive number δ such that the diameter of each $\mathcal{U}_i$ is at least 2δ and that if $\mathcal{V}_i = \{x \in \mathcal{U}_i : \operatorname{dist}(x, \partial\mathcal{U}_i) > \delta\}$ then the sets $\mathcal{V}_i$ also cover X. Then there is a neighborhood $\mathcal{V}$ of X and a collection of functions $\{\varphi_i\}_i$ such that*

- $\operatorname{dist}(X, \partial\mathcal{V}) \geq \delta/3$
- $\varphi_i \in C^\infty_c(\mathcal{U}_i)$ *and* $\operatorname{dist}(\operatorname{supp}\varphi_i, \partial\mathcal{U}_i) \geq \delta/6$
- $0 \leq \varphi_i(x) \leq 1$ *for all* $x \in \mathbf{R}^n$ *and all* i
- $\sum_i \varphi_i(x) = 1$ *for all* $x \in \mathcal{V}$
- *for every multi-index α there is a number C depending on δ and α such that* $\sup_i \max_{x\in\mathcal{U}_i} |D^\alpha\varphi_i(x)| \leq C$.

Proof. Let $\mathcal{V}^1_i$ and $\mathcal{V}^2_i$ be the sets of points in $\mathcal{U}_i$ whose distance to $\partial\mathcal{U}_i$ is greater than $2\delta/3$ and $\delta/3$ respectively, and let $\mathcal{V}^1 = \cup_i\mathcal{V}^1_i$ and $\mathcal{V}^2 = \cup_i\mathcal{V}^2_i$. Then $\operatorname{dist}(X, \partial\mathcal{V}^1)$ and $\operatorname{dist}(\mathcal{V}^1, \partial\mathcal{V}^2)$ are bounded below by $\delta/3$. (To prove the first of these let $x \in \mathcal{V}^1_i$ with $\operatorname{dist}(x, \partial\mathcal{U}_i) = \delta' > \delta$ and check that if ε is small then for $z \in \partial V^1$ the inequality $|x - z| < \delta/3 + \varepsilon$ leads to a contradiction). Thus $\mathcal{V} \equiv \mathcal{V}^1$ satisfies the first conclusion above and there is a function $\chi \in C^\infty_c(\mathcal{V}^2)$ which is identically one on $\mathcal{V}^1$ and which satisfies the third and fifth conditions. Also, there are functions $\varphi'_i \in C^\infty_c(\mathcal{U}_i)$ which are identically one on $\mathcal{V}^2_i$ and which satisfy the second, third, and fifth conditions. Then $\sum_i \varphi'_i(x) \geq 1$ for $x \in \mathcal{V}^2$ and since the collection $\{\mathcal{V}^2_i\}_i$

is locally finite, the reciprocal of this function is C^∞ on $\mathcal{V}^2$. The functions $\varphi_i \equiv \chi\varphi_i' / \left(\sum_i \varphi_i'\right)$ are then easily seen to satisfy all the required conditions. $\square$

Exercises

1.1 Construct a function from a nontrivial interval in $\mathbf{R}$ into a separable Hilbert space which is weakly continuous but not continuous.

1.2 Prove that a weakly measurable mapping of a nontrivial interval in $\mathbf{R}$ into a separable Hilbert space is measurable in the usual sense that inverse images of open sets are measurable. Suggestion: prove that the inverse image of a ball is measurable, then prove that an arbitrary open set is the countable union of balls.

1.3 Let $\mathcal{H}$ and $\mathcal{H}'$ be separable Hilbert spaces with $\mathcal{H}'$ continuously contained in and dense in $\mathcal{H}$. Show that a mapping from a nontrivial interval in $\mathbf{R}$ into $\mathcal{H}'$ is weakly measurable as a mapping into $\mathcal{H}$ if and only if it is weakly measurable as a mapping into $\mathcal{H}'$. Suggestion: apply Theorem A.10(b) in the Appendix.

1.4 Let u be a Lipschitz mapping of an open interval $I \subseteq \mathbf{R}$ into a separable Hilbert space $\mathcal{H}$. Prove that there exists a bounded, weakly measurable mapping $u' : I \to \mathcal{H}$ such that for all t_1, $t_2 \in I$,

$$u(t_2) - u(t_1) = \int_{t_1}^{t_2} u'(t)dt$$

and for almost all $t \in I$, $h^{-1}\big[u(t+h) - u(t)\big] \rightharpoonup u'(t)$ weakly in $\mathcal{H}$ as $h \to 0$. (See also Exercise 2.7 for a related result.)

1.5 Consider the formal problem: find $u : \mathbf{R}^n \times [0,T) \to \mathbf{R}$ such that

$$\begin{cases} u_t = -\Delta^2 u - u + g, \ (x,t) \in \mathbf{R}^n \times (0,T), \\ u(x,t) \to 0 \text{ as } |x| \to \infty, \ t > 0, \\ u(x,0) = \varsigma_0(x), \ x \in \mathbf{R}^n, \end{cases}$$

where Δ^2 is the "biharmonic" operator, which is the Laplacean composed with itself. Derive the weak form of this problem as in section 1.1, identifying the bilinear form A and the spaces $\mathcal{H}_0$ and $\mathcal{H}_1$. Then show that there is a positive constant θ such that $A(u,u) \geq \theta |u|^2_{\mathcal{H}_1}$ for all $u \in \mathcal{H}_1$. Suggestion: make use of the Fourier transform.

1.6 Construct examples showing that neither the segment condition nor the cone condition implies the other.

1.7 Simplify the conditions in Definition 1.7(c) for the case that Ω is bounded and the collection $\{\mathcal{U}_i\}$ is finite.

CHAPTER 2

Differential Equations in Hilbert Space

> Separable Hilbert spaces we can understand; beyond these lies only mysticism.
>
> Anonymous

[The space of almost periodic functions belies the epigraph ([**35**] pg. 327); still, all the spaces occurring here will be separable, and this property will serve us well.]

In this chapter we develop an existence and regularity theory for solutions of the abstract Hilbert space problem introduced in section 1.1. The most basic result, essential for all that follows, is given in Theorem 2.1. In section 2.2 the concept of *weak absolute continuity* of solutions is defined, and corresponding higher order regularity results are proved in subsequent sections in Theorems 2.5, 2.7 and 2.10. The first two of these require a certain compatibility condition between the initial data and the inhomogeneous terms, and the third shows how instantaneous regularization in positive time occurs when this compatibility condition fails. An alternative regularity framework is developed in Theorems 2.11 and 2.12 for problems with symmetry, and in Theorem 2.13 a spectral representation is derived for solutions of problems in which the leading-order terms are symmetric and constant in t, generalizing familiar representations based on Fourier series and the Duhamel integral.

§2.1 Basic Existence and Uniqueness

The problem of interest is formulated as follows. Assume:

(2.1)

- $T \in (0, \infty]$
- *$\mathcal{H}_0$ and $\mathcal{H}_1$ are separable Hilbert spaces with $\mathcal{H}_1$ continuously contained in and dense in $\mathcal{H}_0$ (but see* Remark 1 *below)*
- $A \in L^2_{loc}([0,T); \mathcal{B}(\mathcal{H}_1 \times \mathcal{H}_1))$

- *there is a function* $\lambda \in L^1_{loc}([0,T))$ *and a nonincreasing function* $\theta : [0,T) \to (0,\infty)$ *such that for almost all* $t \in [0,T)$,
$$A(t)(z,z) \geq \theta(t)|z|^2_{\mathcal{H}_1} - \lambda(t)|z|^2_{\mathcal{H}_0}$$
for all $z \in \mathcal{H}_1$.

Then given

$$\varsigma_0 \in \mathcal{H}_0 \text{ and } F \in L^2_{loc}([0,T);\mathcal{H}_1^*) \tag{2.2}$$

the problem is to find $u \in C([0,T);\mathcal{H}_0^{weak})$ whose equivalence class on $[0,T)$ is in $L^2_{loc}([0,T);\mathcal{H}_1)$ and which satisfies

$$\langle u(t),v\rangle_{\mathcal{H}_0} = \langle \varsigma_0, v\rangle_{\mathcal{H}_0} + \int_0^t [-A(s)(u(s),v) + F(s)\cdot v]\, ds \tag{2.3}$$

for all $v \in \mathcal{H}_1$ and all $t \in [0,T)$.

The following is our most basic theorem on existence and uniqueness of solutions:

Theorem 2.1. *Let* $T, \mathcal{H}_0, \mathcal{H}_1$ *and* A *be as in* (2.1) *and* ς_0 *and* F *as in* (2.2).

(a) *There is an element* $u \in C([0,T);\mathcal{H}_0^{weak})$ *whose Lebesgue equivalence class on* $[0,T)$ *is in* $L^2_{loc}([0,T);\mathcal{H}_1)$ *which satisfies* (2.3) *for all* $v \in \mathcal{H}_1$ *and for which the bound*

$$\begin{aligned}\sup_{0\leq t\leq \bar t} |u(t)|^2_{\mathcal{H}_0} &+ \theta(\bar t)\int_0^{\bar t} e^{\int_t^{\bar t} 2\lambda(s)ds}|u(t)|^2_{\mathcal{H}_1}\,dt \\ &\leq e^{\int_0^{\bar t} 2\lambda(t)dt}|\varsigma_0|^2_{\mathcal{H}_0} + \theta(\bar t)^{-1}\int_0^{\bar t} e^{\int_t^{\bar t} 2\lambda(s)ds}|F(t)|^2_{\mathcal{H}_1^*}\,dt\end{aligned} \tag{2.4}$$

holds for all $\bar t \in [0,T)$.

(b) *If in addition* $A \in L^\infty_{loc}([0,T);\mathcal{B}(\mathcal{H}_1\times\mathcal{H}_1))$ *then* $u \in C([0,T);\mathcal{H}_0)$ *(that is,* $\mathcal{H}_0$ *with the norm topology) and the energy equality*

$$\tfrac{1}{2}|u(t)|^2_{\mathcal{H}_0} + \int_0^t A(s)(u(s),u(s))ds = \tfrac{1}{2}|\varsigma_0|^2_{\mathcal{H}_0} + \int_0^t F(s)\cdot u(s)\, ds \tag{2.5}$$

holds for all $t \in [0,T)$. *Solutions of* (2.3) *with* ς_0 *and* F *as in* (2.1) *and* $A \in L^\infty_{loc}([0,T);\mathcal{B}(\mathcal{H}_1\times\mathcal{H}_1))$ *are unique in the regularity class* $C([0,T);\mathcal{H}_0^{weak}) \cap L^2_{loc}([0,T);\mathcal{H}_1)$.

The proof of Theorem 2.1 is given below following several remarks:

Remarks:

(1) There is no loss of generality in assuming that $\mathcal{H}_1$ is dense in $\mathcal{H}_0$: we need only replace $\mathcal{H}_0$ by the completion $\mathcal{H}_0'$ of $\mathcal{H}_1$ with respect to the $\mathcal{H}_0$ norm and ς_0 by its $\mathcal{H}_0$-projection onto $\mathcal{H}_0'$. The solution u of the amended problem will then be a solution of the original problem.

(2) There is no assumption in (2.1) about the sign of λ. In particular, if λ is negative then (2.4) may give important information about the large-time behavior of the solution u.

(3) The application of Theorem 2.1 to the example of section 1.1 is immediate: If $a_{ij} \in L^\infty_{loc}([0,T); L^\infty(\Omega))$ and if for all $\bar{t} \in [0,T)$ there is a positive number $\theta(\bar{t})$ such that for almost all $(x,t) \in [0,\bar{t}] \times \Omega$, $a_{ij}(x,t)\xi^i\xi^j \geq \theta(\bar{t})|\xi|^2$ for all $\xi \in \mathbf{R}^n$, then given $\varsigma_0 \in L^2(\Omega)$ and $f \in L^2_{loc}([0,T); L^2(\Omega))$ there is a unique $u \in C([0,T); L^2(\Omega))$ with $[0,T)$–equivalence class in $L^2_{loc}([0,T); H^1_0(\Omega))$ satisfying (1.2) for every $v \in H^1_0(\Omega)$.

(4) Equality in (2.5) is important in certain applications in which the square of the $\mathcal{H}_0$ norm represents an energy and exact energy balance is of interest. More generally, (2.5) provides rigorous justification for a standard formal "energy estimate" argument familiar in more elementary discussions. For example, for the initial-boundary value problem discussed in section 1.1, the argument consists in multiplying the differential equation in (1.1) by u then integrating over $\Omega \times [0,t]$ and applying the divergence theorem and the boundary condition, all formally. The result is that

$$\begin{aligned}\tfrac{1}{2}|u(\cdot,t)|^2_{L^2(\Omega)} + &\int_0^t\int_\Omega a_{ij}u_{x_i}u_{x_j}\,dxdt \\ &= \tfrac{1}{2}|\varsigma_0|^2_{L^2(\Omega)} + \int_0^t\int_\Omega g(x,s)u(x,s)\,dxds,\end{aligned}$$

which is the specific realization of (2.5) for this example. The two terms on the left may be interpreted as representing respectively the energy in a physical system at time t and the cumulative energy lost to dissipation up to time t, and the two terms on the right as the initial energy and the mechanical work done on the system by an external force g up to time t.

(5) Uniqueness, strong continuity into $\mathcal{H}_0$ and the energy equality (2.5) are not proved for the case that $A \in (L^2 - L^\infty)(\mathcal{B}(\mathcal{H}_1 \times \mathcal{H}_1))$; indeed, the integral on the left in (2.5) may not be defined in this case. These deficits are remedied under certain additional symmetry conditions, however; see Theorem 2.11 below. Notice that, at the very least, the solution u in (a) is strongly continuous from the right at $t = 0$ because by (2.4) and the weak continuity, $\limsup_{t\to 0^+} |u(t)|_{\mathcal{H}_0} \leq |\varsigma_0|_{\mathcal{H}_0} \leq \liminf_{t\to 0^+} |u(t)|_{\mathcal{H}_0}$.

PROOF of Theorem 2.1. Let $\{\varphi_k\}$ be the special orthonormal basis for $\mathcal{H}_0$ constructed in Theorem A.10 in the Appendix: $\varphi_k \in \mathcal{H}_1$ for all k and if $v \in \mathcal{H}_1$ and v^K is its $\mathcal{H}_0$-orthogonal projection onto $\mathcal{H}^K \equiv \operatorname{span}\{\varphi_1, \ldots, \varphi_K\}$, then $|v^K|_{\mathcal{H}_1} \leq 2|v|_{\mathcal{H}_1}$ for all K and $v^K \to v$ in the $\mathcal{H}_1$ norm. An approximate solution $u^K : [0,T) \to \mathcal{H}^K$ is then constructed by requiring that

$$(2.6)\quad \langle u^K(t), v^K\rangle_{\mathcal{H}_0} = \langle \varsigma_0, v^K\rangle_{\mathcal{H}_0} + \int_0^t \left[-A(s)(u^K(s), v^K) + F(s)\cdot v^K\right] ds$$

for all $t \in [0,T)$ and all $v^K \in \mathcal{H}^K$. It suffices that (2.6) holds with $v^K = \varphi_k$ for all $k \le K$, and therefore that $u^K(t) = \sum_1^K a_k(t)\varphi_k$ where $a = (a_1, \dots, a_K)$ satisfies

$$a_k(t) = a_k(0) + \int_0^t [-A(s)(\varphi_i, \varphi_k)a_i + F(s) \cdot \varphi_k] \tag{2.7}$$

for $k = 1, \dots, K$ and $t \in [0,T)$, and where $a_k(0) = \langle \varsigma_0, \varphi_k \rangle_{\mathcal{H}_0}$ (summation over i is understood). The existence of an absolutely continuous solution $a : [0,T) \to \mathbf{R}^K$ of (2.7) is guaranteed by Theorem A.9 in the Appendix. Now recall that an absolutely continuous real or $\mathbf{R}^K$-valued function is differentiable almost everywhere and is the antiderivative of its derivative (see [**34**] pp. 106–107 or [**13**] Corollary 3.33 for example). It follows that, for almost all t, $\dot{a}_k = -A(\varphi_i, \varphi_k)a_i + F \cdot \varphi_k$ and therefore that

$$\begin{aligned} \frac{d}{dt}|u^K|^2_{\mathcal{H}_0} &= -2A(u^K, u^K) + 2F \cdot u^K \\ &\le -2\theta|u^K|^2_{\mathcal{H}_1} + 2\lambda|u^K|^2_{\mathcal{H}_0} + 2F \cdot u^K \\ &\le -\theta|u^K|^2_{\mathcal{H}_1} + 2\lambda|u^K|^2_{\mathcal{H}_0} + \theta^{-1}|F|^2_{\mathcal{H}_1^*}. \end{aligned} \tag{2.8}$$

Multiplying by the integrating factor $\exp\left(-2\int_0^t \lambda(s)ds\right)$ and applying the absolute continuity, we conclude that, for $\bar{t} \in [0,T)$,

$$\begin{aligned} \sup_{0 \le t \le \bar{t}} |u^K(t)|^2_{\mathcal{H}_0} + \theta(\bar{t}) \int_0^{\bar{t}} e^{\int_t^{\bar{t}} 2\lambda(s)ds} |u^K(t)|^2_{\mathcal{H}_1} dt \\ \le e^{\int_0^{\bar{t}} 2\lambda(t)dt} |\varsigma_0|^2_{\mathcal{H}_0} + \theta(\bar{t})^{-1} \int_0^{\bar{t}} e^{\int_t^{\bar{t}} 2\lambda(s)ds} |F(t)|^2_{\mathcal{H}_1^*} dt \end{aligned} \tag{2.9}$$

and therefore that

$$\sup_{0 \le t \le \bar{t}} |u^K(t)|^2_{\mathcal{H}_0} + \int_0^{\bar{t}} |u^K(t)|^2_{\mathcal{H}_1} \le C(\bar{t}) \tag{2.10}$$

where the constant C depends on A, F, ς_0 and $\bar{t}$ but is independent of K. In particular, the sequence u^K is uniformly bounded in $L^2([0,\bar{t}]; \mathcal{H}_1)$ for every $\bar{t} \in (0,T)$ (modulo Lebesgue equivalence class on $[0,T)$) and so has a weakly converging subsequence. A diagonal process then shows that there is a single subsequence $K_l \to \infty$ and an element $u \in L^2_{loc}([0,T); \mathcal{H}_1)$ such that $[u^{K_l}]_{[0,\bar{t}],\mathcal{H}_1} \rightharpoonup u$ weakly in $L^2([0,\bar{t}]; \mathcal{H}_1)$ for every $\bar{t} \in (0,T)$.

We claim that for this same subsequence, $u^{K_l}(t)$ converges weakly in $\mathcal{H}_0$ for *every* $t \in [0,T)$. To see this we fix t and observe that, by the bound in (2.10), $u^{K_l}(t)$ has weak-$\mathcal{H}_0$ subsequential limits; it therefore suffices to show that all such subsequential limits agree. Assume therefore that a further subsequence $u^{K_{l_i}}(t)$ converges weakly in $\mathcal{H}_0$, say to w. Let $v \in \mathcal{H}_1$ and let v^M be its $\mathcal{H}_0$ projection onto $\mathcal{H}^M$. Then $v^M \in \mathcal{H}^{K_{l_i}}$ if $K_{l_i} \ge M$ and so

$$\langle u^{K_{l_i}}(t), v^M \rangle_{\mathcal{H}_0} = \langle \varsigma_0, v^M \rangle_{\mathcal{H}_0} + \int_0^t \left[-A(s)(u^{K_{l_i}}(s), v^M) + F(s) \cdot v^M\right] ds$$

by (2.6). We first let $i \to \infty$ using the weak convergence $u^{K_{l_i}}(t) \rightharpoonup w$ in $\mathcal{H}_0$ and $u^{K_{l_i}} \rightharpoonup u$ in $L^2([0,t];\mathcal{H}_1)$ (and the fact that there is an element $S \in L^2_{loc}([0,T);\mathcal{L}(\mathcal{H}_1))$ such that for all $z, w \in \mathcal{H}_1$ and almost all t, $A(t)(z,w) = \langle z, S(t)w\rangle_{\mathcal{H}_1}$), then let $M \to \infty$ using the strong convergence $v^M \to v$ in $\mathcal{H}_1$. The result is that

$$\langle w, v\rangle_{\mathcal{H}_0} = \langle \varsigma_0, v\rangle_{\mathcal{H}_0} + \int_0^t [-A(s)(u(s), v) + F(s)\cdot v]\, ds. \tag{2.11}$$

The right side here depends on u and on the sequence K_l but not on w or the subsequence K_{l_i}. Thus if $\tilde{w}$ is another such weak-$\mathcal{H}_0$ subsequential limit, then $\langle \tilde{w} - w, v\rangle_{\mathcal{H}_0} = 0$ for all $v \in \mathcal{H}_1$. It follows that $\tilde{w} = w$ because $\mathcal{H}_1$ is dense in $\mathcal{H}_0$. This proves that $u^{K_l}(t)$ converges weakly in $\mathcal{H}_0$, say to $w(t)$, for all $t \in [0,T)$. Observe also that since the right side of (2.11) is continuous in t and $\mathcal{H}_1$ is dense in $\mathcal{H}_0$, the limit w is weakly continuous from $[0,T)$ into $\mathcal{H}_0$.

Recall that $u \in L^2_{loc}([0,T);\mathcal{H}_1)$ is an equivalence class whereas $w \in C([0,T);\mathcal{H}_0^{weak})$ is a function. We will show that u is in fact the Lebesgue equivalence class of w on $[0,T)$. First note that, if $f \in \mathcal{H}_0$ is fixed, then the map $z \in \mathcal{H}_1 \to \langle z, f\rangle_{\mathcal{H}_0}$ is an element of $\mathcal{H}_1^*$, so there exists (a different) $S \in \mathcal{L}(\mathcal{H}_0, \mathcal{H}_1)$ such that $\langle z, Sf\rangle_{\mathcal{H}_1} = \langle z, f\rangle_{\mathcal{H}_0}$ for all $z \in \mathcal{H}_1$. Then since $u(t) \in \mathcal{H}_1$ for almost all t and $u^{K_l}(t) \in \mathcal{H}_1$ for all t, we have that for $t \in [0,T)$ and $f \in L^2([0,t];\mathcal{H}_0)$,

$$\begin{aligned}\int_0^t \langle u, f\rangle_{\mathcal{H}_0} &= \int_0^t \langle u, Sf\rangle_{\mathcal{H}_1} = \lim \int_0^t \langle u^{K_l}, Sf\rangle_{\mathcal{H}_1}\\ &= \lim \int_0^t \langle u^{K_l}, f\rangle_{\mathcal{H}_0} = \int_0^t \langle w, f\rangle_{\mathcal{H}_0},\end{aligned}$$

the last step following from the dominated convergence theorem and the bound (2.10). It follows that $[w]_{[0,T),\mathcal{H}_0} = u$, as claimed. This regularity together with (2.11) shows that u is a solution of (2.3) in the required sense, and (2.4) follows from (2.9). This proves (a).

We now set aside this particular u and instead let $u \in C([0,T);\mathcal{H}_0^{weak})$ with $[0,T)$-equivalence class in $L^2_{loc}([0,T);\mathcal{H}_1)$ be *any* solution of (2.3), and then derive the energy equality (2.5) as an *a priori* consequence of (2.3) under the assumption that $A \in L^\infty_{loc}([0,T);\mathcal{B}(\mathcal{H}_1 \times \mathcal{H}_1))$. Uniqueness of solutions will then follow by the application of (2.5) to the difference between two such solutions, which is a solution of the corresponding problem with $\varsigma_0 = 0$ and $F = 0$. Also, (2.5), once proved, will show that $|u(t)|_{\mathcal{H}_0}$ is continuous in t, so that $u \in C([0,T);\mathcal{H}_0)$, again as an *a priori* consequence of (2.3).

To prove (2.5) we let $u^K(t)$ and ς_0^K denote the $\mathcal{H}_0$-orthogonal projections of $u(t)$ and ς_0 onto $\mathcal{H}^K$. Then repeating the computation in the first line of (2.8) we obtain that $|u^K(t)|_{\mathcal{H}_0}$ is absolutely continuous in t and that for almost all t,

$$\frac{d}{dt}|u^K(t)|^2_{\mathcal{H}_0} = -2A(u, u^K) + 2F \cdot u^K$$

and therefore that

$$\begin{aligned}\tfrac{1}{2}|u^K(t)|^2_{\mathcal{H}_0} + &\int_0^t A(u,u)ds \\ &= \tfrac{1}{2}|\varsigma_0{}^K|^2_{\mathcal{H}_0} + \int_0^t \left[A(u, u-u^K) + F \cdot u^K\right] ds.\end{aligned} \tag{2.12}$$

Recall that, since $u(t) \in \mathcal{H}_1$ for almost all t, $u^K(t) \to u(t)$ a.e. strongly in $\mathcal{H}_1$ by our choice of basis. The integrand on the right in (2.12) therefore converges to $F \cdot u$ for almost all t and is bounded by $C(|u|^2_{\mathcal{H}_1} + |u|_{\mathcal{H}_1}|F|_{\mathcal{H}_1^*})$, where C is independent of K, which is integrable on $[0, t]$ (it is here that the hypothesis that $A \in L^\infty_{loc}$ is applied). The energy equality (2.5) therefore follows in the limit as $K \to \infty$, and the bound in (2.4) is derived as in (2.8). This completes the proof of (b).

□

§2.2 Weak Absolute Continuity

Higher order regularity of solutions typically entails differentiability, and while differentiability of mappings from an interval in **R** into a Banach space can be defined in various ways, our purposes will require that solutions be the antiderivatives in a particular sense of their derivatives. The reader will recall in this regard the familiar fact that a real-valued function of bounded variation is differentiable in the usual sense almost everywhere but need not be the antiderivative of its derivative. The relevant concept for us will be *weak absolute continuity*:

Definition 2.2. *Let $I \subseteq \mathbf{R}$ be a nontrivial interval.*

(a) *A mapping F from I into the dual X^* of a separable Banach space X is weakly absolutely continuous with derivative $F' \in L^1_{loc}(I; X^*)$ if for each $v \in X$ and all $t_1, t_2 \in I$,*

$$F(t_2) \cdot v - F(t_1) \cdot v = \int_{t_1}^{t_2} F'(t) \cdot v\, dt.$$

The vector space of weakly absolutely continuous functions is denoted by $wAC(I; X^*)$ and the set of such functions with derivatives in $L^p_{loc}(I; X^*)$ by $wAC^p(I; X^*)$.

(b) *Let $\mathcal{H}$ and $\mathcal{H}'$ be separable Hilbert spaces. A mapping $A : I \to \mathcal{B}(\mathcal{H} \times \mathcal{H}')$ is weakly absolutely continuous with derivative $A' \in L^1_{loc}(I; \mathcal{B}(\mathcal{H} \times \mathcal{H}'))$ if for each $(z, w) \in \mathcal{H} \times \mathcal{H}'$ and all $t_1, t_2 \in I$,*

$$A(t_2)(z, w) - A(t_1)(z, w) = \int_{t_1}^{t_2} A'(t)(z, w) dt.$$

The vector space of such weakly absolutely continuous mappings is denoted by $wAC(I;\mathcal{B}(\mathcal{H}\times\mathcal{H}'))$ and the space of such maps with $A' \in L^p_{loc}(I;\mathcal{B}(\mathcal{H}\times\mathcal{H}'))$ by $wAC^p(I;\mathcal{B}(\mathcal{H}\times\mathcal{H}'))$.

Remarks:

(1) Since Hilbert spaces can be identified with their duals, part (a) of the definition includes the following statement: if $\mathcal{H}$ is a separable Hilbert space then $u \in wAC^p(I;\mathcal{H})$ with derivative $u' \in L^p_{loc}(I;\mathcal{H})$ if and only if

$$\langle u(t_2), v\rangle_{\mathcal{H}} - \langle u(t_1), v\rangle_{\mathcal{H}} = \int_{t_1}^{t_2} \langle u'(t), v\rangle_{\mathcal{H}} dt$$

for all $v \in \mathcal{H}$ and $t_1, t_2 \in I$.

(2) The notations F' and A' do not reference the underlying Banach or Hilbert spaces, so that some care must be taken when two or more spaces are involved, as in Theorem 2.1. It is an easy exercise to show that if $\mathcal{H}$ and $\mathcal{H}'$ are separable Hilbert spaces with $\mathcal{H}'$ continuously included in $\mathcal{H}$ and if $u \in wAC(I;\mathcal{H}')$ with derivative u', then $u \in wAC(I;\mathcal{H})$ with the same derivative u'. A partial converse is given in Lemma 2.3 below.

(3) If $F \in wAC(I;X^*)$ then $F \in C(I;X^*)$, and if $F \in wAC^p(I;X^*)$ for some $p \in (1,\infty]$ then $F \in C^\lambda(I;X^*)$ with $\lambda = 1 - 1/p$, since for $v \in X$ and $t_1, t_2 \in I$,

$$|F(t_2)\cdot v - F(t_1)\cdot v| \le \left|\int_{t_1}^{t_2} |F'(t)|_{X^*} dt\right| |v|_X$$

which implies that $|F(t_2) - F(t_1)|_{X^*} \le |F'|_{L^p([t_1,t_2];X^*)}|t_2 - t_1|^\lambda$. Similar remarks apply to mappings in $wAC(I;\mathcal{B}(\mathcal{H}\times\mathcal{H}'))$.

(4) Observe that if u is a solution of (2.3) as described in Theorem 2.1 and if $u^* : [0,T) \to \mathcal{H}_1^*$ is defined by $u^*(t)\cdot v = \langle u(t), v\rangle_{\mathcal{H}_0}$, then $u^* \in wAC([0,T);\mathcal{H}_1^*)$ in case (a) and $u^* \in wAC^2([0,T);\mathcal{H}_1^*)$ in case (b), with derivative

$$(u^*)'\cdot v = -A(u,v) + F\cdot v$$

in either case.

(5) If $\mathcal{H}$ is a separable Hilbert space and if $u, v \in wAC(I;\mathcal{H})$, then the inner product $\langle u, v\rangle_{\mathcal{H}}$ is absolutely continuous on I with derivative $\langle u', v\rangle_{\mathcal{H}} + \langle u, v'\rangle_{\mathcal{H}}$. Consequently $|u|^2_{\mathcal{H}}$ is absolutely continuous with derivative $2\langle u', u\rangle_{\mathcal{H}}$. See Exercise 2.6.

The following result, pursuant to Remark 2 above, will be applied in later sections:

Lemma 2.3. *Let $I \subseteq \mathbf{R}$ be a nontrivial interval and $\mathcal{H}$ and $\mathcal{H}'$ separable Hilbert spaces with $\mathcal{H}'$ continuously included in $\mathcal{H}$. Assume that*

$u \in wAC^2(I;\mathcal{H})$ with derivative $u' \in L^2_{loc}(I;\mathcal{H}')$ and that $u(t) \in \mathcal{H}'$ for all $t \in I$. Then $u \in wAC^2(I;\mathcal{H}')$ with the same derivative u'.

PROOF. First, since the hypotheses hold with $\mathcal{H}$ replaced by the $\mathcal{H}$-closure of $\mathcal{H}'$, we can assume that $\mathcal{H}'$ is dense in $\mathcal{H}$. Theorem A.10 in the Appendix then applies to show that there is an orthonormal basis $\{\varphi_k\}_k$ for $\mathcal{H}$ with the following properties: each φ_k is in $\mathcal{H}'$ and if $z \in \mathcal{H}'$ and z^K is its $\mathcal{H}_0$-orthogonal projection onto the span of $\{\varphi_1, \dots, \varphi_K\}$, then $|z^K|_{\mathcal{H}'} \le 2|z|_{\mathcal{H}'}$ for all K and $z^K \to z$ strongly in $\mathcal{H}'$. Thus for $v \in \mathcal{H}'$ and $t_1, t_2 \in I$,

$$\begin{aligned}
\langle u\big|_{t_1}^{t_2}, v\rangle_{\mathcal{H}'} &= \lim \langle u^K\big|_{t_1}^{t_2}, v\rangle_{\mathcal{H}'} = \lim \sum_1^K \langle u\big|_{t_1}^{t_2}, \varphi_k\rangle_{\mathcal{H}} \langle \varphi_k, v\rangle_{\mathcal{H}'} \\
&= \lim \sum_1^K \int_{t_1}^{t_2} \langle u'(s), \varphi_k\rangle_{\mathcal{H}} \langle \varphi_k, v\rangle_{\mathcal{H}'} ds \\
&= \lim \int_{t_1}^{t_2} \langle (u')^K(s), v\rangle_{\mathcal{H}'} ds = \int_{t_1}^{t_2} \langle u'(s), v\rangle_{\mathcal{H}'} ds,
\end{aligned}$$

where the dominated convergence theorem and the fact that $|(u')^K|_{\mathcal{H}'} \le 2|u'|_{\mathcal{H}'}$ a.e. have been applied in the last step.

□

Compositions of mappings such as the integrand $t \to A(t)(u(t), \cdot)$ in (2.3) will arise frequently and it will therefore be necessary to consider their absolute continuity. The following abstract product rule will be applied in later sections:

Lemma 2.4. *Let $I \subseteq \mathbf{R}$ be a nontrivial interval and $\mathcal{H}$ and $\mathcal{H}'$ separable Hilbert spaces with $\mathcal{H}'$ continuously included in $\mathcal{H}$, and let X be a third separable Hilbert space. Assume that*

- $u \in wAC^2(I;\mathcal{H})$ with derivative $u' \in L^2_{loc}(I;\mathcal{H})$
- $u(t) \in \mathcal{H}'$ for every $t \in I$ and $|u(\cdot)|_{\mathcal{H}'} \in L^2_{loc}(I)$
- $B \in L^2_{loc}(I;\mathcal{B}(\mathcal{H} \times X))$
- *the restriction $B\big|_{\mathcal{H}' \times X}$ of B to $\mathcal{H}' \times X$ is in $wAC^2(I;\mathcal{B}(\mathcal{H}' \times X))$ with derivative B'.*

Then the map $t \in I \to B(t)(u(t), \cdot) \in X^*$ is in $wAC(I;X^*)$ with derivative $B(t)(u'(t), \cdot) + B'(t)(u(t), \cdot)$.

PROOF. First, we can assume that $\mathcal{H}'$ is dense in $\mathcal{H}$. To prove this let $\tilde{\mathcal{H}}$ be the $\mathcal{H}$-closure of $\mathcal{H}'$ so that $u(t) \in \tilde{\mathcal{H}}$ for all t. It follows that if $v \in \tilde{\mathcal{H}}^\perp$ then $\langle u'(t), v\rangle_{\mathcal{H}} = 0$ for almost all t. Applying separability, we then find that $u'(t) \in \tilde{\mathcal{H}}$ for almost all t depending on the specific representative. The hypotheses of the lemma are therefore satisfied with $\mathcal{H}$ replaced by $\tilde{\mathcal{H}}$, which we rename $\mathcal{H}$ and in which $\mathcal{H}'$ is dense, as required. Now let $\{\varphi_k\}_k \subset \mathcal{H}'$ be the $\mathcal{H}$-orthonormal basis of Theorem A.10(b) and for general $v \in \mathcal{H}$ let v^K denote the $\mathcal{H}$-orthogonal projection of v onto the span of $\{\varphi_1, \ldots, \varphi_K\}$. Then $u(t)^K \to u(t)$ strongly in $\mathcal{H}'$ for all t and $u'(t)^K \to u'(t)$ strongly in $\mathcal{H}$ for almost all t. Thus for $v \in X$ and $t_1, t_2 \in I$,

$$\begin{aligned} B(\cdot)(u(\cdot), v)\Big|_{t_1}^{t_2} &= \lim B(\cdot)(u^K(\cdot), v)\Big|_{t_1}^{t_2} = \lim \sum_1^K \langle u(\cdot), \varphi_k\rangle_{\mathcal{H}} B(\cdot)(\varphi_k, v)\Big|_{t_1}^{t_2} \\ &= \lim \sum_1^K \int_{t_1}^{t_2} \left[\langle u', \varphi_k\rangle_{\mathcal{H}} B(\varphi_k, v) + \langle u, \varphi_k\rangle_{\mathcal{H}} B'(\varphi_k, v)\right] dt \\ &= \lim \int_{t_1}^{t_2} \left[B((u')^K, v) + B'(u^K, v)\right] dt \\ &= \int_{t_1}^{t_2} \left[B(u', v) + B'(u, v)\right] dt \end{aligned}$$

as required. We have applied here the absolute continuity of the product of two absolutely continuous functions of t in the third equality and the dominated convergence theorem in the last. To justify the latter, observe that the integrand in the second-last integral is bounded a.e. by

$$\left(\|B\|_{\mathcal{H}\times X} |u'|_{\mathcal{H}} + \|B'\|_{\mathcal{H}'\times X} |u|_{\mathcal{H}'}\right) |v|_X,$$

which is independent of K and whose summands are products of elements of $L^2([t_1, t_2])$.

□

A Leibnitz rule for higher order absolute continuity can be derived by iterating the product rule. We give the statement for the case that the three Hilbert spaces in Lemma 2.4 are the same:

Corollary. *Let $I \subseteq \mathbf{R}$ be a nontrivial interval, $\mathcal{H}$ a separable Hilbert space and $m \geq 1$. Suppose that $B = B^{(0)}, \ldots, B^{(m-1)} \in wAC^2(I; \mathcal{B}(\mathcal{H} \times \mathcal{H}))$ are given with respective derivatives the equivalence classes of $B^{(1)}, \ldots, B^{(m)}$, and $u = u^{(0)}, \ldots, u^{(m-1)} \in wAC^2(I; \mathcal{H})$ are given with respective derivatives the equivalence classes of $u^{(1)}, \ldots, u^{(m)}$ (thus $B^{(m)}$ and $u^{(m)}$ are themselves equivalence classes). Let $v \in \mathcal{H}$ and define $h(t) = B(t)(u(t), v)$ and*

$$h^{(j)}(t) = \sum_{i=0}^{j} \binom{j}{i} B^{(i)}(t)(u^{(j-i)}(t), v)$$

for $j = 0, \ldots, m$. Then $h^{(0)}, \ldots, h^{(m-1)}$ are absolutely continuous on I with respective derivatives the equivalence classes of $h^{(1)}, \ldots, h^{(m)}$, and if $m \geq 2$ then $h^{(0)}, \ldots, h^{(m-2)}$ are continuously differentiable on I (with one-sided derivatives at endpoints in I, if any).

Proof. The $m = 1$ case is included in Lemma 2.4 and the general case is proved by induction in the usual way. Continuous differentiability follows from the second remark following Definition 2.2. □

We now consider the conditions under which the solution u of the problem (2.3) described in Theorem 2.1 is in $wAC^2([0,T);\mathcal{H}_0)$, say with derivative $u' \in L^2_{loc}([0,T);\mathcal{H}_0)$. If so, then higher order regularity can be considered by induction provided that u' has a specific representative $u^{(1)} \in C([0,T);\mathcal{H}_0)$ with equivalence class in $L^2_{loc}([0,T);\mathcal{H}_1)$ solving an initial value problem of the same type with $u^{(1)}(0) \equiv \varsigma_1$, which is to be determined. To explore this we proceed formally from (2.3), anticipating that for all $v \in \mathcal{H}_1$,

$$\langle u^{(1)}(t), v\rangle_{\mathcal{H}_0} = -A(t)(u(t), v) + F(t)\cdot v, \tag{2.13}$$

which may be considered the "strong form" of (2.3). This suggests that $u^{(1)}(0) = \varsigma_1$ where

$$\langle \varsigma_1, v\rangle_{\mathcal{H}_0} = -A(0)(\varsigma_0, v) + F(0)\cdot v, \tag{2.14}$$

which in the applications of interest expresses a necessary compatibility between the initial value ς_0 and boundary conditions satisfied by elements of $\mathcal{H}_1$ (see the detailed discussion in section 3.1). Formally differentiating (2.13) and applying the product rule and integrating, we arrive at

$$\begin{aligned}\langle u^{(1)}(t), v\rangle_{\mathcal{H}_0} = \langle \varsigma_1, v\rangle_{\mathcal{H}_0} + \int_0^t \big[&- A(s)(u^{(1)}(s), v) \\ &- A'(u(s), v) + F'(s)\cdot v\big]\,ds,\end{aligned} \tag{2.15}$$

which is of the type considered in (2.3) with ς_0 and $F(t)$ replaced by ς_1 and $F'(t) - A'(t)(u(t),\cdot)$. We will see in Theorem 2.5 below that if

- $A \in wAC^2([0,T);\mathcal{B}(\mathcal{H}_1 \times \mathcal{H}_1))$
- $F \in wAC^2([0,T);\mathcal{H}_1^*)$
- $(\varsigma_0, \varsigma_1) \in \mathcal{H}_1 \times \mathcal{H}_0$ satisfies (2.14)

then indeed $u \in wAC^2([0,T);\mathcal{H}_0)$ with derivative u' having a representative $u^{(1)} \in C([0,T);\mathcal{H}_0)$ satisfying (2.13)–(2.15). This is the $m = 1$ case of the following theorem. We point out, however, that, while satisfactory in applications to differential equations whose coefficients are highly regular, this theorem is suboptimal because the absolute continuity hypotheses on A are overly strong, as we will see in the next section. The more satisfactory higher-order result is proved in Theorem 2.7, the following being a necessary starting point:

Theorem 2.5. *Let T, $\mathcal{H}_0$, $\mathcal{H}_1$ and A satisfy the hypotheses in* (2.1) *and let $m \geq 1$. Assume that $A = A^{(0)}, \ldots, A^{(m-1)} \in wAC^2([0,T);\mathcal{B}(\mathcal{H}_1 \times \mathcal{H}_1))$ are* *given with respective derivatives*

$$[A^{(1)}],\dots,[A^{(m-1)}],\ A^{(m)}\in L^2_{loc}([0,T);\mathcal{B}(\mathcal{H}_1\times\mathcal{H}_1))$$

($[\cdot]$ denotes equivalence class on $[0,T)$). Then given $F=F^{(0)},\dots,F^{(m-1)}\in wAC^2([0,T);\mathcal{H}_1^*)$ with respective derivatives $[F^{(1)}],\dots,[F^{(m-1)}]$, $F^{(m)}\in L^2_{loc}([0,T);\mathcal{H}_1^*)$ and given $(\varsigma_0,\dots,\varsigma_m)\in\mathcal{H}_1\times\cdots\times\mathcal{H}_1\times\mathcal{H}_0$ satisfying

$$\langle\varsigma_{j+1},v\rangle_{\mathcal{H}_0}=-\sum_{i=0}^{j}\binom{j}{i}A^{(i)}(0)(\varsigma_{j-i},v)+F^{(j)}(0)\cdot v \tag{2.16}$$

for $j=0,\dots,m-1$ and all $v\in\mathcal{H}_1$, the corresponding solution u of the problem (2.3) *described in* Theorem 2.1(b) *satisfies the following:*

(a) *u is m times continuously differentiable in the following sense: there are mappings $u=u^{(0)},\dots,u^{(m-1)}\in wAC^2([0,T);\mathcal{H}_1)$ and $u^{(m)}\in C([0,T);\mathcal{H}_0)$ such that the derivative of $u^{(j)}$ is $[u^{(j+1)}]$ for $j\le m-1$.*

(b) *The equivalence classes of the linear functionals*

$$G_j\equiv-\sum_{i=1}^{j}\binom{j}{i}A^{(i)}(u^{(j-i)},\cdot)+F^{(j)} \tag{2.17}$$

are in $L^2_{loc}([0,T);\mathcal{H}_1^*)$ for $j=1,\dots,m$; and if $m\ge 2$ then $G_1,\dots,G_{m-1}\in wAC^2([0,T);\mathcal{H}_1^*)$ with respective derivatives the equivalence classes of $G_2,\dots,G_m$.

(c) *For $1\le j\le m$, $u^{(j)}$ is the unique solution of the problem*

$$\langle u^{(j)}(t),v\rangle_{\mathcal{H}_0}=\langle\varsigma_j,v\rangle_{\mathcal{H}_0}+\int_0^t\Big[-A(s)(u^{(j)}(s),v)+G_j(s)\cdot v\Big]\,ds \tag{2.18}$$

in the sense of Theorem 2.1(b); *and for $0\le j\le m-1$ the strong forms*

$$\langle u^{(j+1)}(t),v\rangle_{\mathcal{H}_0}=-A(t)(u^{(j)}(t),v)+G_j(t)\cdot v \tag{2.19}$$

hold for all $v\in\mathcal{H}_1$ and all $t\in[0,T)$ (where $G_0=F$).

(d) *Given $\bar t\in(0,T)$ there is a constant C depending on upper bounds for m, $\bar t$, $\theta(\bar t)^{-1}$, the norms of λ and the $A^{(j)}$ on $[0,\bar t]$ occurring in the above hypotheses and on the norm of the inclusion $\mathcal{H}_1\subseteq\mathcal{H}_0$, but independent of F and of $(\varsigma_0,\dots,\varsigma_m)$, such that*

$$\sum_{j=0}^{m-1}\sup_{0\le t\le\bar t}|u^{(j)}(t)|^2_{\mathcal{H}_1}+\sup_{0\le t\le\bar t}|u^{(m)}(t)|^2_{\mathcal{H}_0}+\int_0^{\bar t}|u^{(m)}|^2_{\mathcal{H}_1}dt$$
$$\le C(\bar t)\Big(\sum_{j=0}^{m}|\varsigma_j|^2_{\mathcal{H}_0}+\sum_{j=0}^{m-1}\sup_{0\le t\le\bar t}|F^{(j)}(t)|^2_{\mathcal{H}_1^*}+\int_0^{\bar t}|F^{(m)}|^2_{\mathcal{H}_1^*}dt\Big).$$

PROOF. The proof is by induction on m. First note that the fourth hypothesis in (2.1) can be combined with the absolute continuity assumption on A

to show that there are positive functions θ_0 and λ_0 on $[0,T)$ such that for $\bar{t} \in [0,T)$,

$$A(t)(z,z) \geq \theta_0(\bar{t})|z|^2_{\mathcal{H}_1} - \lambda_0(\bar{t})|z|^2_{\mathcal{H}_0} \tag{2.20}$$

for all $z \in \mathcal{H}_1$ and $t \in [0,\bar{t}]$ (see Exercise 2.9).

For the case $m=1$ we let $\mathcal{H}^K$ and u^K be as in the first half of the proof of Theorem 2.1 except that $u^K(0) \equiv \tilde{u}_0^K$ is chosen differently. Specifically, since $\tilde{A} \equiv A(0) + \lambda_0(0)\langle\cdot,\cdot\rangle_{\mathcal{H}_0}$ is coercive on $\mathcal{H}_1 \times \mathcal{H}_1$, we can define $\tilde{u}_0^K \in \mathcal{H}^K$ by the requirement that

$$\tilde{A}(\tilde{u}_0^K, v^K) = -\langle \varsigma_1, v^K\rangle_{\mathcal{H}_0} + \lambda_0\langle \Pi^K \varsigma_0, v^K\rangle_{\mathcal{H}_0} + F(0)\cdot v^K \tag{2.21}$$

for all $v^K \in \mathcal{H}^K$, where Π^K is the $\mathcal{H}_0$-orthogonal projection onto $\mathcal{H}^K$. We claim that $u_0^K \to \varsigma_0$ in $\mathcal{H}_1$ as $K \to \infty$. To prove this we subtract (2.14) from (2.21) to obtain that $\tilde{A}(\tilde{u}_0^K - \varsigma_0, v^K) = 0$ for all $v^K \in \mathcal{H}^K$. Then taking $v^K = \Pi^K\varsigma_0 - \tilde{u}_0^K$ we compute that for a generic constant C independent of K,

$$\begin{aligned} C^{-1}|\tilde{u}_0^K - \varsigma_0|^2_{\mathcal{H}_1} &\leq \tilde{A}(\tilde{u}_0^K - \varsigma_0, \tilde{u}_0^K - \varsigma_0 + v^K) \\ &= A(\tilde{u}_0^K - \varsigma_0, \Pi^K\varsigma_0 - \varsigma_0) \leq C|\tilde{u}_0^K - \varsigma_0|_{\mathcal{H}_1}|\Pi^K\varsigma_0 - \varsigma_0|_{\mathcal{H}_1}, \end{aligned}$$

so that $|\tilde{u}_0^K - \varsigma_0|_{\mathcal{H}_1} \leq C|\Pi^K\varsigma_0 - \varsigma_0|_{\mathcal{H}_1}$, which goes to zero as $K \to \infty$ by Theorem A.10(b), as claimed. Next we prove that $u_t^K(0)$ converges to ς_1 in $\mathcal{H}_0$ as $K \to \infty$. First note that u^K is continuously differentiable on $[0,T)$ by (2.7) and our hypotheses on A and F. Thus

$$\langle u_t^K(t), v^K\rangle_{\mathcal{H}_0} = -A(t)(u^K(t), v^K) + F(t)\cdot v^K \tag{2.22}$$

for all $v^K \in \mathcal{H}^K$ and $t \in [0,T)$. Setting $t=0$ and substituting from (2.21) we find that

$$\langle u_t^K(0), v^K\rangle_{\mathcal{H}_0} = \langle \varsigma_1, v^K\rangle_{\mathcal{H}_0} + \lambda_0\langle \tilde{u}_0^K - \Pi^K\varsigma_0, v^K\rangle_{\mathcal{H}_0}.$$

This shows that $u_t^K(0) = \Pi^K\varsigma_1 + \lambda_0(\tilde{u}_0^K - \Pi^K\varsigma_0)$, which converges to ς_1 in $\mathcal{H}_0$ as $K \to \infty$, as claimed.

Now recall from the first half of the proof of Theorem 2.1 that (2.6)–(2.10) hold, that for a particular subsequence K_l, $u^{K_l}(t)$ converges to $u(t)$ weakly in $\mathcal{H}_0$ for every t and that $[u^{K_l}]$ converges to $[u]$ weakly in $L^2_{loc}([0,T);\mathcal{H}_1)$. We will prove convergence of the derivatives u_t^K, bounds for which are obtained as follows. Since the right side of (2.22) is absolutely continuous in t, we can apply the product rule Lemma 2.4 to obtain that

$$\langle u_{tt}^K(t), v^K\rangle_{\mathcal{H}_0} = -A(u_t^K, v^K) - A^{(1)}(u^K, v^K) + F^{(1)}\cdot v^K \tag{2.23}$$

for almost all t. Putting $v^K = u_t^K(t) \in \mathcal{H}^K$ for such t and applying the absolute continuity, we get that

$$\tfrac{1}{2}|u_t^K(t)|^2_{\mathcal{H}_0} = \tfrac{1}{2}|u_t^K(0)|^2_{\mathcal{H}_0} + \int_0^t \big[-A(u_t^K, u_t^K) - A^{(1)}(u^K, u_t^K) + F^{(1)}\cdot u_t^K\big].$$

Applying (2.10) and Gronwall's inequality we then obtain

$$\sup_{0\le t\le \bar t} |u_t^K(t)|^2_{\mathcal{H}_0} + \int_0^{\bar t} |u_t^K|^2_{\mathcal{H}_1} dt \le C(\bar t)\Big[|\varsigma_0|^2_{\mathcal{H}_0} + |\varsigma_1|^2_{\mathcal{H}_0} + \int_0^{\bar t} \big(\|A^{(1)}\|^2_{\mathcal{H}_1\times\mathcal{H}_1}|u^K|^2_{\mathcal{H}_1} + |F^{(1)}|^2_{\mathcal{H}_1^*}\big)dt\Big]$$

for $\bar t \in (0,T)$, where $C(\bar t)$ is a generic constant as described in (d). We also have from (2.20) and (2.22) that

$$|u^K(t)|^2_{\mathcal{H}_1} \le C(\bar t)\left[|u^K|^2_{\mathcal{H}_0} + |u_t^K|^2_{\mathcal{H}_0} + |F(t)|^2_{\mathcal{H}_1^*}\right]$$

for every $t \in [0,\bar t]$. Combining the last two inequalities and applying the bound in (2.10) and Gronwall's inequality again, we conclude that for $\bar t \in (0,T)$ there is a constant $\tilde C(\bar t)$ depending on the quantities listed in (d) and also on ς_0, ς_1 and F, but which is independent of K, such that

$$\sup_{0\le t\le \bar t} |u^K(t)|^2_{\mathcal{H}_1} + \sup_{0\le t\le \bar t} |u_t^K(t)|^2_{\mathcal{H}_0} + \int_0^{\bar t} |u_t^K|^2_{\mathcal{H}_1} dt \le \tilde C(\bar t). \tag{2.24}$$

We can now repeat the compactness argument following (2.10) in the proof of Theorem 2.1 to show that there is a further subsequence, still denoted by K_l, and there is a function $z \in C([0,T);\mathcal{H}_0^{weak})$ whose equivalence class $[z]$ is in $L^2_{loc}([0,T),\mathcal{H}_1)$ such that $u_t^{K_l}(t) \rightharpoonup z(t)$ weakly in $\mathcal{H}_0$ for every $t\in[0,T)$ and $[u_t^{K_l}] \rightharpoonup [z]$ weakly in $L^2_{loc}([0,T),\mathcal{H}_1)$. Observe also that the uniform pointwise bound in (2.24) improves the weak $\mathcal{H}_0$-convergence $u^K(t) \rightharpoonup u(t)$ to weak $\mathcal{H}_1$-convergence. Taking limits in the equation

$$\langle u^{K_l}(t), v\rangle_{\mathcal{H}_1} = \langle \varsigma_0^{K_l}, v\rangle_{\mathcal{H}_1} + \int_0^t \langle u_t^{K_l}(s), v\rangle_{\mathcal{H}_1} ds, \ \ v\in\mathcal{H}_1,$$

we then obtain that $u \in wAC^2([0,T),\mathcal{H}_1)$ with derivative the equivalence class of z, which we therefore designate $u^{(1)}$. This proves the regularity statement in (a) for u. The statement for $u^{(1)}$ will follow by the application of Theorem 2.1 to (2.18) once we prove (b) and (c).

First, we have that $G_1(t) = -A^{(1)}(u(t),\cdot) + F^{(1)}(t)$, whose equivalence class is in $L^2_{loc}([0,T),\mathcal{H}_1^*)$ by our hypotheses on A and F and the above regularity of u; this proves (b). To prove (c) we integrate (2.23) to get

$$\langle u_t^K(t), v^K\rangle_{\mathcal{H}_0} = \langle u_t^K(0), v^K\rangle_{\mathcal{H}_0} + \int_0^t \big[-A(u_t^K, v^K) - A^{(1)}(u^K, v^K) + F^{(1)}\cdot v^K\big]ds,$$

then take the limit as $K = K_l \to \infty$ to obtain that

$$\langle u^{(1)}, v\rangle_{\mathcal{H}_0} = \langle \varsigma_1, v\rangle_{\mathcal{H}_0} + \int_0^t \Big[-A(u^{(1)}, v) + G_1\cdot v\Big]ds. \tag{2.25}$$

To justify the last step, say for the $A^{(1)}$ term, we observe that

$$A^{(1)}(t)(u^{K_l}(t), v^{K_l}) \to A^{(1)}(t)(u(t), v)$$

for almost all t because $u^{K_l}(t) \rightharpoonup u(t)$ weakly in $\mathcal{H}_1$ for all t, $v^{K_l} \to v$ strongly in $\mathcal{H}_1$, and $A^{(1)}(t) \in \mathcal{B}(\mathcal{H}_1 \times \mathcal{H}_1)$ for almost all t. Also,

$$|A^{(1)}(u^{K_l}, v^{K_l})| \leq \|A^{(1)}\|_{\mathcal{H}_1\times\mathcal{H}_1} |u^{K_l}|_{\mathcal{H}_1} |v^{K_l}|_{\mathcal{H}_1} \leq \tilde{C}(\bar{t}) \|A^{(1)}\|_{\mathcal{H}_1\times\mathcal{H}_1} |v|_{\mathcal{H}_1},$$

which is integrable on $[0,t]$ and is independent of K_l; convergence can therefore be passed under the integral. This proves that $u^{(1)}$ is a solution of (2.25) in the sense of Theorem 2.1(b), which therefore applies to show that $u^{(1)} \in C([0,T); \mathcal{H}_0)$, thus completing the proof of (a) and the first assertion in (c). Finally we take the limit as $K = K_l \to \infty$ in (2.22) to obtain the other assertion in (c), and (d) follows immediately from (2.24). This completes the proof for the case $m = 1$.

Now assume that $m \geq 2$ and that the theorem holds with m replaced by $m-1$. We have only to show that the $m = 1$ result applies to the problem (2.18) satisfied by $u^{(m-1)}$. To do this we need to check the compatibility condition $\langle \varsigma_m, \cdot \rangle_{\mathcal{H}_0} = -A(0)(\varsigma_{m-1}, \cdot) + G_{m-1}(0)$ and the regularity conditions $G_{m-1} \in wAC^2([0,T), \mathcal{H}_1^*)$ with $(G_{m-1})' = G_m \in L^2_{loc}([0,T), \mathcal{H}_1^*)$. The first is immediate because $u^{(l)}(0) = \varsigma_l$ for $1 \leq l \leq m-1$. For the second we check only the absolute continuity of terms $A^{(i)}(u^{(m-1-i)}, \cdot)$ for $1 \leq i \leq m-1$, everything else being similar. For such i, $A^{(i)} \in wAC^2([0,T); \mathcal{B}(\mathcal{H}_1 \times \mathcal{H}_1))$ by hypothesis, and since $m - i - 1 \leq m - 2$, $u^{(m-i-1)} \in wAC^2([0,T); \mathcal{H}_1)$ by induction. Lemma 2.4 (with $\mathcal{H} = \mathcal{H}' = X = \mathcal{H}_1$) therefore applies to show that the term in question is in $wAC([0,T); \mathcal{H}_1^*)$ with derivative $A^{(i+1)}(u^{(m-1-i)}, \cdot) + A^{(i)}(u^{(m-i)}, \cdot)$, whose $\mathcal{H}_1^*$ norm is bounded locally on $[0,T)$ by

$$\begin{aligned} &\|A^{(i+1)}\|_{\mathcal{H}_1\times\mathcal{H}_1} |u^{(m-1-i)}|_{\mathcal{H}_1} + \|A^{(i)}\|_{\mathcal{H}_1\times\mathcal{H}_1} |u^{(m-i)}|_{\mathcal{H}_1} \\ &\qquad \in L^2 \cdot L^\infty + L^\infty \cdot L^2 \subseteq L^2. \end{aligned}$$

This shows that $A^{(i)}(u^{(m-1-i)}, \cdot) \in wAC^2([0,T); \mathcal{B}(\mathcal{H}_1 \times \mathcal{H}_1))$, and the relation $(G_{m-1})' = G_m$ then follows easily. This completes the induction step and the proof of the theorem.

□

§2.3 Higher Order Regularity

In this section we improve the result in Theorem 2.5 by giving more refined regularity information about the solution u and by weakening the hypotheses on the bilinear form A. The complete statement is given in Theorem 2.7 below and its proof in section 2.4. The following preliminary discussion will motivate the rather lengthy technical analysis to follow. First recall the Lax-Milgram theorem: if $\mathcal{H}$ is a Hilbert space and B is a bounded, coercive bilinear form on $\mathcal{H}$, that is, there is a positive constant θ such that $B(z,z) \geq \theta |z|^2_{\mathcal{H}}$ for all $z \in \mathcal{H}$, then there is an $S \in \mathcal{L}(\mathcal{H}^*, \mathcal{H})$ such that $B(SG, v) = G \cdot v$ for all $v \in \mathcal{H}$ and $G \in \mathcal{H}^*$. The operator S is injective, so if $\mathcal{V}_0$ is a proper subspace of $\mathcal{H}^*$ then $S(\mathcal{V}_0)$ is a proper subspace of $\mathcal{H}$. If $S(\mathcal{V}_0)$ is contained in a proper subspace $\mathcal{H}_2$ of $\mathcal{H}$ and inclusion in $\mathcal{H}_2$ is regarded as a regularity restriction, then the obvious statement that $G \in \mathcal{V}_0$ implies $SG \in \mathcal{H}_2$ can be interpreted as a regularity theorem.

To apply this observation to the problem at hand, suppose that $\mathcal{H}_0$, $\mathcal{H}_1$, A, F and u are as in the $m = 1$ case of Theorem 2.5 and that $A(t)$ is coercive for every t, so that the operators $S(t) \in \mathcal{L}(\mathcal{H}_1^*, \mathcal{H}_1)$ are defined as above. Assume further that $S(t)(\mathcal{V}_0) \subseteq \mathcal{H}_2$ for a particular subspace $\mathcal{V}_0 \subseteq \mathcal{H}_1^*$ containing the set $\{f^* \equiv \langle f, \cdot \rangle_{\mathcal{H}_0} : f \in \mathcal{H}_0\}$ and for a particular subspace $\mathcal{H}_2$ of $\mathcal{H}_1$. We can then deduce from the equation

$$A(u(t), v) = F(t) \cdot v - \langle u^{(1)}(t), v \rangle_{\mathcal{H}_0},$$

which is the $j = 0$ case of (2.19), that $u(t) \in \mathcal{H}_2$ provided that $F(t) \in \mathcal{V}_0$.

While significant in itself, the above result has an important implication for our hypotheses on A. Recall that the analysis of Theorem 2.5 required the weak absolute continuity of $A(t)(u(t), \cdot)$ as a mapping into $\mathcal{H}_1^*$, and this in turn required the weak absolute continuity of A as a mapping into $\mathcal{B}(\mathcal{H}_1 \times \mathcal{H}_1)$. Now, if it is known that $u(t) \in \mathcal{H}_2$ for every t, then the milder condition that the *restriction* $A\big|_{\mathcal{H}_2 \times \mathcal{H}_1}$ be weakly absolutely continuous as a mapping into $\mathcal{B}(\mathcal{H}_2 \times \mathcal{H}_1)$ should suffice.

Let us examine this point in the context of the example of section 1.1 in which $\mathcal{H}_0 = L^2(\Omega)$ and $\mathcal{H}_1 = H_0^1(\Omega)$ for a bounded open set $\Omega \subset \mathbf{R}^n$ and A is given by (1.3). We can reasonably anticipate that if A is to be weakly absolutely continuous as a mapping into $\mathcal{B}(H_0^1 \times H_0^1)$ (abbreviating $H_0^1(\Omega) = H_0^1$, etc.), then its derivative $A^{(1)}$ should be given by

$$A^{(1)}(t)(u, v) = \int_\Omega (D_t a_{ij}) u_{x_i} v_{x_j} \, dx$$

with $D_t a_{ij} \in L^2_{loc}([0,T); L^\infty)$. On the other hand, we also expect on the basis of standard elliptic theory (see Theorem 4.2) that under reasonable hypotheses on a_{ij} and with $\mathcal{V}_0 = \{f^* : f \in L^2\}$, we can take $\mathcal{H}_2 = H^2 \cap H_0^1$. If so, then $u(t) \in H^2$ for all t, so that $u_{x_i}(t) \in L^p$ for all $p \in [2, p^*]$ (if $n \geq 3$) for a certain p^* determined by n (Theorem A.3(a)). The weaker condition that the restriction of A be weakly absolutely continuous as a map into $\mathcal{B}((H^2 \cap H_0^1) \times H_0^1))$ will therefore hold if $D_t a_{ij} \in L^2_{loc}([0,T); L^q)$ for any particular q satisfying $q^{-1} + p^{-1} = \frac{1}{2}$ for some $p \in [2, p^*]$. This is a clear improvement on the above assumption required for the application of Theorem 2.5 that $D_t a_{ij} \in L^2_{loc}([0,T); L^\infty)$.

The reader will no doubt have detected a circularity in the above reasoning, however: for the specific example discussed above, the weak absolute continuity of the restricted bilinear form will suffice for the proof of Theorem 2.5 only if it is known that $u(t) \in H^2 \cap H_0^1$ for every t. But this fact was derived from (2.19), which is a *conclusion* of Theorem 2.5. We will circumvent this difficulty by replacing A by approximate bilinear forms A_ε for small $\varepsilon > 0$, applying Theorem 2.5 to obtain corresponding approximate solutions u_ε with the desired regularity properties, and then showing that this regularity persists in the limit as $\varepsilon \to 0$. The analysis is straightforward but rather technical. For example, the compatibility conditions (2.16) for A_ε and F will differ from those for A and F; it will therefore be necessary to construct nearby compatible initial data for the perturbed problem.

Before stating the main theorem we introduce approximate bilinear forms, constructed by a standard mollifying procedure:

Definition 2.6. *Fix a C^∞ function $h : \mathbf{R} \to [0, \infty)$ whose support is contained in $[-1, 1]$ and whose integral over $\mathbf{R}$ equals one, and let $h_\varepsilon(t) = \varepsilon^{-1}h(t/\varepsilon)$ for $\varepsilon > 0$. Let $\mathcal{H}$ and $\mathcal{H}'$ be separable Hilbert spaces, $T \in (0, \infty]$ and $B \in L^1_{loc}([0, T); \mathcal{B}(\mathcal{H} \times \mathcal{H}'))$. Then for $\varepsilon > 0$ and $t \in [0, T - 2\varepsilon)$ define the approximation B_ε by*

$$B_\varepsilon(t)(z, w) = \int_{|s-t|<\varepsilon} h_\varepsilon(t - s)B(s + \varepsilon)(z, w)ds$$

for $(z, w) \in \mathcal{H} \times \mathcal{H}'$.

We leave to the reader to check that if

$$(B_\varepsilon)^{(j)}(z, w) \equiv \int_{|s-t|<\varepsilon} h_\varepsilon^{(j)}(t - s)B(s + \varepsilon)(z, w)ds$$

for $j \geq 0$, then $(B_\varepsilon)^{(j)} \in wAC^2([0, T - 2\varepsilon); \mathcal{B}(\mathcal{H} \times \mathcal{H}'))$ with derivative $(B_\varepsilon)^{(j+1)}$.

We can now state the basic theorem on higher order regularity of solutions of (2.3). Several important remarks follow, especially Remark 3 concerning hypothesis (b).

Theorem 2.7. *Let $m \geq 1$ and assume that separable Hilbert spaces $\mathcal{H}_0 \supseteq \mathcal{H}_1 \ldots \supseteq \mathcal{H}_{2m+1}$ and $\mathcal{V}_{-1} = \mathcal{H}_1^* \supseteq \mathcal{V}_0 \supseteq \mathcal{V}_1 \ldots \mathcal{V}_{2m-1}$ are given in which all inclusions are continuous, $\mathcal{H}_1$ is dense in $\mathcal{H}_0$, and the mapping $*$ that takes $w \in \mathcal{H}_0$ to $w^* \in \mathcal{H}_1^*$ defined by $w^*v = \langle w, v\rangle_{\mathcal{H}_0}$ is in $\mathcal{L}(\mathcal{H}_j, \mathcal{V}_j)$ for $j = 0, \ldots, 2m - 1$.*

$$\begin{array}{ccccccccccc}
\mathcal{V}_{-1} = \mathcal{H}_1^* & \supseteq & \mathcal{V}_0 & \supseteq & \mathcal{V}_1 & \ldots & \supseteq \mathcal{V}_{2m-1} & & & & \\
 & & \uparrow * & & \uparrow * & & \uparrow * & & & & \\
 & & \mathcal{H}_0 & \supseteq & \mathcal{H}_1 & \ldots & \supseteq \mathcal{H}_{2m-1} & \supseteq & \mathcal{H}_{2m} & \supseteq & \mathcal{H}_{2m+1}
\end{array}$$

Let $T \in (0, \infty]$ and let $A : [0, T) \to \mathcal{B}(\mathcal{H}_1 \times \mathcal{H}_1)$ be weakly measurable, and assume that there are parameters $\delta \in (0, 1]$, $\beta \in L^2_{loc}([0, T))$ and for every $\bar{t} \in [0, T)$ positive numbers $\theta(\bar{t})$, $\varepsilon_0(\bar{t})$, $C_A(\bar{t})$ and a nonnegative number $\lambda(\bar{t})$ such that the following hold:

(a) *For all* $z, w \in \mathcal{H}_1$ *and all* $t \in [0, \bar{t}]$,

$$|A(t)(z,w)| \leq C_A(\bar{t})|z|_{\mathcal{H}_1}|w|_{\mathcal{H}_1}$$

and

$$A(t)(z,z) \geq \theta(\bar{t})|z|^2_{\mathcal{H}_1} - \lambda(\bar{t})|z|^2_{\mathcal{H}_0}.$$

(b) *For* $j = -1, \ldots, 2m-1$, $\varepsilon \in (0, \varepsilon_0(\bar{t})]$, $t \in [0, \bar{t}]$ *and* $\lambda' \geq \lambda(\bar{t})$*: if* $z \in \mathcal{H}_1$ *and* $G \in \mathcal{V}_j$ *satisfy*

$$A_\varepsilon(t)(z,v) + \lambda' \langle z, v \rangle_{\mathcal{H}_0} = G \cdot v$$

for all $v \in \mathcal{H}_1$*, where* A_ε *is the approximation defined above in* Definition 2.6*, then* $z \in \mathcal{H}_{j+2}$ *and*

$$|z|_{\mathcal{H}_{j+2}} \leq C_A(\bar{t})|G|_{\mathcal{V}_j}$$

(*see* Remark 3 *below concerning this hypothesis*).

(c) *For* $i = 0, \ldots, m$ *there are weakly measurable maps* $A^{(i)} : [0,T) \to \mathcal{B}(\mathcal{H}_{2i+1} \times \mathcal{H}_1)$ *with* $A^{(0)} = A$ *such that:*

(*i*) *for* $i = 1, \ldots, m$, $t \in [0, \bar{t}]$ *and all* $(z, w) \in \mathcal{H}_{2i+1} \times \mathcal{H}_1$,

$$|A^{(i)}(t)(z,w)| \leq (\beta(t)|z|_{\mathcal{H}_{2i}})^\delta |z|^{(1-\delta)}_{\mathcal{H}_{2i+1}}|w|_{\mathcal{H}_1};$$

(*ii*) *for* $i = 0, \ldots, m-1$,

$$A^{(i)}|_{\mathcal{H}_{2i+2}\times\mathcal{H}_1} \in C([0,T); \mathcal{B}(\mathcal{H}_{2i+2} \times \mathcal{H}_1));$$

(*iii*) *for* $i = 0, \ldots, m-1$, $A^{(i)}|_{\mathcal{H}_{2i+3}\times\mathcal{H}_1} \in wAC^2([0,T); \mathcal{B}(\mathcal{H}_{2i+3} \times \mathcal{H}_1))$ *with derivative* $[A^{(i+1)}]$*, where* $[\cdot]$ *denotes Lebesgue equivalence class on* $[0,T)$*;*

(*iv*) *if* $m \geq 2$ *then for* $i = 1, \ldots, m-1$ *and* $l = 0, \ldots, 2m-2i-1$*: if* $z \in \mathcal{H}_{2i+l+2}$ *then* $A^{(i)}(t)(z, \cdot) \in \mathcal{V}_l$*; and if* $t \leq \bar{t} < T$*, then*

$$|A^{(i)}(t)(z,\cdot)|_{\mathcal{V}_l} \leq C_A(\bar{t})|z|^\delta_{2i+l}|z|^{1-\delta}_{2i+l+2}.$$

Then given $F = F^{(0)}, \ldots, F^{(m-1)} \in wAC^2([0,T); \mathcal{H}_1^*)$ *with respective derivatives* $[F^{(1)}], \ldots, [F^{(m-1)}], F^{(m)} \in L^2_{loc}([0,T); \mathcal{H}_1^*)$ *satisfying*

$$[F^{(j)}] \in L^\infty_{loc}([0,T); \mathcal{V}_{2m-2j-2}) \tag{2.26}$$

for $j = 0, \ldots, m-1$ *and*

$$[F^{(j)}] \in L^2_{loc}([0,T); \mathcal{V}_{2m-2j-1}) \tag{2.27}$$

for $j = 0, \ldots, m$*, and given* $(\varsigma_0, \ldots, \varsigma_m) \in (\mathcal{H}_{2m} \times \cdots \mathcal{H}_2 \times \mathcal{H}_0)$ *satisfying the compatibility conditions* (2.16)*, the corresponding unique solution* u *of the problem* (2.3) *constructed in* Theorem 2.1(b) *satisfies the following:*

(d) *For* $j = 0, \dots, m$ *there are weakly measurable maps* $u^{(j)} : [0,T) \to \mathcal{H}_{2m-2j+1}$, *with* $u^{(0)} = u$, *such that:*

(*i*) *for* $j = 0 \dots, m$, $u^{(j)} \in C([0,T); \mathcal{H}^{weak}_{2m-2j})$ *with equivalence class in* $L^2_{loc}([0,T); \mathcal{H}_{2m-2j+1})$, *and* $u^{(m)}$ *is continuous into* $\mathcal{H}_0$ *in the norm topology;*

(*ii*) *for* $j = 0, \dots, m-1$,

$$u^{(j)} \in Lip_{loc}([0,T); \mathcal{H}_{2m-2j-2}) \cap wAC^2([0,T); \mathcal{H}_{2m-2j-1})$$

with $\mathcal{H}_{2m-2j-1}$*-derivative* $[u^{(j+1)}]$;

(*iii*) $(u^{(m)})^* \in wAC^2([0,T); \mathcal{H}_1^*)$ *with derivative the equivalence class of* $-A(u^{(m)}, \cdot) + G_m$, *where* G_m *is as in* (2.17).

(e) *Parts* (b) *and* (c) *of* Theorem 2.5 *hold as stated.*

(f) *Given* $\bar{t} \in [0,T)$ *there is a constant* $C(\bar{t})$ *depending on upper bounds for* $C_A(\bar{t})$, $\theta(\bar{t})^{-1}$, $\lambda(\bar{t})$, δ^{-1} *and the norms of the inclusions* $\mathcal{H}_{j+1} \subseteq \mathcal{H}_j$, $\mathcal{V}_{j+1} \subseteq \mathcal{V}_j$ *and the* $*$*-operators* $\mathcal{H}_j \to \mathcal{V}_j$, *but independent of* β, F *and* $(\varsigma_0, \dots, \varsigma_m)$, *such that*

$$\begin{aligned}
&\sum_{j=0}^{m} \Big(\sup_{0 \le t \le \bar{t}} |u^{(j)}(t)|^2_{\mathcal{H}_{2m-2j}} + \int_0^{\bar{t}} |u^{(j)}|^2_{\mathcal{H}_{2m-2j+1}} dt \Big) \\
&\qquad\qquad + \int_0^{\bar{t}} |(u^{(m)})^*|^2_{\mathcal{H}_1^*} \\
&\le C(\bar{t}) e^{\int_0^{\bar{t}} \beta^2} \Big[\sum_{j=0}^{m} |\varsigma_j|^2_{\mathcal{H}_{2m-2j}} + \sum_{j=0}^{m-1} \sup_{0 \le t \le \bar{t}} |F^{(j)}(t)|^2_{\mathcal{V}_{2m-2j-2}} \\
&\qquad\qquad + \sum_{j=0}^{m} \int_0^{\bar{t}} |F^{(j)}|^2_{\mathcal{V}_{2m-2j-1}} \Big].
\end{aligned} \tag{2.28}$$

(g) *Statement* (d*i*) *can be strengthened to* $u^{(j)} \in C([0,T); \mathcal{H}_{2m-2j})$, *that is, continuity in the norm topology, under the additional hypotheses that* $F^{(j)} \in C([0,T); \mathcal{V}_{\mathcal{H}_{2m-2j-2}})$ *for* $j = 0, \dots, m-1$; *that* $A^{(i)}(z, \cdot) \in C([0,T); \mathcal{V}_l)$ *for* $i = 0, \dots, m-1$, $l = 0, \dots, 2m-2i-2$ *and* $z \in \mathcal{H}_{2i+l+2}$; *and that* A *satisfies the ellipticity property described in* (b), *that is, if* $A(z,v) + \lambda' \langle z, v \rangle_{\mathcal{H}_0} = G \cdot v$ *for all* $v \in \mathcal{H}_1$, *where* $z \in \mathcal{H}_1$ *and* $G \in \mathcal{V}_j$, *then* $z \in \mathcal{H}_{j+2}$ *and* $|z|_{\mathcal{H}_{j+2}} \le C_A(\bar{t}) |G|_{\mathcal{V}_j}$.

The proof of Theorem 2.7 is given in the next section. We conclude with several remarks:

Remarks:

(1) The special case that $\mathcal{V}_j = \mathcal{H}_1^*$ for $j \ge -1$ and $\mathcal{H} = \mathcal{H}_1$ for $j \ge 1$ is included in the hypotheses of the theorem. Theorem 2.5 is therefore a special case of Theorem 2.7.

(2) Typical realizations of the Hilbert space array described in the hypotheses, and the examples of most interest in Chapters 5 and 6, are those in which $\mathcal{H}_0 = [L^2(\Omega)]^N$, where Ω is an open set in $\mathbf{R}^n$ and $N \geq 1$, $\mathcal{H}_1$ is a closed subspace of $[H^1(\Omega)]^N$ whose elements satisfy given "essential" boundary conditions, $\mathcal{H}_j = H^j(\Omega) \cap \mathcal{H}_1$, $j \geq 2$, and $\mathcal{V}_j$ is the set of dual elements F given by

$$F \cdot v = \int_\Omega (g \cdot v + G : \nabla_x v)\, dx + \int_{\partial\Omega} \Gamma \cdot v\, d\sigma, \; v \in \mathcal{H}_1$$

where $g \in [H^j(\Omega)]^N$, $G \in [H^{j+1}(\Omega)]^{N\times n}$, Γ satisfies a certain H^{j+1} regularity condition, and σ denotes surface measure on $\partial\Omega$ (the boundary integral is omitted if $\Omega = \mathbf{R}^n$). See Definition 5.1 and the discussion which follows.

(3) The awkwardness of hypothesis (b) is mitigated somewhat by the fact that, for the applications of interest in Chapter 5, this requirement is satisfied automatically–that is, the required conditions for the elliptic regularity hypothesis (b) are inherited by A_ε from those assumed for A nearly by inspection. On the other hand, it can be proved that if A satisfies (b), then so too does A_ε provided that for each $\bar{t} \in [0,T)$ there is a constant C and a continuous function $\omega(\cdot;\bar{t}) : [0,\infty) \to [0,\infty)$ with $\omega(0;\bar{t}) = 0$ such that $|A(t)(z,\cdot)|_{\mathcal{V}_j} \leq C|z|_{\mathcal{H}_{j+2}}$ and

$$|(A(t) - A(s))(z,\cdot)|_{\mathcal{V}_j} \leq \omega(|t-s|;\bar{t})|z|_{\mathcal{H}_{j+2}}$$

for $t, s \in [0,\bar{t}]$, $j = 0,\ldots,2m-1$ and $z \in \mathcal{H}_{j+2}$; see Exercise 2.10. This hypothesis is not required elsewhere and so is not made here.

(4) Observe that the compatibility condition (2.16) makes sense here because $F^{(j)}$ and $A^{(i)}(\varsigma_{j-i},\cdot)$ have definite values at $t=0$: for $F^{(j)}$ by hypothesis (since $j \leq m-1$), and for $A^{(i)}(\varsigma_{j-i},\cdot)$ by hypothesis (cii) and the fact that $\varsigma_{j-i} \in \mathcal{H}_{2m-2j+2i} \subseteq \mathcal{H}_{2i+2}$.

(5) More explicit bounds for the solution and its derivatives can be obtained if necessary by applying the results (2.4) and (2.5) of Theorem 2.1 to the equation (2.18) satisfied by $u^{(j)}$, first for $j=0$, then proceeding inductively via hypothesis (b) and Gronwall's inequality. See the application in the proof of Theorem 6.14 for example.

§2.4 Proof of Theorem 2.7

We begin with two preliminary lemmas, the first showing that the approximations $A_\varepsilon^{(i)}$ inherit the essential properties of $A^{(i)}$ and converge to $A^{(i)}$ in a suitable sense as $\varepsilon \to 0$:

Lemma 2.8. *Assume that the hypotheses and notations of* Theorem 2.7 *are in force and let* $A^{(i)}_\varepsilon = (A^{(i)})_\varepsilon$ *be the approximations to* $A^{(i)}$ *defined in* Definition 2.6. *Then:*

(a) $A^{(0)}_\varepsilon, \ldots, A^{(m)}_\varepsilon$ *satisfy hypotheses* (a) and (c) *of* Theorem 2.7 *with* T, $C_A(\bar{t})$, $\lambda(\bar{t})$ *and* $\theta(\bar{t})$ *replaced respectively by* $T - 2\varepsilon$, $C_A(\bar{t}+2\varepsilon)$, $\lambda(\bar{t}+2\varepsilon)$ *and* $\theta(\bar{t}+2\varepsilon)$, *and* $\beta(t)$ *replaced by* $\beta_\varepsilon(t) = \int_{|s-t|\le\varepsilon} h_\varepsilon(t-s)\beta(s+\varepsilon)ds$.

(b) *For* $i = 0, \ldots, m$ *and* $\bar{t} \in [0,T)$: *if a sequence* z_{ε_k} *converges weakly in* $L^2([0,\bar{t}];\mathcal{H}_{2i+1})$ *to* z *as* $\varepsilon_k \to 0$, *then for all* $v \in \mathcal{H}_1$,

$$\int_0^{\bar{t}} A^{(i)}_{\varepsilon_k}(t)(z_{\varepsilon_k}(t), v)\, dt \to \int_0^{\bar{t}} A^{(i)}(t)(z(t), v)\, dt$$

as $\varepsilon_k \to 0$.

Proof. The proofs that the $A^{(i)}_\varepsilon$ satisfy (a) and (c*i*)–(c*iii*) of Theorem 2.7 are straightforward: for example, that of (c*iii*) requires only Fubini's theorem and shows in addition that for $i \le m-1$, $A^{(i)}_\varepsilon$ is weakly absolutely continuous as a mapping into $\mathcal{B}(\mathcal{H}_{2i+3}\times\mathcal{H}_1)$ with derivative $[A^{(i+1)}_\varepsilon]$. Thus in this context, differentiation and mollifying commute.

To prove that the $A^{(i)}_\varepsilon$ satisfy property (c*iv*) of Theorem 2.7 we let i and l be as in the statement, $z \in \mathcal{H}_{2i+l+2}$, $v \in \mathcal{H}_1$ and $t \le \bar{t} < T - 2\varepsilon$. Then since $A^{(i)}(s+\varepsilon)(z,v)$ is a continuous function of s by hypothesis, the integral defining $A^{(i)}_\varepsilon(t)(z,v)$ is the limit of Riemann sums in which Δs is a positive increment and $s_k = k\Delta s$:

$$\begin{aligned} A^{(i)}_\varepsilon(t)(z,v) &= \int_{|s|\le\varepsilon} h_\varepsilon(t-s)A^{(i)}(s+\varepsilon)(z,v)ds \\ &= \lim_{\Delta s\to 0} \sum_k h_\varepsilon(t-s_k)A^{(i)}(s_k+\varepsilon)(z,v)\Delta s \\ &= \lim_{\Delta s\to 0} \Big[\sum_k h_\varepsilon(t-s_k)A^{(i)}(s_k+\varepsilon)(z,\cdot)\Delta s\Big]\cdot v. \end{aligned}$$

This shows that the linear functionals $G_{\Delta s}$ in the brackets are weak-$*$ convergent to $A^{(i)}_\varepsilon(t)(z,\cdot)$ in $\mathcal{H}_1^*$ as $\Delta s \to 0$. Also, $G_{\Delta s} \in \mathcal{V}_l$ by property (c*iv*) assumed for $A^{(i)}$, and

$$\begin{aligned} |G_{\Delta s}|_{\mathcal{V}_l} &\le \Big(\sum_k h_\varepsilon(t-s_k)\Delta s\Big)C_A(\bar{t}+2\varepsilon)|z|^\delta_{\mathcal{V}_{2i+l}}|z|^{1-\delta}_{\mathcal{V}_{2i+l+2}} \\ &\to C_A(\bar{t}+2\varepsilon)|z|^\delta_{\mathcal{V}_{2i+l}}|z|^{1-\delta}_{\mathcal{V}_{2i+l+2}} \end{aligned}$$

as $\Delta s \to 0$. These facts suffice to show that $A^{(i)}(t)(z,\cdot) \in \mathcal{V}_l$ with norm bounded as in (c*iv*).

Next we prove (b) of the lemma. Thus let i, $\bar{t}$, z_{ε_k} and z be as in the statement, where without loss of generality $\bar{t} + 2\varepsilon_k \le \bar{t}' < T$ for all k. Then for $v \in \mathcal{H}_1$,

$$\begin{aligned}\int_0^{\bar{t}} A^{(i)}_{\varepsilon_k}(t)(z_{\varepsilon_k}(t),v)dt - \int_0^{\bar{t}} A^{(i)}(t)(z(t),v)dt \\ = \int_0^{\bar{t}} (A^{(i)}_{\varepsilon_k} - A^{(i)})(z_{\varepsilon_k},v)dt + \int_0^{\bar{t}} A^{(i)}(z_{\varepsilon_k} - z,v)dt.\end{aligned} \tag{2.29}$$

For the second term on the right recall from (a) and (ci) of Theorem 2.7 that $[A^{(i)}] \in L^2_{loc}([0,T);\mathcal{B}(\mathcal{H}_{2i+1}\times\mathcal{H}_1))$, so that for almost all t there is a map $S(t) \in \mathcal{L}(\mathcal{H}_{2i+1},\mathcal{H}_1)$ such that $A^{(i)}(t)(w,v) = \langle S(t)w,v\rangle_{\mathcal{H}_1}$ for all $(w,v) \in \mathcal{H}_{2i+1}\times\mathcal{H}_1$, and $\|S(t)\|_{\mathcal{L}(\mathcal{H}_{2i+1},\mathcal{H}_1)} = \|A^{(i)}(t)\|_{\mathcal{H}_{2i+1}\times\mathcal{H}_1} \in L^2_{loc}([0,T))$. It follows that $[S^*v] \in L^2_{loc}([0,T);\mathcal{H}_{2i+1})$ and therefore that the second integral on the right in (2.29) is the inner product $\langle z_{\varepsilon_k} - z, S^*v\rangle_{L^2([0,\bar{t}];\mathcal{H}_{2i+1})}$, which goes to zero by the weak convergence of z_{ε_k} to z.

For the first integral on the right side of (2.29) we fix $v \in \mathcal{H}_1$ and define weak-$*$ measurable mappings G, $G_{\varepsilon_k} : [0,\bar{t}'] \to \mathcal{H}^*_{2i+1}$ (different from and unrelated to the previous G and $G_{\Delta s}$) by $G(t)\cdot w = A^{(i)}(t)(w,v)$, and similarly for G_{ε_k}. The equivalence classes of G and G_{ε_k} are then in $L^2_{loc}([0,T);\mathcal{H}^*_{2i+1})$ and

$$G_{\varepsilon_k}(t)\cdot w = \int_{|s-t|\le\varepsilon_k} h_{\varepsilon_k}(t-s)G(s+\varepsilon_k)\cdot w\,ds \tag{2.30}$$

for $w \in \mathcal{H}_{2i+1}$ and all t. The first integral on the right side of (2.29) is then bounded by

$$\int_0^{\bar{t}} |G_{\varepsilon_k} - G|_{\mathcal{H}^*_{2i+1}} |z_{\varepsilon_k}|_{\mathcal{H}_{2i+1}} |v|_{\mathcal{H}_1} dt \le C\Big(\int_0^{\bar{t}} |G_{\varepsilon_k} - G|^2_{\mathcal{H}^*_{2i+1}} dt\Big)^{1/2},$$

where C is independent of k (note that the weakly convergent sequence $\{z_{\varepsilon_k}\}$ is bounded in $L^2([0,\bar{t}];\mathcal{H}_{2i+1})$). To show that this goes to zero as $\varepsilon_k \to 0$ we let $\{\varphi_j\}$ be an orthonormal basis for $\mathcal{H}_{2i+1}$ and apply the fact that for any $g \in \mathcal{H}^*_{2i+1}$, $|g|^2_{\mathcal{H}^*_{2i+1}} = \sum_j (g\cdot\varphi_j)^2$, as the reader can easily check. Then if $J < J'$,

$$\begin{aligned}\int_0^{\bar{t}} \sum_1^{J'} |(G_{\varepsilon_k} - G)\cdot\varphi_j|^2 dt &\le \sum_1^{J} \int_0^{\bar{t}} |(G_{\varepsilon_k} - G)\cdot\varphi_j|^2 dt \\ &\quad + \int_0^{\bar{t}} \sum_{J+1}^{J'} \big(|G_{\varepsilon_k}\cdot\varphi_j|^2 + |G\cdot\varphi_j|^2\big) dt \\ &\le \sum_1^{J} \int_0^{\bar{t}} |(G_{\varepsilon_k} - G)\cdot\varphi_j|^2 dt + 2\int_0^{\bar{t}'} \sum_{J+1}^{\infty} |G\cdot\varphi_j|^2 dt\end{aligned}$$

by (2.30). The second term on the right goes to zero as $J \to \infty$ by the dominated convergence theorem because the integrand goes to zero pointwise in t and is bounded by $|G(\cdot)|^2_{\mathcal{H}^*_{2i+1}}$, which is integrable. Then for fixed J the first term on the right goes to zero as $\varepsilon_k \to 0$ by standard properties of mollifiers of real-valued functions and by (2.30), which shows that $G_{\varepsilon_k} \cdot \varphi_j$ is the mollifier of the real-valued function $G \cdot \varphi_j$, whose equivalence class is in $L^2([0,t])$. This shows that the first integral on the right side of (2.29) goes to zero as $\varepsilon_k \to 0$ and completes the proof of (b).

□

The second preliminary result, which concerns the algebraic solvability of the systems (2.16) and (2.19), will be applied at two crucial points in the proof of Theorem 2.7:

Lemma 2.9. *Let $\mathcal{H}_j$ and $\mathcal{V}_j$ be the Hilbert spaces of* Theorem 2.7 *and $B \in \mathcal{B}(\mathcal{H}_1 \times \mathcal{H}_1)$. Assume that there are numbers $C_B > 0$, $\theta_B > 0$, $\lambda_B \geq 0$ and $\delta_B \in (0,1]$ such that the following hold:*

(a) *$B(z,z) \geq \theta_B |z|^2_{\mathcal{H}_1} - \lambda_B |z|^2_{\mathcal{H}_0}$ for all $z \in \mathcal{H}_1$.*

(b) *For $j = -1, \ldots, 2m-1$ and $\lambda' \geq \lambda_B$: if for some $z \in \mathcal{H}_1$ and $G \in \mathcal{V}_j$, $B(z,v) + \lambda' \langle z, v \rangle_{\mathcal{H}_0} = G \cdot v$ for all $v \in \mathcal{H}_1$, then $z \in \mathcal{H}_{j+2}$ and $|z|_{\mathcal{H}_{j+2}} \leq C_B |G|_{\mathcal{V}_j}$.*

(c) *If $m \geq 2$ then for $i = 1, \ldots, m-1$ there are bilinear forms $B^{(i)} \in \mathcal{B}(\mathcal{H}_{2i+1} \times \mathcal{H}_1)$ such that if $0 \leq l \leq 2m - 2i - 1$ and $z \in \mathcal{H}_{2i+l+2}$, then $B^{(i)}(z, \cdot) \in \mathcal{V}_l$ and*

$$|B^{(i)}(z,\cdot)|_{\mathcal{V}_l} \leq C_B |z|^{\delta_B}_{\mathcal{H}_{2i+l}} |z|^{1-\delta_B}_{\mathcal{H}_{2i+l+2}}.$$

Let σ be either 0 or 1. Then given

$$f = (f_0 \ldots, f_{m-1}) \in Y_\sigma \equiv \mathcal{V}_{2m-2+\sigma} \times \cdots \times \mathcal{V}_\sigma$$

and $w_m \in \mathcal{H}_\sigma$, there is a unique

$$w = (w_0, \ldots, w_{m-1}) \in X_\sigma \equiv \mathcal{H}_{2m+\sigma} \times \cdots \times \mathcal{H}_{2+\sigma}$$

such that for $j = 0, \ldots, m-1$,

$$\langle w_{j+1}, v \rangle_{\mathcal{H}_0} + \sum_{i=0}^{j} \binom{j}{i} B^{(i)}(w_{j-i}, v) = f_j \cdot v \tag{2.31}$$

for all $v \in \mathcal{H}_1$, where $B^{(0)} = B$. Furthermore, there is a constant C depending on upper bounds for m, C_B, θ_B^{-1}, λ_B, $\delta_B{}^{-1}$ and the norms of the inclusions and the $*$-maps in the Hilbert space array in the statement of Theorem 2.7 *but independent of w_m and the f_j, such that*

$$\sum_{j=0}^{m-1} |w_j|_{\mathcal{H}_{2m-2j+\sigma}} \leq C\Big[|w_m|_{\mathcal{H}_\sigma} + \sum_{j=0}^{m-1} |f_j|_{\mathcal{V}_{2m-2j-2+\sigma}}\Big]. \tag{2.32}$$

Finally, if every $B^{(i)} \in \mathcal{B}(\mathcal{H}_1 \times \mathcal{H}_1)$, then the solution w is unique in the space $[\mathcal{H}_1]^m$.

PROOF. First, it suffices to prove that there is a number $\lambda \geq \lambda_B$ determined by the same quantities as the constant C in (2.32) such that the conclusions of the lemma hold with B replaced by $B_\lambda \equiv B + \lambda\langle\cdot,\cdot\rangle_{\mathcal{H}_0}$ and $B^{(1)}, \ldots, B^{(m-1)}$ unchanged. The proof of this assertion is a discrete version of the change of variables in Exercise 2.1. Specifically, we define for given $\lambda \in \mathbf{R}$ a map $\mathcal{T}_\lambda : X_\sigma \to X_\sigma$ by

$$(\mathcal{T}_\lambda(w))_j = \sum_{i=0}^{j} \binom{j}{i} \lambda^i w_{j-i}$$

for $j = 0, \ldots, m-1$. Abusing notation slightly, we also denote by $\mathcal{T}_\lambda$ the map from Y_σ to Y_σ defined by exactly the same formula. We then claim first that the composition $\mathcal{T}_\lambda \circ \mathcal{T}_{-\lambda}$ is the identity map and second that w satisfies the equations (2.31) if and only $\mathcal{T}_{-\lambda} w$ satisfies the corresponding system in which f is replaced by $\mathcal{T}_{-\lambda} f$, w_m by $\sum_{i=0}^{m} \binom{m}{i} (-\lambda)^i w_{m-i}$, B by B_λ, and $B^{(1)}, \ldots, B^{(m-1)}$ unchanged. The proofs of these assertions involve little more than algebraic manipulations of binomial coefficients and so are omitted. Once proven, these facts allow us to change variables, solve the amended problem, and then invert the change of variables to recover a solution of the original system (2.31). Observe that hypotheses (b) and (c) will hold for the amended problem with the same constant C_B independently of $\lambda \geq \lambda_B$ and that

$$B_\lambda(z,z) \geq \theta_B |z|^2_{\mathcal{H}_1} + \lambda |z|^2_{\mathcal{H}_0}, \quad z \in \mathcal{H}_1, \tag{2.33}$$

(we have taken the now inconsequential constant λ_B to be zero). In what follows λ will be a free parameter to be chosen later and C will denote a generic positive constant as described in the conclusion of the lemma but which is independent of λ.

Given w_m and f as in the statement we proceed to solve (2.31) by iteration. Thus define $S : X_\sigma \to X_\sigma$ by $z = Sw$, where for $j = 0, \ldots, m-1$ and all $v \in \mathcal{H}_1$,

$$\langle z_{j+1}, v\rangle_{\mathcal{H}_0} + B_\lambda(z_j, v) = -\sum_{i=1}^{j} \binom{j}{i} B^{(i)}(w_{j-i}, v) + f_j \cdot v,$$

where $z_m = w_m$ and the sum is zero for $j = 0$. The existence of a unique solution $z \in X_\sigma$ follows easily by induction, starting with $j = m-1$, from hypothesis (b) and the observation that $B^{(i)}(w_{j-i}, \cdot)$ and f_j are in $\mathcal{V}_{2m-2j-2+\sigma}$ by (c) and our assumptions on f_j.

The next step is to construct a metric on X_σ in which S is contractive. To do this we let $w', w'' \in X_\sigma$, $z' = S(w')$ and $z'' = S(w'')$, and $w = w'' - w'$ and $z = z'' - z'$. Then for $j = 0, \ldots, m-1$,

$$\langle z_{j+1}, \cdot\rangle_{\mathcal{H}_0} + B_\lambda(z_j, \cdot) = -\sum_{i-1}^{j} \binom{j}{i} B^{(i)}(w_{j-i}, \cdot) \tag{2.34}$$

where $z_m = 0$ and the right side is zero if $j = 0$. Two preliminary bounds are required. We claim first that

$$\sum_{i=0}^{m-1} |z_i|_{\mathcal{H}_\sigma} \leq C\lambda^{(\sigma-1)/2} M \tag{2.35}$$

where

$$M = \sum_{i=0}^{m-1} |w_i|_{\mathcal{H}_{2m-2i}}$$

(which hints at contractiveness for $\sigma = 0$ and λ large). To prove this we apply (2.33) in (2.34) to get that

$$\theta_B |z_j|^2_{\mathcal{H}_1} + \lambda |z_j|^2_{\mathcal{H}_0} \leq |\langle z_{j+1}, z_j\rangle_{\mathcal{H}_0}| + \sum_{i=1}^{j} \binom{j}{i} |B^{(i)}(w_{j-i}, z_j)|. \tag{2.36}$$

For the $B^{(i)}$ term on the right we have $1 \leq i \leq j \leq m-1$ so that if $l = 2m - 2j - 2$, then $0 \leq l \leq 2m - 2i - 1$ as in hypothesis (c), and $2i + l + 2 = 2m - 2(j - i)$. Therefore

$$\begin{aligned} |B^{(i)}(w_{j-i}, z_j)| &\leq C|B^{(i)}(w_{j-i}, \cdot)|_{\mathcal{V}_l} |z_j|_{\mathcal{H}_1} \\ &\leq C|w_{j-i}|_{\mathcal{H}_{2m-2(j-i)}} |z_j|_{\mathcal{H}_1} \leq C\eta^{-1} M^2 + \eta |z_j|^2_{\mathcal{H}_1} \end{aligned}$$

for arbitrary $\eta > 0$. Applying this bound in (2.36) we thus obtain

$$|z_j|^2_{\mathcal{H}_1} + \lambda |z_j|^2_{\mathcal{H}_0} \leq C\big(|z_{j+1}|^2_{\mathcal{H}_0} + M^2\big).$$

The $\sigma = 0$ case of (2.35) now follows by induction on j starting with $j = m-1$ and the fact that $z_m = 0$, and the $\sigma = 1$ case follows immediately from the $\sigma = 0$ case. This proves (2.35).

Next we define norms

$$\langle v\rangle^{(j)}_\sigma \equiv \sum_{i=0}^{j\wedge m-1} |v_i|_{\mathcal{H}_{2j-2i+\sigma}}$$

for $j = 0, \ldots, m$ and $v = (v_0, \ldots, v_{m-1}) \in X_\sigma$ (the upper limit here is the smaller of j and $m - 1$). We claim that for $j = 1, \ldots, m$ and z and w as in (2.34),

$$\langle z\rangle^{(j)}_\sigma \leq C\left[\lambda^{(\sigma-1)/2} M + (\langle w\rangle^{(j-1)}_\sigma)^{\delta_B} (\langle w\rangle^{(j)}_\sigma)^{1-\delta_B}\right] \tag{2.37}$$

for a new constant C, still independent of λ. The proof is a double induction, as follows. First, as a notational convenience set $\langle w\rangle^{(-1)}_\sigma = 0$ so that (2.37) holds for $j = 0$ by (2.35). Next fix $j \geq 1$ and assume that (2.37) holds with $j - 1$ in place of j. We will show that $|z_{j-k}|_{\mathcal{H}_{2k+\sigma}}$ is bounded by the right side of (2.37) by induction on $k = 0, \ldots, j$. This is true for $k = 0$ by (2.35), and if it holds for a given $k \leq j-1$, then by hypotheses (b) and (c) applied in (2.34) with j replaced by $j-(k+1)$ and with $l = 2k+\sigma$,

$$|z_{j-(k+1)}|_{\mathcal{H}_{2(k+1)+\sigma}} \leq C\Big[|z_{j-k}|_{\mathcal{H}_{2k+\sigma}} + \sum_{i=1}^{j-(k+1)} |B^{(i)}(w_{j-(k+1)-i},\cdot)|_{\mathcal{V}_{2k+\sigma}}\Big]$$
$$\leq C\big[|z_{j-k}|_{\mathcal{H}_{2k+\sigma}} + |w_{(j-1)-(k+i)}|^{\delta_B}_{\mathcal{H}_{2(k+i)+\sigma}} |w_{j-(k+1+i)}|^{1-\delta_B}_{\mathcal{H}_{2(k+1+i)+\sigma}}\big],$$

which by the induction hypothesis on k is bounded by the right side of (2.37). This completes the induction on k and proves that $\langle z\rangle_\sigma^{(j)}$ is bounded by the quantity on the right side of (2.37); and this in turn completes the induction on j, which proves (2.37) for all j.

We now fix the constant C in (2.37), which we recall is independent of λ, and construct a norm on X_σ in which S is contractive. Given any positive numbers $c_0, \ldots, c_m$ we have from (2.37) and our convention that $\langle w\rangle_\sigma^{(-1)} = 0$ that for the same z and w,

$$\sum_{j=0}^{m} c_j\langle z\rangle_\sigma^{(j)} \leq C\lambda^{(\sigma-1)/2} M \sum_j c_j$$
$$+C\sum_{j=1}^{m}\Big(\frac{c_j}{c_{j-1}}\Big)^{\delta_B} \big(c_{j-1}\langle w\rangle_\sigma^{(j-1)}\big)^{\delta_B}\big(c_j\langle w\rangle_\sigma^{(j)}\big)^{1-\delta_B}$$
$$\leq C\Big(\lambda^{(\sigma-1)/2}c_m^{-1}\sum_j c_j\Big)c_m\langle w\rangle_0^{(m)} + C\max_j\Big(\frac{c_j}{c_{j-1}}\Big)^{\delta_B}\sum_{j=0}^{m} c_j\langle w\rangle_\sigma^{(j)}$$

since $M = \langle w\rangle_0^{(m)}$. The coefficient of the second sum on the right can be made arbitrarily small by choosing the ratios c_j/c_{j-1} sufficiently small depending on C and δ_B, which are independent of λ. Then for the case $\sigma = 0$ the coefficient of the term $c_m\langle w\rangle_0^{(m)}$ can be made arbitrarily small by choosing λ sufficiently large. We can therefore choose the c_j's and λ so that if $\|z\|_\sigma$ is defined to be the sum on the left above, then

$$\|z\|_0 \leq \tfrac{1}{2}\|w\|_0 \text{ and } \|z\|_1 \leq C\|w\|_0 + \tfrac{1}{2}\|w\|_1$$

for another constant C. The first bound here shows that S is a contraction on X_0, which proves the existence and uniqueness assertion for $\sigma = 0$; and the two bounds taken together show that S is a contraction on X_1 in the norm $\|\cdot\|_0 + \eta\|\cdot\|_1$ for η sufficiently small, which proves existence and uniqueness for $\sigma = 1$.

To prove the bound in (2.32) we let $\tilde{z} = S(0)$, so that

$$\langle \tilde{z}_{j+1}, \cdot\rangle_{\mathcal{H}_0} + B_\lambda(\tilde{z}_j, \cdot) = f_j$$

for $j = 0, \ldots, m-1$, where $\tilde{z}_m = w_m$. It follows easily by induction on j starting with $j = m-1$ that each $|\tilde{z}_j|_{\mathcal{H}_{2m-2j+\sigma}}$ is bounded by the right side of (2.32), hence so is $\|\tilde{z}\|_\sigma$. Then if w is the fixed point of S and $\|\cdot\|$ is either of the contractive norms constructed above, $\|w\| \le \|S(w) - S(0)\| + \|\tilde{z}\|$, so that $\|w\|$ is bounded by $C\|\tilde{z}\|$ and therefore also by the right side of (2.32).

Finally, to prove the last statement of the lemma we consider the Hilbert space array in which each $\mathcal{V}_j$ is replaced by $\mathcal{H}_1^*$ and $\mathcal{H}_j$ is replaced by $\mathcal{H}_1$ for $j \ge 1$ and then check that the other hypotheses of the lemma still hold. The above analysis therefore applies to show that solutions in $[\mathcal{H}_1]^m$ are unique.

□

We can now apply the results of Theorem 2.5 and Lemmas 2.8 and 2.9 to prove Theorem 2.7. Briefly, we replace A by the approximations A_ε of Definition 2.6 and the initial data ς_0 by corresponding approximate data ς_0^ε to obtain approximate solutions u_ε via Theorem 2.5. Lemma 2.9 will then show that the u_ε satisfy the required regularity independent of ε, and Lemma 2.8(b) will enable us to extract a limiting solution which retains this regularity.

PROOF of Theorem 2.7. We let A_ε be the approximations to A defined in Definition 2.6 and leave to the reader to check that Lemma 2.8 guarantees first that each A_ε satisfies the hypotheses of Theorem 2.5 and second that the hypotheses of Lemma 2.9 hold with $B^{(i)} = A_\varepsilon^{(i)}(0)$. Consequently the given data ς_m and $F^{(0)}(0), \ldots, F^{(m-1)}(0)$ uniquely determine approximate data $(\varsigma_0^\varepsilon, \ldots, \varsigma_{m-1}^\varepsilon) \in \mathcal{H}_{2m} \times \ldots \times \mathcal{H}_2$ satisfying the system of equations

$$\text{(2.38)} \qquad \langle \varsigma_{j+1}^\varepsilon, v\rangle_{\mathcal{H}_0} = -\sum_{i=0}^{j} \binom{j}{i} A_\varepsilon^{(i)}(0)(\varsigma_{j-i}^\varepsilon, v) + F^{(j)}(0)\cdot v$$

for $j = 0, \ldots, m-1$ and all $v \in \mathcal{H}_1$, where $\varsigma_m^\varepsilon = \varsigma_m$. Lemma 2.9 also shows that $(\varsigma_0^\varepsilon, \ldots, \varsigma_{m-1}^\varepsilon)$ is bounded in $\mathcal{H}_{2m} \times \ldots \times \mathcal{H}_2$ independently of ε. There is therefore a sequence $\varepsilon \to 0$ such that ς_j^ε converges weakly in $\mathcal{H}_{2m-2j}$, say to $\tilde{\varsigma}_j$, for each $j = 0, \ldots, m-1$. We claim that $\tilde{\varsigma}_j = \varsigma_j$ for these j. To prove this let $\varepsilon \to 0$ in (2.38) above: convergence of the term on the left is obvious and for the summand on the right we have that at $t = 0$,

$$A_\varepsilon^{(i)}(\varsigma_{j-i}^\varepsilon, v) = (A_\varepsilon^{(i)} - A^{(i)})(\varsigma_{j-i}^\varepsilon, v) + A^{(i)}(\varsigma_{j-i}^\varepsilon, v).$$

The second term on the right converges to $A^{(i)}(\tilde{\varsigma}_{j-i}, v)$ because $\varsigma_{j-i}^\varepsilon$ converges weakly in $\mathcal{H}_{2m-2j+2i} \subseteq \mathcal{H}_{2i+2}$ and $A^{(i)}(0) \in \mathcal{B}(\mathcal{H}_{2i+2} \times \mathcal{H}_1)$; and the first term on the right is bounded by $\|A_\varepsilon^{(i)}(0) - A^{(i)}(0)\|_{\mathcal{H}_{2i+2}\times\mathcal{H}_1} |\varsigma_{j-i}^\varepsilon|_{\mathcal{H}_{2i+2}}$, which goes to zero because $A^{(i)} \in C([0,T]; \mathcal{B}(\mathcal{H}_{2i+2} \times \mathcal{H}_1))$. This proves that $(\tilde{\varsigma}_0, \ldots, \tilde{\varsigma}_{m-1})$ satisfies (2.16), which is the compatibility system satisfied by $(\varsigma_0, \ldots, \varsigma_{m-1})$. The uniqueness assertion in Lemma 2.9 therefore applies to show that $\tilde{\varsigma}_j = \varsigma_j$ for each j, as claimed. Thus

$$\text{(2.39)} \qquad \varsigma_j^\varepsilon \rightharpoonup \varsigma_j \text{ weakly in } \mathcal{H}_{2m-2j}$$

for $j = 0, \ldots, m-1$.

We have now checked the hypotheses of Theorem 2.5 for the problem

$$\langle u_\varepsilon(t), v\rangle_{\mathcal{H}_0} = \langle \varsigma_0^\varepsilon, v\rangle_{\mathcal{H}_0} + \int_0^t [-A_\varepsilon(u_\varepsilon(s), v) + F(s)\cdot v]\, ds;$$

there is therefore a unique solution u_ε satisfying its conclusions. In particular, $u_\varepsilon = u_\varepsilon^{(0)}, \ldots, u_\varepsilon^{(m-1)} \in wAC^2([0,T);\mathcal{H}_1)$ with respective derivatives $[u_\varepsilon^{(1)}], \ldots, [u_\varepsilon^{(m)}]$ and

$$\langle u_\varepsilon^{(j+1)}(t), v\rangle_{\mathcal{H}_0} = -\sum_{i=0}^{j} \binom{j}{i} A_\varepsilon^{(i)}(t)(u_\varepsilon^{(j-i)}(t), v) + F^{(j)}(t)\cdot v$$

for each $t \in [0,T)$ and $j = 0, \ldots, m-1$. This is a special case of the system (2.31) of Lemma 2.9, and since $u_\varepsilon^{(m)}(t)$ is in $\mathcal{H}_\sigma$ and $(F^{(0)}(t), \ldots, F^{(m-1)})(t) \in (\mathcal{V}_{2m-2+\sigma} \times \ldots \times \mathcal{V}_\sigma)$ for all t if $\sigma = 0$ and almost all t if $\sigma = 1$, we conclude from Lemma 2.9 that $u_\varepsilon, \ldots, u_\varepsilon^{(m-1)} \in \mathcal{H}_{2m} \times \ldots, \mathcal{H}_2$ for all t and in $\mathcal{H}_{2m+1} \times \ldots \mathcal{H}_3$ for almost all t. Also, (2.32) shows that for each $\bar{t} \in (0,T)$ there is a positive constant $C(\bar{t})$ as described in Theorem 2.7(f) which is independent of ε such that

$$\begin{aligned}\text{(2.40)}\quad &\sum_{j=0}^{m-1} \Big[\sup_{0\le t\le \bar{t}} |u_\varepsilon^{(j)}(t)|^2_{\mathcal{H}_{2m-2j}} + \int_0^{\bar{t}} |u_\varepsilon^{(j)}|^2_{\mathcal{H}_{2m-2j+1}} dt\Big] \\ &\qquad \le C(\bar{t})\Big[\sup_{0\le t\le \bar{t}} |u_\varepsilon^{(m)}(t)|^2_{\mathcal{H}_0} + \int_0^{\bar{t}} |u_\varepsilon^{(m)}|^2_{\mathcal{H}_1} dt + M_{F,\{\varsigma_j\}}(\bar{t})\Big]\end{aligned}$$

where $M_{F,\{\varsigma_j\}}(\bar{t})$ is the quantity in brackets on the right side of (2.28). To bound the $u_\varepsilon^{(m)}$ terms on the right side of (2.40) we apply the conclusion (2.4) of Theorem 2.1 to the problem (2.18) with $j = m$ and each $A^{(i)}$ replaced by $A_\varepsilon^{(i)}$, and then apply hypothesis Theorem 2.7(ci), which by Lemma 2.8(a) is satisfied by $A_\varepsilon^{(i)}$:

$$\begin{aligned}&\sup_{0\le t\le \bar{t}} |u_\varepsilon^{(m)}(t)|^2_{\mathcal{H}_0} + \int_0^{\bar{t}} |u_\varepsilon^{(m)}|^2_{\mathcal{H}_1} \\ &\quad \le C(\bar{t})\Big[M_{F,\{\varsigma_j\}}(\bar{t}) + \sum_{i=1}^{m} \int_0^{\bar{t}} |A_\varepsilon^{(i)}(u_\varepsilon^{(m-i)}, \cdot)|^2_{\mathcal{V}_{-1}}) dt\Big] \\ &\quad \le C(\bar{t})\Big[M_{F,\{\varsigma_j\}}(\bar{t}) + \sum_{i=1}^{m} \int_0^{\bar{t}} (\beta_\varepsilon |u_\varepsilon^{(m-i)}|_{\mathcal{H}_{2i}})^{2\delta} |u^{(m-i)}|^{2(1-\delta)}_{\mathcal{H}_{2i+1}} dt\Big].\end{aligned}$$

Substituting this into (2.40) we then obtain that, for arbitrary $\eta > 0$,

$$\sum_{j=0}^{m}\Big[\sup_{0\le t\le \bar t}|u_\varepsilon^{(j)}(t)|^2_{\mathcal{H}_{2m-2j}} + \int_0^{\bar t}|u_\varepsilon^{(j)}|^2_{\mathcal{H}_{2m-2j+1}}dt\Big]$$
$$\le C\Big[M_{F,\{\varsigma_j\}}(\bar t) + \sum_{i=1}^{m}\int_0^{\bar t}(\eta^{-1}\beta_\varepsilon^2|u_\varepsilon^{(m-i)}|^2_{\mathcal{H}_{2i}} + \eta|u_\varepsilon^{(m-i)}|^2_{\mathcal{H}_{2i+1}})dt\Big]$$

(observe that the $j = m$ term is now included on the left). The second term in the integrand on the right can be absorbed into the left side by taking η sufficiently small, and the first term can then be eliminated by Gronwall's inequality. The result is that for $\bar t \in (0, T - \varepsilon)$ there is a constant $C(\bar t)$ as described in Theorem 2.7(f) which is independent of β and ε such that

$$\sum_{j=0}^{m}\Big[\sup_{0\le t\le \bar t}|u_\varepsilon^{(j)}(t)|^2_{\mathcal{H}_{2m-2j}} + \int_0^{\bar t}|u_\varepsilon^{(j)}|^2_{\mathcal{H}_{2m-2j+1}}dt\Big] \le C(\bar t)e^{\int_0^{\bar t+\varepsilon}\beta_\varepsilon^2}M_{F,\{\varsigma_j\}}(\bar t). \tag{2.41}$$

It follows that there is a further subsequence $\varepsilon_l \to 0$ such that for $j = 0, \ldots, m$, $[u_{\varepsilon_l}^{(j)}]$ converges weakly to some z_j in $L^2([0,\bar t]; \mathcal{H}_{2m-2j+1})$ for every $\bar t \in [0, T)$. We claim that, for these j, for all t, and for this same subsequence, $u_{\varepsilon_l}^{(j)}(t)$ converges weakly in $\mathcal{H}_{2m-2j}$, say to $\tilde z_j(t)$. The argument is essentially the same as in the proof of Theorem 2.1: the pointwise bound in (2.41) shows that, for given t fixed, the sequence $u_{\varepsilon_l}^{(j)}(t)$ has weak $\mathcal{H}_{2m-2j}$ subsequential limits, and all such limits are shown to be the same. The latter follows because for $v \in \mathcal{H}_1$,

$$\begin{aligned}\langle u_{\varepsilon_l}^{(j)}(t), v\rangle_{\mathcal{H}_0} &= \langle \varsigma_j^{\varepsilon_l}, v\rangle_{\mathcal{H}_0}\\ &+ \int_0^t\Big[-A_{\varepsilon_l}(t)(u_{\varepsilon_l}^{(j)}(t), v) - \sum_{i=1}^{j}\binom{j}{i}A_{\varepsilon_l}^{(i)}(u_{\varepsilon_l}^{(j-i)}, v) + F^{(j)}\cdot v\Big]ds\end{aligned} \tag{2.42}$$

(the sum is absent if $j = 0$). The right side here converges as $\varepsilon_l \to 0$ to

$$\langle \varsigma_j, v\rangle_{\mathcal{H}_0} + \int_0^t\Big[-A(t)(z_j(t), v) - \sum_{i=1}^{j}\binom{j}{i}A^{(i)}(z_{j-i}, v) + F^{(j)}\cdot v\Big]ds$$

by (2.39) and Lemma 2.8(b), and this therefore holds for every further subsequence ε_{l_i} as well. Consequently the difference w between any two such subsequential limits satisfies $\langle w, v\rangle_{\mathcal{H}_0} = 0$ for all $v \in \mathcal{H}_1$, and therefore that $w = 0$ by the density of $\mathcal{H}_1$ in $\mathcal{H}_0$. This proves that $u_{\varepsilon_l}^{(j)}(t)$ converges weakly in $\mathcal{H}_{2m-2j}$, say to $\tilde z_j(t)$, as claimed. We can then show exactly as in the proof of Theorem 2.1 that $[\tilde{z}_j] = z_j$, so that from (2.42)

$$\begin{aligned}\langle z_j(t), v\rangle_{\mathcal{H}_0} = \langle \varsigma_j, v\rangle_{\mathcal{H}_0} + \int_0^t \Big[&- A(t)(z_j(t), v) \\ &- \sum_{i=1}^{j} \binom{j}{i} A^{(i)}(z_{j-i}, v) + F^{(j)} \cdot v \Big] ds\end{aligned} \tag{2.43}$$

for all $v \in \mathcal{H}_1$, $t \in [0, T)$ and $j = 0, \ldots, m$ (again, the sum is absent if $j = 0$). From this we conclude three things: first that $z_0 = u$ is the unique solution of the problem (2.3); second that $z_m \in C([0, T); \mathcal{H}_0)$; and third that $z_j^* \equiv \langle z_j, \cdot\rangle_{\mathcal{H}_0} \in C([0, T); \mathcal{H}_1^*)$, which, together with the uniform $\mathcal{H}_{2m-2j}$ bound in (2.41) shows that z_j is continuous into $\mathcal{H}_{2m-2j}^{weak}$ for $j = 0, \ldots, m-1$. On the other hand, we can also take limits in the equation

$$\langle u_{\varepsilon_l}^{(j)}(t), v\rangle_{\mathcal{H}_1} = \langle \varsigma_j^{\varepsilon_l}, v\rangle_{\mathcal{H}_1} + \int_0^t \langle u_{\varepsilon_l}^{(j+1)}, v\rangle_{\mathcal{H}_1} ds$$

to find that $z_j \in wAC^2([0, T); \mathcal{H}_1)$ with derivative z_{j+1}, and then by Lemma 2.3 that $z_j \in wAC^2([0, T); \mathcal{H}_{2m-2j-1})$ with derivative z_{j+1}. We then relabel $z_j = u^{(j)}$ so that $u^{(j)} \in C([0, T); \mathcal{H}_{2m-2j}^{weak})$ (with strong continuity for $j = m$) and $[u^{(j)}] \in L_{loc}^2([0, T); \mathcal{H}_{2m-2j+1})$, and leave to the reader the task of checking the remaining details for the assertions in (d)–(f).

Finally we prove (g) by induction on j, starting with $j = m$. The strong continuity $u^{(m)} \in C([0, T); \mathcal{H}_0)$ has already been proved. Assume therefore that $u^{(j+1)} \in C([0, T); \mathcal{H}_{2m-2j-2})$ for a fixed $j \in \{0, \ldots, m-1\}$, let $t_0, t \in [0, T)$ and write

$$\begin{aligned}A(t)(u^{(j)}(t) - u^{(j)}(t_0), \cdot) &= -\langle u^{(j+1)}(t) - u^{(j+1)}(t_0), \cdot\rangle_{\mathcal{H}_0} \\ &+ \big[F^{(j)}(t) - F^{(j)}(t_0)\big] - \sum_{i=1}^{j} \binom{j}{i} A^{(i)}(t)(u^{(j-i)}(t) - u^{(j-i)}(t_0), \cdot) \\ &- \sum_{i=0}^{j} \binom{j}{i} (A^{(i)}(t) - A^{(i)}(t_0))(u^{(j-i)})(t_0), \cdot).\end{aligned}$$

Let $l = 2m - 2j - 2$, so that for i as in the above sums, $0 \le l \le 2m - 2i - 2$, consistent with the hypotheses in (civ) and (g). The first two terms on the right then go to zero in $\mathcal{V}_l$ as $t \to t_0$ by induction on j and by the assumption on $F^{(j)}$. The summand in the third term on the right goes to zero in $\mathcal{V}_l$ by hypothesis (civ) and by the conclusions already proved in (d) that $u^{(j-i)}$ is a Lipschitz map into $\mathcal{H}_{2m-2j+2i-2} = \mathcal{H}_{2i+l}$ and is a bounded map into $\mathcal{H}_{2m-2j+2i} = \mathcal{H}_{2i+l+2}$. Finally, the summand in the last sum on the right goes to zero by the second hypothesis in (g) and the latter fact that $u^{(j-i)}(t_0) \in \mathcal{H}_{2i+l+2}$. This shows that the right side above goes to zero in $\mathcal{V}_l$ as $t \to t_0$ and therefore that $u^{(j)}(t) \to u^{(j)}(t_0)$ in $\mathcal{H}_{l+2} = \mathcal{H}_{2m-2j}$ by the ellipticity property of $A(t)$ assumed in (g). This proves that $u^{(j)} \in C([0,T);\mathcal{H}_{2m-2j})$, thereby completing the induction, the proof of (g) and the proof of Theorem 2.7.

□

§2.5 Incompatible Data and Initial Layer Regularization

In this section we consider problems (2.3) in which the regularity conditions on A in Theorem 2.7 are satisfied but the compatibility conditions (2.16) are not. It is reasonable to ask in this situation whether solutions attain the regularity described in Theorem 2.7 instantaneously in time, so that the conclusions of the theorem hold locally in $(0,T)$ if not in $[0,T)$.

Consider the case $m=1$ in which A and F satisfy the $m=1$ hypotheses of Theorem 2.7, $\varsigma_0 \in \mathcal{H}_0$, but no assumption is made concerning compatibility. There is then a unique solution $u \in C([0,T);\mathcal{H}_0)$ to the problem (2.3) as described in Theorem 2.1, but Theorem 2.7 does not apply. However, for the new unknown function $w(t) = tu(t)$ we find that, for all $v \in \mathcal{H}_1$,

$$(2.44) \qquad \langle w(t), v\rangle_{\mathcal{H}_0} = \int_0^t \big[- A(s)(w(s),v) + \tilde{F}(s)\cdot v\big]\,ds$$

where $\tilde{F}(t) = tF(t) + \langle u,\cdot\rangle_{\mathcal{H}_0}$. The $m=1$ compatibility condition (2.16) for this system is then satisfied with $(\varsigma_0,\varsigma_1)$ replaced by $(0,\varsigma_0)$, and the conclusions of Theorem 2.7 therefore apply to w. Corresponding regularity statements for u then follow from the relation $u(t) = w(t)/t$.

Observe that the above argument requires only that F satisfy the hypotheses of Theorem 2.1 and that $tF(t)$, not $F(t)$, satisfy those of Theorem 2.7. In fact, any algebraic blowup of the norms of F at $t=0^+$ can be accommodated by taking $w(t) = t^{1+\delta}u(t)$ for δ sufficiently large. (We will require either that $\delta = 0$ or $\delta > \frac{1}{2}$, this stipulation insuring that $t^\delta\langle u(t),\cdot\rangle_{\mathcal{H}_0}$ is in $wAC^2([0,T);\mathcal{V}_{-1})$ as required in Theorem 2.7(c).) The equation (2.44) for w will then hold with $\tilde{F}(t) = t^{1+\delta}F(t) + (1+\delta)t^\delta\langle u(t),\cdot\rangle_{\mathcal{H}_0}$, and the argument can be repeated under the following assumptions (in which $(0,T)$ and $[0,T)$ are carefully distinguished):

- $F \in wAC^2((0,T);\mathcal{V}_{-1})$ with $F^{(1)} \in L^2_{loc}((0,T);\mathcal{V}_{-1})$
- $t^{1+\delta}F^{(1)} \in L^2_{loc}([0,T);\mathcal{V}_{-1})$
- $t^{1+\delta}[F] \in L^\infty_{loc}([0,T);\mathcal{V}_0) \cap L^2_{loc}([0,T);\mathcal{V}_1)$
- there is a $\gamma_1 \in \mathcal{H}_0$ such for all $v \in \mathcal{H}_1$,

$$(2.45) \qquad \lim_{t\to 0^+} t^{1+\delta}F(t)\cdot v = \langle \gamma_1, v\rangle_{\mathcal{H}_0}.$$

The second and third of these are the hypotheses (2.26) and (2.27) of Theorem 2.7 imposed on $\tilde{F}$, and the last shows that the compatibility condition for (2.44) is satisfied with $(\varsigma_0,\varsigma_1)$ replaced by $(0,\gamma)$, where γ is either γ_1 or $\gamma_1+\varsigma_0$ according as $\delta > \frac{1}{2}$ or $\delta = 0$. Theorem 2.7 therefore applies with

$m = 1$ and shows in particular that $w(t) \rightharpoonup 0$ weakly in $\mathcal{H}_2$ and $w^{(1)}(t) \to \gamma$ strongly in $\mathcal{H}_0$ as $t \to 0^+$. Then since $w^{(1)}(t) = t^{1+\delta}u^{(1)}(t) + (1+\delta)t^{\delta}u(t)$,

- $t^{1+\delta}u(t) \rightharpoonup 0$ weakly in $\mathcal{H}_2$
- $t^{1+\delta}u^{(1)}(t) \to \gamma_1$ strongly in $\mathcal{H}_0$,

which give controlled rates of blowup of $|u(t)|_{\mathcal{H}_2}$ and $|u^{(1)}(t)|_{\mathcal{H}_0}$ as $t \to 0^+$.

These results are generalized to arbitrary order in the following theorem in which the four itemized hypotheses above correspond to the rather technical hypotheses (b) and (c) with $m = 1$ and the two itemized conclusions above are generalized in (e).

Theorem 2.10. *Let $m \geq 1$ and $\mathcal{H}_j$, $\mathcal{V}_j$ and A satisfy the hypotheses of* Theorem 2.7. *Assume the following:*

(a) *$F \in L^2_{loc}([0,T); \mathcal{V}_{-1})$ and $\varsigma_0 \in \mathcal{H}_0$ (as in* Theorem 2.1*).*

(b) *$F^{(0)} = F, \ldots, F^{(m-1)} \in wAC^2((0,T); \mathcal{V}_{-1})$ are given with respective derivatives $[F^{(1)}]_{(0,T)}, \ldots, [F^{(m-1)}]_{(0,T)}$, $F^{(m)} \in L^2_{loc}((0,T); \mathcal{V}_{-1})$.*

(c) *There is a $\delta \in \{0\} \cup (\frac{1}{2}, \infty)$, such that:*

(i) *for $j = 0, \ldots, m$ and $k = j, \ldots, m$,*

$$t^{k+\delta}F^{(j)} \in L^2_{loc}([0,T); \mathcal{V}_{2k-2j-1});$$

(ii) *for $j = 0, \ldots, m-1$ and $k = j+1, \ldots, m$,*

$$t^{k+\delta}F^{(j)} \in L^\infty_{loc}([0,T); \mathcal{V}_{2k-2j-2});$$

(iii) *for $j = 0, \ldots, m-1$,*

$$\lim_{t \to 0} t^{j+1+\delta}F^{(j)}(t) \cdot v = \langle \gamma_{j+1}, v \rangle_{\mathcal{H}_0}$$

for some $\gamma_{j+1} \in \mathcal{H}_0$ and all $v \in \mathcal{H}_1$.

Then the unique solution u of (2.3) *satisfies the following, in addition to the conclusions of* Theorem 2.1:

(d) *The regularity statements in* Theorem 2.7(d) *and the strong forms* (2.19) *hold with $[0,T)$ replaced by $(0,T)$; and the weak forms* (2.18) *hold with $[0,t] \subset [0,T)$ replaced by $[t_1, t_2] \subset (0,T)$.*

(e) *For $k = 1, \ldots, m$,*

$$t^{k+\delta}u^{(j)}(t) \begin{cases} \rightharpoonup 0 & \text{weakly in } \mathcal{H}_{2k-2j},\ k = j+1, \ldots, m \\ \to \gamma_j & \text{strongly in } \mathcal{H}_0,\ k = j \end{cases}$$

as $t \to 0^+$.

(f) *Given* $\bar{t} \in [0, T)$ *there is a constant* $C(\bar{t})$ *as described in* Theorem 2.7(f) *such that*

$$\sum_{k=0}^{m} \sum_{j=0}^{k} \Big[\sup_{0 \le t \le \bar{t}} t^{2(k+\delta)} |u^{(j)}(t)|^2_{\mathcal{H}_{2k-2j}} + \int_0^{\bar{t}} t^{2(k+\delta)} |u^{(j)}(t)|^2_{\mathcal{H}_{2k-2j+1}} dt \Big] + \int_0^{\bar{t}} t^{2(m+\delta)} |((u^{(m)})^*)'|^2_{\mathcal{V}_{-1}} dt$$

$$\le C(\bar{t}) e^{\int_0^{\bar{t}} \beta^2} \Big[|\varsigma_0|^2_{\mathcal{H}_0} + \sum_{j=1}^{m} |\gamma_j|^2_{\mathcal{H}_0} + M_F \Big]$$

where β *is as in hypothesis* (c*i*) *of* Theorem 2.7 *and* M_F *bounds the sum of the squares of the norms of* $t^{k+\delta} F^{(j)}$ *on* $[0, \bar{t}]$ *in hypotheses* (c*i*) *and* (c*ii*) *of the present theorem.*

(g) *The regularity statement that* $u^{(j)} \in C((0,T); \mathcal{H}^{weak}_{2m-2j})$ *asserted in* (d) *above can be strengthened to* $u^{(j)} \in C((0,T); \mathcal{H}_{2m-2j})$, *that is, continuity in the norm topology, under the additional hypotheses that* $F^{(j)} \in C((0,T); \mathcal{V}_{\mathcal{H}_{2m-2j-2}})$ *for* $j = 0, \dots, m-1$ *and that* A *satisfies the additional hypotheses described in* Theorem 2.7(g).

PROOF. We leave to the reader to check the details of the discussion preceding the statement of the theorem for the case that $m = 1$. Assuming then that the theorem holds with $m - 1 \ge 1$ in place of m we let $w(t) \equiv t^{m+\delta} u(t)$ and compute that

$$\langle w(t), v \rangle_{\mathcal{H}_0} = \int_0^t \big[-A(w, v) + \tilde{F} \cdot v \big] ds \tag{2.46}$$

for all $v \in \mathcal{H}_1$, where $\tilde{F}(t) = t^{m+\delta} F(t) + (m+\delta) t^{m+\delta-1} u(t)^*$ and $u(t)^* = \langle u(t), \cdot \rangle_{\mathcal{H}_0}$. We will check that Theorem 2.7 applies to w. First, the hypotheses in (c*i*) and (c*ii*) and the induction hypotheses on u show that the $\tilde{F}^{(j)}$ satisfy the L^2 and L^∞ bounds required in (2.26) and (2.27). Next, to show that the absolute continuity in hypothesis (b) extends to $t = 0$ and to identify $\tilde{F}^{(j)}(0)$ we let $0 < \tau \le t < T$, $j \le m - 1$ and $v \in \mathcal{H}_1$. Then by (b),

$$\big(\tilde{F}^{(j)}(t) - \tilde{F}^{(j)}(\tau)\big) \cdot v = \int_\tau^t \tilde{F}^{(j+1)}(s) \cdot v \, ds$$

for all $v \in \mathcal{H}_1$. Hypothesis (c*i*) with $k = m$ shows that the integral on the right converges as $\tau \to 0$. Concerning the left side we note that $\tilde{F}^{(j)}(\tau)$ is a linear combination of terms $\tau^{m+\delta-j+i-1} (u^{(i)}(\tau))^*$ and $\tau^{m+\delta-j+i} F^{(i)}(\tau)$ for $i = 0 \dots, j$. Hypothesis (c*iii*) and the induction hypotheses on u show that for $j < m - 1$ these terms all go to zero in $\mathcal{V}_0$ as $\tau \to 0^+$ and that $\tilde{F}^{(m-1)}(\tau) \rightharpoonup \gamma^* \equiv \langle \gamma, \cdot \rangle_{\mathcal{H}_0}$ weak-$*$ in $\mathcal{H}_1^*$ for some $\gamma \in \mathcal{H}_0$ determined by ς_0 and the γ_j, but which need not be identified explicitly. Thus $\tilde{F}^{(j)} \in wAC^2([0,T); \mathcal{V}_{-1})$ with value $\tilde{F}^{(j)}(0) = 0$ for $j < m - 1$ and $\tilde{F}^{(m-1)}(0) = \gamma^*$.

It follows that the compatibility conditions (2.16) for (2.46) are satisfied with $(\varsigma_0, \ldots, \varsigma_m)$ replaced by $(0, \ldots, \gamma)$. Theorem 2.7 therefore applies to w and most of the corresponding conclusions for $u(t) = t^{-(m+\delta)}w(t)$ then follow easily.

One exception is the second conclusion in (e) that $t^{m+\delta}u^{(m)}(t) \to \gamma_m$ strongly in $\mathcal{H}_0$ as $t \to 0^+$. To prove this we note first that $w^{(m)}(t) = (t^{m+\delta}u(t))^{(m)} \to \gamma$ strongly in $\mathcal{H}_0$ as $t \to 0^+$ by Theorem 2.7. Since terms $t^{i+\delta}u^{(i)}(t)$ for $i < m$ in its Leibniz expansion have strong $\mathcal{H}_0$ limits by the induction hypothesis, so too does the term $t^{m+\delta}u^{(m)}(t)$. To identify its limit we recall from the paragraph above that $\tilde{F}^{(m-1)}(t)$ and $w^{(m)}(t)^* = \langle w^{(m)}(t), \cdot\rangle_{\mathcal{H}_0}$ have the same limit γ^* in $\mathcal{V}_{-1}$ as $t \to 0$. Since

$$w^{(m)}(t) = \left[t^{m+\delta}u^{(1)}(t) + (m+\delta)t^{m+\delta-1}u(t)\right]^{(m-1)}$$

and

$$\tilde{F}^{(m-1)}(t) = \left[t^{m+\delta}F(t) + (m+\delta)t^{m+\delta-1}u(t)^*\right]^{(m-1)}$$

we conclude that $\left[t^{m+\delta}\left(u^{(1)}(t)^* - F(t)\right)\right]^{(m-1)}$ goes to zero weak-$*$ in $\mathcal{V}_{-1}$ as $t \to 0^+$. But terms $\left[t^{i+1+\delta}\left(u^{(i+1)}(t)^* - F^{(i)}(t)\right)\right]$ in its Leibniz expansion go to zero by the induction hypothesis for $i \leq m-2$, hence so too does the term with $i = m-1$. This together with hypothesis $(ciii)$ shows that $\langle t^{m+\delta}u^{(m)}(t) - \gamma_m, v\rangle_{\mathcal{H}_0} \to 0$ for all $v \in \mathcal{H}_1$ and therefore that $t^{m+\delta}u^{(m)}(t) \to \gamma_m$ strongly in $\mathcal{H}_0$, since the strong limit is known to exist and $\mathcal{H}_1$ is dense in $\mathcal{H}_0$.

□

§2.6 Systems with Symmetry

In this section we examine problems in which the bilinear form A in (2.3) is symmetric or nearly so; that is, for which $A(u,v) - A(v,u)$ is either zero for all u and v or is of lower order in a certain sense. This symmetry leads to a regularity result not included in either Theorem 2.1 or Theorem 2.7: for the problem

$$\langle u(t), v\rangle_{\mathcal{H}_0} = \langle \varsigma_0, v\rangle_{\mathcal{H}_0} + \int_0^t \left[-A(s)(u(s),v) + \langle g(s), v\rangle_{\mathcal{H}_0}\right] ds \tag{2.47}$$

with $g \in L^2_{loc}([0,T); \mathcal{H}_0)$ and $\varsigma_0 \in \mathcal{H}_1$, there exists a unique solution $u \in wAC^2([0,T); \mathcal{H}_0) \cap C([0,T); \mathcal{H}_1^{weak})$. A corresponding higher-order regularity theory for nearly symmetric problems, parallel to that of Theorem 2.7, can then be developed, but this requires considerable effort, similar to the extensive analysis of section 2.4, and will not be included here. On the other hand this improvement in regularity can often be obtained in particular problems rather easily by applying the results in Theorems 2.11 and 2.12 below to the strong forms (2.18) with $j > 0$. See Theorem 6.14 for an application in which this idea is applied to a large class of quasilinear systems in two and three space dimensions.

It is instructive to first consider the example in section 1.1 of an initial-boundary value problem for a scalar unknown $u : [0,T) \to H_0^1(\Omega)$, where Ω is an open set in $\mathbf{R}^n$ and the differential equation includes a lower order term:

$$\begin{cases} u_t = (a_{ij}(x,t)u_{x_i})_{x_j} + b_i(x,t)u_{x_i} + g(x,t), \ (x,t) \in \Omega \times (0,T), \\ u(x,t) = 0, \ (x,t) \in \partial\Omega \times [0,T), \\ u(x,0) = \varsigma_0(x), \ x \in \Omega. \end{cases}$$

The following purely formal argument parallels that of Remark 4 following Theorem 2.1: if we multiply the differential equation by u_t and integrate over $\Omega \times [0,t]$, taking u_t to be zero on $\partial\Omega$, we obtain

$$\int_0^t \int_\Omega u_t^2 dxdt = \int_0^t \int_\Omega \big(- a_{ij}u_{x_i}u_{tx_j} + b_i u_{x_i} u_t + g u_t \big) dxdt.$$

If the symmetry condition $a_{ij} = a_{ji}$ holds for all i and j, then

$$a_{ij}u_{x_i}u_{tx_j} = \tfrac{1}{2}\big(a_{ij}u_{x_i}u_{x_j}\big)_t - \tfrac{1}{2}(a_{ij})_t u_{x_i}u_{x_j}$$

so that

$$\int_0^t \int_\Omega u_t^2 dxdt + \tfrac{1}{2} \int_\Omega a_{ij}u_{x_i}u_{x_j}dx \Big|_0^t$$
$$= \int_0^t \int_\Omega \big[- \tfrac{1}{2}(a_{ij})_t u_{x_i}u_{x_j} + b_i u_{x_i} u_t + g u_t \big] dxdt.$$

It follows that under suitable hypotheses on the coefficients a_{ij} and b_i,

$$\int_0^{\bar{t}} \int_\Omega u_t^2 dxdt + \sup_{0 \le t \le \bar{t}} |u(t)|_{H_0^1}^2$$
$$\le C(\bar{t}) \Big[|\varsigma_0|_{H_0^1}^2 + \int_0^{\bar{t}} \int_\Omega (|\nabla_x u|^2 + |g|^2) dxdt \Big],$$

which is stronger than the minimal regularity result of Theorem 2.1 but weaker than the $m = 1$ case of Theorem 2.7. Observe that the bilinear form $A(u,v) = \int_\Omega (a_{ij}u_{x_i}v_{x_j} + b_i u_{x_i} v) dx$ in this example is not symmetric, but rather is the sum $A = A_{sym} + B$ of a symmetric part A_{sym} and a lower-order term B satisfying $|B(u,v)| \le C|u|_{H_0^1}|v|_{L^2}$ for bounded b_i.

The following two theorems give both a rigorous justification for the above formal argument as well as a generalization to the abstract problem (2.47). In Theorem 2.11 we assume that ς_0 is in $\mathcal{H}_1$ as in the example, and in Theorem 2.12 we consider the case that $\varsigma_0 \in \mathcal{H}_0$ and an initial-layer regularization occurs. Several remarks follow the statements of the theorems and the proofs are given at the end of the section.

Theorem 2.11. *Let $\mathcal{H}_0$, $\mathcal{H}_1$ and T be as in* Theorem 2.1 *and assume the following:*

(a) *A_{sym}, B and $A = A_{sym} + B$ are weakly measurable mappings from $[0,T)$ to $\mathcal{B}(\mathcal{H}_1 \times \mathcal{H}_1)$.*

(b) *A_{sym} satisfies*

(*i*) *$A_{sym} \in wAC^2([0,T);\mathcal{B}(\mathcal{H}_1 \times \mathcal{H}_1))$;*

(*ii*) *for every $\bar{t} \in [0,T)$ there are positive numbers $\theta(\bar{t})$ and $C_A(\bar{t})$ and a nonnegative number $\lambda(\bar{t})$ such that for all $t \in [0,\bar{t}]$ and z, $w \in \mathcal{H}_1$,*

$$|A_{sym}(t)(z,w)| \le C_A(\bar{t})|z|_{\mathcal{H}_1}|w|_{\mathcal{H}_1},$$

$$\int_0^{\bar{t}} \|A'_{sym}\|^2_{\mathcal{B}(\mathcal{H}_1\times\mathcal{H}_1)}dt \le C_A(\bar{t})$$

and

$$A_{sym}(t)(z,z) \ge \theta(\bar{t})|z|^2_{\mathcal{H}_1} - \lambda(\bar{t})|z|^2_{\mathcal{H}_0};$$

(*iii*) *for all $t \in [0,T)$ and w, $z \in \mathcal{H}_1$, $A_{sym}(t)(z,w) = A_{sym}(t)(w,z)$.*

(c) *There is a function $\tilde{\beta} \in L^2_{loc}([0,T))$ such that for almost all $t \in [0,T)$,*

$$|B(t)(z,w)| \le \tilde{\beta}(t)|z|_{\mathcal{H}_1}|w|_{\mathcal{H}_0}$$

for all z, $w \in \mathcal{H}_1$.

Then given $\varsigma_0 \in \mathcal{H}_1$ and $g \in L^2_{loc}([0,T);\mathcal{H}_0)$ there is a unique element

$$u \in wAC^2([0,T);\mathcal{H}_0) \cap C([0,T);\mathcal{H}_1^{weak}) \tag{2.48}$$

which satisfies (2.47) *for all $v \in \mathcal{H}_1$. In addition, given specific representatives of g and the derivative u' of u, there is a null set $E \subset [0,T)$ such that for all $t \notin E$ and all $v \in \mathcal{H}_1$,*

$$\langle u'(t), v\rangle_{\mathcal{H}_0} = -A(u(t),v) + \langle g(t), v\rangle_{\mathcal{H}_0}. \tag{2.49}$$

The energy equality

$$\tfrac{1}{2}|u(t)|^2_{\mathcal{H}_0} + \int_0^t A(s)(u(s),u(s))ds = \tfrac{1}{2}|\varsigma_0|^2_{\mathcal{H}_0} + \int_0^t \langle g(s), u(s)\rangle_{\mathcal{H}_0}\, ds \tag{2.50}$$

holds for $t \in [0,T)$, and given $\bar{t} \in [0,T)$ there is a constant $C(\bar{t})$ depending only on upper bounds for the norm of the inclusion $\mathcal{H}_1 \subseteq \mathcal{H}_0$ and on $\bar{t}$, $C_A(\bar{t})$, $\theta^{-1}(\bar{t})$, $\lambda(\bar{t})$ and $|\tilde{\beta}|_{L^2([0,\bar{t}])}$ such that

$$\sup_{0\le t\le \bar{t}} |u(t)|^2_{\mathcal{H}_1} + \int_0^{\bar{t}} |u'(t)|^2_{\mathcal{H}_0}dt \le C(\bar{t})\Big[|\varsigma_0|^2_{\mathcal{H}_1} + \int_0^{\bar{t}} |g(t)|^2_{\mathcal{H}_0}dt\Big]. \tag{2.51}$$

If ς_0 is in $\mathcal{H}_0$ but not in $\mathcal{H}_1$, the solution u attains the regularity in (2.48) instantaneously in positive time, this effect being similar to the behavior described in Theorem 2.10. The precise result is as follows:

Theorem 2.12. *Let $\mathcal{H}_0$, $\mathcal{H}_1, T$, A, A_{sym} and B be as in* Theorem 2.11. *Then given $\varsigma_0 \in \mathcal{H}_0$ and g satisfying*

$$t^{\delta/2}g(t) \in L^2_{loc}([0,T);\mathcal{H}_0) \text{ and } g^* \in L^2_{loc}([0,T);\mathcal{H}_1^*) \tag{2.52}$$

for some $\delta \geq 1$, there is a unique u in the regularity class

$$C([0,T);\mathcal{H}_0) \cap wAC^2((0,T);\mathcal{H}_0) \cap C((0,T);\mathcal{H}_1^{weak}) \tag{2.53}$$

with equivalence class on $[0,T)$ in $L^2_{loc}([0,T);\mathcal{H}_1)$ which satisfies (2.47) for all $v \in \mathcal{H}_1$. The strong form (2.49) holds for almost all $t \in [0,T)$ and the energy equality (2.50) holds for all $t \in [0,T)$, just as in Theorem 2.11; *and given $\bar{t} \in [0,T))$ there is a constant $C(\bar{t})$ as described in that theorem such that*

$$\begin{aligned} &\sup_{0<t\leq\bar{t}} t^\delta |u(t)|^2_{\mathcal{H}_1} + \int_0^{\bar{t}} t^\delta |u'(t)|^2_{\mathcal{H}_0} dt \\ &\qquad \leq C(\bar{t})\Big[|\varsigma_0|^2_{\mathcal{H}_0} + \int_0^{\bar{t}} \left(|g^*(t)|^2_{\mathcal{H}_1^*} + t^\delta |g(t)|^2_{\mathcal{H}_0} dt\right)\Big]. \end{aligned} \tag{2.54}$$

Remarks:

(1) In the case that A, A_{sym}, and B are independent of t, the continuity into $\mathcal{H}_1^{weak}$ in these theorems can be improved to strong continuity: $u \in C([0,T);\mathcal{H}_1)$ and $u \in C((0,T);\mathcal{H}_1)$ respectively. See Corollary 2.14 in section 2.7 below.

(2) Observe that the assumption made in Theorem 2.1(b) that A is locally bounded on $[0,T)$ is not satisfied here. Consequently neither the energy equality (2.50) nor the uniqueness of solutions is known from Theorem 2.1. The proofs given below will instead depend on the improved regularity resulting from the symmetry hypothesis. In particular, the uniqueness proof applies the known existence results to an "adjoint" problem to deduce that the difference between two solutions with the same data is orthogonal to a dense set. This adjoint argument, familiar in other areas of analysis, does not appear elsewhere in the text.

(3) It is of some interest to compare the two conditions in (2.52), say for the example at the beginning of this section with $n \geq 3$. In this case there is a $p^* \in (2,\infty)$ such that $H_0^1(\Omega) \subseteq L^p(\Omega)$ for all $p \in [2,p^*]$ (Theorem A.3). Then for $v \in \mathcal{H}_1 = H_0^1(\Omega)$,

$$|g^*(t)\cdot v| = \Big|\int_\Omega g(x,t)v(x)dx\Big| \leq C|g(\cdot,t)|_{L^q}|v|_{H_0^1}$$

for any $q \in [q^*,2]$, where q^* is the Hölder conjugate of p^*. The second requirement in (2.52) is therefore satisfied if $g \in L^2_{loc}([0,T);L^q(\Omega))$ for some such q, and this condition is independent of the first condition that $t^\delta g \in L^2_{loc}([0,T);L^2(\Omega))$.

(4) The bound (2.54) in Theorem 2.12 shows that if g is zero and $\varsigma_0 \in \mathcal{H}_0$, then the solution $u(t)$ regularizes into $\mathcal{H}_1$ at least at the rate $|u(t)|_{\mathcal{H}_1} \leq Ct^{-1/2}$ as t increases from 0. The following familiar example provides an illustration and can also be applied to show that this rate is optimal in general. Let $\mathcal{H}_0 = L^2(\mathbf{R}^n)$ and $\mathcal{H}_1 = H^1(\mathbf{R}^n)$ and let u be the solution of the heat equation with initial data $\varsigma_0 \in L^2(\mathbf{R}^n)$. Then u is given by $\hat{u}(\xi, t) = e^{-|\xi|^2 t}\hat{\varsigma}_0(\xi)$, modulo constants, where ˆ denotes Fourier transform with respect to the spatial variable x. Thus by the Plancherel theorem,

$$
\begin{aligned}
|u(\cdot,t)|_{H^1}^2 &= \int_{\mathbf{R}^n} (1+|\xi|^2)|\hat{u}(\xi,t)|^2 d\xi \\
&= \int_{\mathbf{R}^n} (1+|\xi|^2)e^{-2|\xi|^2 t}|\hat{\varsigma}_0(\xi)|^2 d\xi \leq C(1+t^{-1})|\varsigma_0|_{L^2(\mathbf{R}^n)}^2,
\end{aligned}
$$

consistent with (2.54). This computation can also be applied with various specific initial functions ς_0 to show that $t^{-1/2}$ cannot be replaced by any more favorable rate; see Exercise 2.12.

PROOFS of Theorems 2.11 and 2.12. First observe that the hypotheses (2.1) of Theorem 2.1(a) hold so that the details of its proof apply: if $\{\varphi_k\}_k$ is the orthonormal basis for $\mathcal{H}_0$ from Theorem A.10(b) and $u^K \in \text{span}\{\varphi_1, \dots, \varphi_K\}$ is the approximate solution define by

$$
\begin{aligned}
\langle u^K(t), v^K\rangle_{\mathcal{H}_0} = \langle \varsigma_0, v^K\rangle_{\mathcal{H}_0} + \int_0^t \big[&- A(s)(u^K(s), v^K) \\
&+ \langle g(s), v^K\rangle_{\mathcal{H}_0}\big] ds
\end{aligned}
\tag{2.55}
$$

for all $v^K \in \mathcal{H}^K$ and all t, then the bound (2.10) holds independently of K and there is a subsequence $K \to \infty$ and a solution $u \in C([0,T); \mathcal{H}_0^{weak})$ with equivalence class in $L^2_{loc}([0,T); \mathcal{H}_1)$ to (2.47) such that $u^K(t) \rightharpoonup u(t)$ weakly in $\mathcal{H}_0$ for every t and, modulo equivalence classes, $u^K \rightharpoonup u$ weakly in $L^2([0,\bar{t}]; \mathcal{H}_1)$ for every $\bar{t}$.

To prove the bounds in (2.51) and (2.54) we first note that by Theorem A.9 the projections $\langle u^K(t), \varphi_k\rangle_{\mathcal{H}_0}$ are absolutely continuous functions of t. We can therefore differentiate with respect to t in (2.55) and then take $v^K = u_t^K$ to obtain that for almost all t,

$$
\begin{aligned}
|u_t^K|_{\mathcal{H}_0}^2 &= - A(u^K, u_t^K) + \langle g, u_t^K\rangle_{\mathcal{H}_0} \\
&= - A_{sym}(u^K, u_t^K) - B(u^K, u_t^K) + \langle g, u_t^K\rangle_{\mathcal{H}_0}.
\end{aligned}
\tag{2.56}
$$

Writing

$$
A_{sym}(u^K, u^K) = A_{sym}(\varphi_k, \varphi_j)\langle u^K, \varphi_k\rangle_{\mathcal{H}_0}\langle u^K, \varphi_j\rangle_{\mathcal{H}_0}
$$

and differentiating with respect to t and applying the symmetry assumption, we find that

$$
A_{sym}(u^K, u_t^K) = \tfrac{1}{2}D_t\big[A_{sym}(u^K, u^K)\big] - \tfrac{1}{2}A'_{sym}(u^K, u^K)
$$

and then substituting into (2.56) that

$$\begin{aligned}|u_t^K|^2_{\mathcal{H}_0}+\tfrac{1}{2}D_t\big[A_{sym}(u^K,u^K)\big]&\\ &= \tfrac{1}{2}A'_{sym}(u^K,u^K)-B(u^K,u_t^K)+\langle g(s),u_t^K\rangle_{\mathcal{H}_0}\\ &\le \tfrac{1}{2}\|A'_{sym}\|_{\mathcal{B}(\mathcal{H}_1\times\mathcal{H}_1)}|u^K|^2_{\mathcal{H}_1}+\tilde\beta|u^K|_{\mathcal{H}_1}|u_t^K|_{\mathcal{H}_0}+|g|_{\mathcal{H}_0}|u_t^K|_{\mathcal{H}_0}.\end{aligned}$$

We absorb the $|u_t^K|_{\mathcal{H}_0}$ terms on the right and then integrate with respect to t and apply absolute continuity, Gronwall's inequality and the bound (2.10) from the proof of Theorem 2.1(a). The result is that for every $\bar t\in[0,T)$ there is a constant $C(\bar t)$ as described in Theorem 2.11 such that

$$\sup_{0\le t\le \bar t}|u^K(t)|^2_{\mathcal{H}_1}+\int_0^{\bar t}|u_t^K|^2_{\mathcal{H}_0}\le C(\bar t)\Big[|\varsigma_0|^2_{\mathcal{H}_1}+\int_0^{\bar t}|g(t)|^2_{\mathcal{H}_0}dt\Big].$$

This together with the known weak convergence $u^K(t)\rightharpoonup u(t)$ in $\mathcal{H}_0$ shows that $u^K(t)\rightharpoonup u(t)$ weakly in $\mathcal{H}_1$ for every $t\in[0,T)$ and also that u_t^K converges weakly in $L^2([0,\bar t];\mathcal{H}_0)$ for every $\bar t\in[0,T)$. The statements in (2.48), (2.49) and (2.51) then follow easily.

For the proof of the corresponding statements in Theorem 2.12 we multiply (2.56) by t^δ and then proceed as above. The result is that

$$\sup_{0<t\le\bar t}t^\delta|u^K(t)|^2_{\mathcal{H}_1}+\int_0^{\bar t}t^\delta|u_t^K|^2_{\mathcal{H}_0}\le C(\bar t)\Big[|\varsigma_0|^2_{\mathcal{H}_0}+\int_0^{\bar t}(t^\delta|g(t)|^2_{\mathcal{H}_0}+|g^*|^2_{\mathcal{H}_1^*})dt\Big].$$

The bounds and regularity statements in (2.52), (2.53) and (2.54) then follow as before. (The strong continuity at $t=0$ in (2.53) follows from Remark 5 following the statement of Theorem 2.1.)

To prove the energy equality (2.50) we apply the absolute continuity of u already established (on $[0,T)$ if $\varsigma_0\in\mathcal{H}_1$ and on $(0,T)$ if $\varsigma_0\in\mathcal{H}_0$) together with Exercise 2.6 to obtain that $\frac{1}{2}|u(t)|^2_{\mathcal{H}_0}$ is absolutely continuous with derivative $\langle u',u\rangle_{\mathcal{H}_0}$, which by (2.49) equals $-A(u,u)+\langle g,u\rangle_{\mathcal{H}_0}$. If $\varsigma_0\in\mathcal{H}_1$ we integrate over $[0,t]$; otherwise we integrate over an interval $[\tau,t]\subseteq(0,T)$ and then let $\tau\to 0$ applying the strong continuity of u from the right at $t=0$, which was noted in the paragraph above.

To prove the uniqueness statements in Theorems 2.11 and 2.12 we suppose that u is an element of the regularity class (2.53) solving (2.47) with ς_0 and g both zero. Let A denote a specific representative of its equivalence class and let $\{\varphi_k\}_k$ be the orthonormal basis for $\mathcal{H}_0$ described in Theorem A.10(b), say with $\varepsilon=1$. Then for almost all t,

$$\frac{d}{dt}\langle u(t),\varphi_k\rangle_{\mathcal{H}_0}=-A(t)(u(t),\varphi_k) \tag{2.57}$$

for all k. Next fix $t_0\in(0,T)$ and define $A^*(t)(z,w)=A(t_0-t)(w,z)$ for $z,w\in\mathcal{H}_1$. Then A^* satisfies the hypotheses in Theorems 2.11 and 2.12 with $[0,T)$ replaced by $[0,t_0)$ and with $\theta(\bar t),C_A(\bar t),\lambda(\bar t)$, and $|\tilde\beta|_{L^2(0,\bar t)}$ bounded independently of $\bar t\in[0,t_0)$. Therefore given $\psi\in\mathcal{H}_1$ there is a solution z to the problem

$$\langle z(t), v\rangle_{\mathcal{H}_0} = \langle \psi, v\rangle_{\mathcal{H}_0} - \int_0^t A^*(s)(z(s), v)ds$$

for $v \in \mathcal{H}_1$ and $t \in [0, t_0)$ satisfying (2.48)–(2.51) with $[0, T)$ replaced by $[0, t_0)$. Also, since the right side of (2.51) is independent of $\bar{t} \in [0, t_0)$, the domain of absolute continuity of z extends to $[0, t_0]$. Changing variables, we obtain that for $t \in [0, t_0]$ and all $v \in \mathcal{H}_1$,

$$\langle z(t_0 - t), v\rangle_{\mathcal{H}_0} = \langle \psi, v\rangle_{\mathcal{H}_0} - \int_t^{t_0} A(s)(v, z(t_0 - s))ds,$$

and therefore for almost all t,

$$\frac{d}{dt}\langle z(t_0 - t), \varphi_k\rangle_{\mathcal{H}_0} = A(t)(\varphi_k, z(t_0 - t)) \tag{2.58}$$

for all k. Now multiply (2.57) by $\langle z(t_0 - t), \varphi_k\rangle_{\mathcal{H}_0}$ and (2.58) by $\langle u(t), \varphi_k\rangle_{\mathcal{H}_0}$ and sum over $k = 1, \ldots, K$; the result is that

$$\frac{d}{dt}\langle u^K(t), z^K(t_0 - t)\rangle_{\mathcal{H}_0} = -A(t)(u(t), z^K(t_0 - t)) + A(t)(u^K(t), z(t_0 - t))$$

where u^K and z^K now denote $\mathcal{H}_0$-orthogonal projections onto the span of $\{\varphi_1, \ldots, \varphi_K\}$ (different from the approximations u^Kin (2.55)). It follows from the absolute continuity and the fact that $u^K(0) = 0$ that

$$\langle u^K(t_0), \psi\rangle_{\mathcal{H}_0} = \int_0^{t_0} \big[- A(u, z^K) + A(u^K, z)\big].$$

Now recall the special properties of u^K and z^K described in Theorem A.10(b): $u^K(t) \to u(t)$ in $\mathcal{H}_1$ for all t and $|u^K(t)|_{\mathcal{H}_1} \leq 2|u(t)|_{\mathcal{H}_1}$; similarly for z^K. Since $\|A\|_{\mathcal{B}(\mathcal{H}_1 \times \mathcal{H}_1)} \in L^2$ and $|u|_{\mathcal{H}_1}$, $|z|_{\mathcal{H}_1} \in L^\infty$ on $[0, t_0]$, we can let $K \to \infty$ and apply the dominated convergence theorem on the right to conclude that $\langle u(t_0), \psi\rangle_{\mathcal{H}_0} = 0$ for all $\psi \in \mathcal{H}_1$. It follows that $u(t_0)$ is zero because $\mathcal{H}_1$ is dense in $\mathcal{H}_0$. This proves the uniqueness statements in both Theorems 2.11 and 2.12.

□

§2.7 Spectral Representations

In this section we continue the study of problems with symmetry, now under the additional assumption that A_{sym} is constant in t. This allows for explicit representations of solutions in terms of a certain spectral measure associated to A_{sym}, and in the case that $\mathcal{H}_1$ is compactly contained in $\mathcal{H}_0$, a more familiar representation in terms of Fourier series and the Duhamel integral.

Specifically, suppose that A_{sym} satisfies the hypotheses in (b) of Theorem 2.11 with A_{sym}, θ and λ constant in t. Then $A_0 \equiv A_{sym} + \lambda\langle\cdot,\cdot\rangle_{\mathcal{H}_0}$ is symmetric and $A_0(z, z) \geq \theta|z|^2_{\mathcal{H}_1}$ for all $z \in \mathcal{H}_1$, so that A_0 is an $\mathcal{H}_1$-equivalent inner product on $\mathcal{H}_1$. There is therefore an operator $S \in \mathcal{L}(\mathcal{H}_0, \mathcal{H}_1)$ such that

$$A_0(Sf, v) = \langle f, v \rangle_{\mathcal{H}_0} \tag{2.59}$$

for all $f \in \mathcal{H}_0$ and all $v \in \mathcal{H}_1$. Since S is symmetric as an element of $\mathcal{L}(\mathcal{H}_0)$ there is a corresponding spectral measure E, which is a countably additive mapping of the Borel sets in $\mathbf{R}$ into the set of orthogonal projections onto subspaces of $\mathcal{H}_0$. (A concise account of the elements of spectral theory is given in Halmos [**21**]. Readers unfamiliar with this theory may consider the more restrictive case that $\mathcal{H}_1$ is *compactly* contained in $\mathcal{H}_0$ so that the spectral measure is identified in terms of eigenvalues and eigenvectors. The following paragraph can then be omitted and readers may proceed directly to the paragraph after next in which (2.64) and (2.65) appear.)

It is easily checked that as consequence of the coerciveness of S and the assumption that $\mathcal{H}_1$ is dense in $\mathcal{H}_0$,

$$E(V) = E(V \cap (0, L]) \tag{2.60}$$

for Borel sets V, where $L = \|S\|_{\mathcal{L}(\mathcal{H}_0)}$. Recall also the following standard facts: If $\psi : (0, L]) \to \mathbf{R}$ is bounded and Borel-measurable, then the integral $\int_{\mathbf{R}} \psi(\mu) dE(\mu) = \int_{(0,L]} \psi(\mu) dE(\mu)$ is defined and is a self-adjoint operator on $\mathcal{H}_0$; the integral of a product of such functions is the composition of the integrals; and

$$\begin{cases} \int_{(0,L]} dE = I \text{ (the identify map on } \mathcal{H}_0) \\ \int_{(0,L]} \mu \, dE(\mu) = S. \end{cases} \tag{2.61}$$

In addition, for z and w in $\mathcal{H}_0$ there is a regular (nonnegative) Borel measure $E_{(z,w)}$ such that for ψ as above,

$$\int_{(0,L]} \psi(\mu) dE_{(z,w)} = \langle \Big(\int_{(0,L]} \psi(\mu) dE \Big) z, w \rangle_{\mathcal{H}_0}, \tag{2.62}$$

and in particular $\int_{(0,L]} dE_{(z,z)} = |z|^2_{\mathcal{H}_0}$. The following less standard fact is proved in Theorem A.10: if $z \in \mathcal{H}_0$, then $z \in \mathcal{H}_1$ if and only if $\int_{(0,L]} \mu^{-1} dE_{(z,z)}$ is finite, in which case

$$\int_{(0,L]} \mu^{-1} dE_{(z,z)} = A_0(z, z) \geq \theta |z|^2_{\mathcal{H}_1}. \tag{2.63}$$

If $\mathcal{H}_0$ is compactly included in $\mathcal{H}_0$, then S is a compact symmetric operator in $\mathcal{L}(\mathcal{H}_0)$ and there is therefore an orthonormal basis $\{\varphi_k\}_1^\infty$ of eigenvectors of S corresponding to positive eigenvalues $\mu_1 \geq \mu_2 \ldots \to 0$ (see

Evans [**12**] or Gilbarg and Trudinger [**19**], for example). In this case the spectral measure E is given by

$$E(V) = \sum_{\{j:\mu_j \in V\}} P_j \equiv \sum_{\{j:\mu_j \in V\}} \langle \cdot, \varphi_j \rangle_{\mathcal{H}_0} \varphi_j \tag{2.64}$$

for Borel sets V, and

$$\int \psi(\mu) dE(\mu) = \sum_j \psi(\mu_j) P_j \tag{2.65}$$

for bounded measurable ψ. Notice that ψ and φ agree E-almost everywhere if and only if they agree at each μ_j. (Readers unfamiliar with the spectral theory for noncompact operators may proceed making the above substitutions throughout.)

The following gives an explicit representation of the solution to the problem (2.47) in terms of the spectral measure E:

Theorem 2.13. *Let $\mathcal{H}_0$ and $\mathcal{H}_1$ be separable Hilbert spaces with $\mathcal{H}_1$ dense and continuously contained in $\mathcal{H}_0$, let $T \in (0, \infty]$ and assume that $A = A_{sym} + B$ satisfies the following:*

(a) *$A_{sym} \in \mathcal{B}(\mathcal{H}_1 \times \mathcal{H}_1)$ is symmetric and constant in t and there are constants $\theta > 0$ and $\lambda \geq 0$ such that*

$$A_{sym}(z, z) \geq \theta |z|^2_{\mathcal{H}_1} - \lambda |z|^2_{\mathcal{H}_0}$$

for all $z \in \mathcal{H}_1$.

(b) *$B : [0, T) \to \mathcal{B}(\mathcal{H}_1 \times \mathcal{H}_0)$ is weakly measurable and there is a $\tilde{\beta} \in L^2_{loc}([0, T))$ such that for almost all t,*

$$|B(t)(z, w)| \leq \tilde{\beta}(t) |z|_{\mathcal{H}_1} |w|_{\mathcal{H}_0}$$

for all $(z, w) \in \mathcal{H}_1 \times \mathcal{H}_0$.

Let E be the spectral measure described above and $\tilde{B}(t) \in \mathcal{L}(\mathcal{H}_1, \mathcal{H}_0)$ the unique operator satisfying $\langle \tilde{B}(t) z, w \rangle_{\mathcal{H}_0} = B(t)(z, w)$ for all $(z, w) \in \mathcal{H}_1 \times \mathcal{H}_0$. Then given $\varsigma_0 \in \mathcal{H}_0$ and $g \in L^2_{loc}([0, T); \mathcal{H}_0)$, the unique solution u of the problem (2.47) satisfies the representation

$$\begin{aligned} u(t) &= e^{\lambda t} \Big(\int e^{-t/\mu} dE(\mu) \Big) \varsigma_0 \\ &\quad + \int_0^t \Big[e^{\lambda(t-s)} \Big(\int e^{(s-t)/\mu} dE(\mu) \Big) \Big(g(s) + \tilde{B}(s) u(s) \Big) \Big] ds. \end{aligned} \tag{2.66}$$

If $\mathcal{H}_1$ is compactly contained in $\mathcal{H}_0$ then E is given by (2.64) *and* (2.66) *takes the form*

$$u(t) = e^{\lambda t}\sum_{j=1}^{\infty} e^{-t/\mu_j}\langle\varsigma_0, \varphi_j\rangle_{\mathcal{H}_0}\varphi_j + \int_0^t e^{\lambda(t-s)}\Big[\sum_{j=1}^{\infty}\big(e^{(s-t)/\mu_j}\langle g(s) + \tilde{B}(s)u(s), \varphi_j\rangle_{\mathcal{H}_0}\varphi_j\Big]ds. \tag{2.67}$$

The two series in (2.67) *converge in $\mathcal{H}_0$ uniformly on $[0, \bar{t}]$ for all $\bar{t} \in [0, T)$ and the first converges in $\mathcal{H}_1$ uniformly on $[\tau, \bar{t}]$ for all $0 < \tau < \bar{t} < T$.*

Of course, (2.66) and (2.67) give explicit formulas for u in terms of ς_0 and g only when $B = 0$. Still, important information can be deduced from these representations in the general case because the term $\tilde{B}u$ is of lower order. One example is the strong continuity described in Remark 1 following the statement of Theorem 2.12:

Corollary 2.14. *Let all the hypotheses and notations in* Theorem 2.13 *be in force. Then $u \in C((0, T); \mathcal{H}_1)$, and if $\varsigma_0 \in \mathcal{H}_1$ then $u \in C([0, T); \mathcal{H}_1)$.*

Before proceeding to the proofs of Theorem 2.13 and Corollary 2.14 we derive the following general bounds for $dE \times ds$ integrals such those in (2.66), in both $\mathcal{H}_0$ and $\mathcal{H}_1$:

Lemma 2.15. *Let $\mathcal{H}_0$, $\mathcal{H}_1$, T and S be as in* Theorem 2.13 *with E the spectral measure of S and $L = \|S\|_{\mathcal{L}(\mathcal{H}_0)}$. Suppose that $\psi \in C([0, T) \times [0, L])$, $g \in L^2_{loc}([0, T); \mathcal{H}_0)$, $t \in [0, T)$ and $\sigma \in \{0, 1\}$. Then*

$$\Big|\int_0^t \Big(\int \psi(s, \mu)dE(\mu)\Big)g(s)ds\Big|^2_{\mathcal{H}_\sigma} \leq C_\sigma\Big[\sup_{\mu\in(0,L]}\int_0^t \mu^{-\sigma}\psi(s, \mu)^2 ds\Big]\int_0^t |g(s)|^2_{\mathcal{H}_0} ds \tag{2.68}$$

where $C_0 = 1$, $C_1 = \theta^{-1}$ and θ is as in hypothesis (a) *of* Theorem 2.13.

Proof. We can assume that the quantity on the right in (2.68) is finite. Let $\chi : \mathbf{R} \to \mathbf{R}$ be C^∞ and nondecreasing with $\chi(\mu) = 0$ for $\mu \leq 1$ and $\chi(\mu) = 1$ for $\mu \geq 2$, and let $\chi_\varepsilon(\mu) = \chi(\mu/\varepsilon)$ for $\varepsilon > 0$. Denote by $v \in \mathcal{H}_0$ the integral with respect to s on the left above and $z(s) \in \mathcal{H}_0$ its integrand. Then omitting the domain $(0, L]$ and the indicator $dE(\mu)$ in integrals with respect to E and applying Definition 1.5 and (2.62), we compute that

$$\begin{aligned}
\int_{(0,L]} \chi_\varepsilon(\mu)\mu^{-\sigma} dE_{(v,v)} &= \langle \int_0^t z(s)ds, \Big(\int \chi_\varepsilon(\mu)\mu^{-\sigma}\Big)v\rangle_{\mathcal{H}_0} \\
&= \int_0^t \langle g(s), \Big(\int \chi_\varepsilon(\mu)\mu^{-\sigma}\psi(s,\mu)\Big)v\rangle_{\mathcal{H}_0} ds \\
&\leq \Big(\int_0^t |g|^2_{\mathcal{H}_0}\Big)^{1/2}\Big(\int_0^t \Big[\int_{(0,L]} \chi_\varepsilon(\mu)^2\mu^{-2\sigma}\psi(s,\mu)^2 dE_{(v,v)}\Big]ds\Big)^{1/2} \\
&= \Big(\int_0^t |g|^2_{\mathcal{H}_0}\Big)^{1/2}\Big(\int_{(0,L]} \chi_\varepsilon(\mu)^2\mu^{-\sigma}\Big[\int_0^t \mu^{-\sigma}\psi(s,\mu)^2 ds\Big]dE_{(v,v)}\Big)^{1/2} \\
&\leq \gamma_\sigma^{1/2}\Big(\int_0^t |g|^2_{\mathcal{H}_0}\Big)^{1/2}\Big(\int_{(0,L]} \chi_\varepsilon(\mu)\mu^{-\sigma} dE_{(v,v)}\Big)^{1/2}
\end{aligned}$$

where γ_σ is the sup in (2.68). Thus

$$\int_{(0,L]} \chi_\varepsilon(\mu)\mu^{-\sigma} dE_{(v,v)} \leq \gamma_\sigma \int_0^t |g|^2_{\mathcal{H}_0}$$

and the result follows by letting $\varepsilon \to 0$ on the left and applying (2.62) for $\sigma = 0$ and (2.63) for $\sigma = 1$.

□

PROOF of Theorem 2.13. We first project the solution $u(t)$ away from $\mu = 0$ as follows. Let χ_ε be as in the proof of Lemma 2.15 and $P_\varepsilon = \int \chi_\varepsilon(\mu)dE(\mu)$. Then $P_{\varepsilon/2}P_\varepsilon = P_\varepsilon$ and if $v \in \mathcal{H}_0$ then $P_\varepsilon v \in \mathcal{H}_1$. Setting $\tilde{g} = g + \tilde{B}u \in L^2_{loc}([0,T);\mathcal{H}_0)$ we then have that for $v \in \mathcal{H}_0$,

$$\begin{aligned}
\langle P_\varepsilon u(t), v\rangle_{\mathcal{H}_0} &= \langle u(t), P_\varepsilon v\rangle_{\mathcal{H}_0} \\
&= \langle P_\varepsilon \varsigma_0, v\rangle_{\mathcal{H}_0} + \int_0^t \big[-A_0(u, P_\varepsilon v) + \lambda\langle P_\varepsilon u, v\rangle_{\mathcal{H}_0} + \langle P_\varepsilon \tilde{g}, v\rangle_{\mathcal{H}_0}\big].
\end{aligned} \tag{2.69}$$

However, $\int \mu\, dE(\mu) = S$ where S is as in (2.59), so that

$$A_0(u, P_\varepsilon v) = A_0(u, S\Big(\int \mu^{-1}\chi_\varepsilon\Big)v) = \langle\Big(\int \mu^{-1}\chi_{\varepsilon/2}\Big)P_\varepsilon u, v\rangle_{\mathcal{H}_0}.$$

Substituting into (2.69) and applying the fact that $\mathcal{H}_1$ is dense in $\mathcal{H}_0$ we conclude that $P_\varepsilon u \in wAC([0,T);\mathcal{H}_0)$ with derivative

$$(P_\varepsilon u)' = -\Big(\int \mu^{-1}\chi_{\varepsilon/2}\Big)P_\varepsilon u + \lambda P_\varepsilon u + P_\varepsilon \tilde{g}.$$

We may regard this as a linear differential equation for the unknown $P_\varepsilon u$ for which the operator

$$M_\varepsilon(t) \equiv e^{-\lambda t}\int e^{t/\mu}\chi_\varepsilon(\mu)dE(\mu)$$

functions as an integrating factor; that is, $M_\varepsilon P_\varepsilon u$ is absolutely continuous with derivative $M_\varepsilon P_\varepsilon \tilde{g}$ (the proof is straightforward and is given in Theorem A.11). Thus for $t \in [0,T)$,

$$M_\varepsilon(t)P_\varepsilon u(t) = P_\varepsilon^2 \varsigma_0 + \int_0^t M_\varepsilon(s)P_\varepsilon\tilde{g}(s)ds.$$

Applying the operator $Q_\varepsilon(t) \equiv e^{\lambda t}\int e^{-t/\mu}dE$ to both sides we then obtain

$$P_\varepsilon^2 u(t) = Q_\varepsilon(t)P_\varepsilon^2\varsigma_0 + Q_\varepsilon(t)\int_0^t M_\varepsilon(s)P_\varepsilon\tilde{g}(s)ds. \tag{2.70}$$

We now let $\varepsilon \to 0$ as follows. The square of the $\mathcal{H}_0$ norm of the difference between the ς_0 terms in (2.70) and (2.66) is

$$e^{2\lambda t}\int_{(0,L]} e^{-2t/\mu}(1-\chi_\varepsilon^2(\mu))^2 dE_{(\varsigma_0,\varsigma_0)} \le e^{2\lambda t}\int_{(0,L]}(1-\chi_\varepsilon^2(\mu))^2 dE_{(\varsigma_0,\varsigma_0)}$$

which goes to zero uniformly on $[0,\bar{t}]$ for every $\bar{t} \in [0,T)$. Also, by (2.63) the square of the A_0 norm of this same difference is bounded by

$$e^{2\lambda t}\int_{(0,L]} \mu^{-1}e^{-2t/\mu}(1-\chi_\varepsilon^2(\mu))^2 dE_{(\varsigma_0,\varsigma_0)}$$
$$\le (e^{2\lambda t}t^{-1}\sup_{s\ge 0} se^{-2s})\int_{(0,L]}(1-\chi_\varepsilon^2(\mu))^2 dE_{(\varsigma_0,\varsigma_0)},$$

which goes to zero uniformly on $[\tau,\bar{t}]$ for $0<\tau<\bar{t}<T$. Similar considerations apply to the term on the left side of (2.70). Finally, the difference between the integral on the right and the corresponding term in (2.66) can be written

$$e^{\lambda t}\int_0^t \Big[\int e^{(s-t)/\mu}(\chi_\varepsilon(\mu)^2-1)dE\Big]e^{-\lambda s}\tilde{g}(s)ds,$$

so that the square of its $\mathcal{H}_0$ norm is bounded by

$$e^{2\lambda t}\Big[\sup_\mu \int_0^t e^{2(s-t)/\mu}(\chi_\varepsilon(\mu)^2-1)^2 ds\Big]\int_0^t e^{-2\lambda s}|\tilde{g}(s)|_{\mathcal{H}_0}^2 ds$$

by Lemma 2.15. The sup here may be taken over $\mu \in [0,2\varepsilon]$ and therefore goes to zero as $\varepsilon \to 0$ by direct computation. This proves (2.66) as well as the statements concerning the convergence of the two series in (2.67), and completes the proof of Theorem 2.13.

□

PROOF of Corollary 2.14. First, since the continuity of $e^{\lambda t}$ is not in doubt, we may assume that $\lambda = 0$ (or see Exercise 2.1). We can also assume that $\varsigma_0 \in \mathcal{H}_1$, otherwise the initial time can be taken to be $\tau > 0$. Let S be as in (2.59), E its spectral measure and $L = \|S\|_{\mathcal{L}(\mathcal{H}_0)}$.

We let $0 \le t_1 \le t_2 \le \bar{t} < T$ and prove that $u(t_2) \to u(t_1)$ in $\mathcal{H}_1$ as $t_2 \to t_1$. Again let $\tilde{g} = f + \tilde{B}u$ so that $|\tilde{g}|_{L^2([0,\bar{t}];\mathcal{H}_0)}$ and $\sup_{t\in[0,\bar{t}]} |u(t)|_{\mathcal{H}_1}$ are bounded by a positive constant $C(\bar{t})$. If $z(t)$ is the first term on the right side of (2.66) then by (2.63),

$$\theta|z(t_2) - z(t_1)|^2_{\mathcal{H}_1} \le \int_{(0,L]} \mu^{-1}(e^{-t_2/\mu} - e^{-t_1/\mu})^2 dE_{(\varsigma_0,\varsigma_0)},$$

which goes to zero as $|t_2 - t_1| \to 0$ because the integrand goes to zero for every μ and is bounded by μ^{-1}, which is integrable with respect to the measure $E_{(\varsigma_0,\varsigma_0)}$ by (2.63).

Next let $w(t)$ be the second term on the right side of (2.66) and let χ_ε and P_ε be as in the proof of Theorem 2.13. Then

$$\begin{aligned} w(t_2) - w(t_1) &= \int_{t_1}^{t_2} \Big(\int e^{(s-t_2)/\mu}\Big)\tilde{g}(s)ds \\ &+ \int_0^{t_1} \Big(\int \big[e^{(s-t_2)/\mu} - e^{(s-t_1)/\mu}\big]\chi_\varepsilon(\mu)\Big)\tilde{g}(s)ds \\ &+ \int_0^{t_1} \Big(\int \big[e^{(s-t_2)/\mu} - e^{(s-t_1)/\mu}\big]\Big)(P_\varepsilon\tilde{g}(s) - \tilde{g}(s))ds. \end{aligned} \tag{2.71}$$

The square of the $\mathcal{H}_1$ norm of the first term on the right is bounded by

$$\theta^{-1}\Big(\sup_{\mu\in(0,L]} \int_{t_1}^{t_2} \mu^{-1}e^{2(s-t_2)/\mu}ds\Big) \int_{t_1}^{t_2} |\tilde{g}(s)|^2_{\mathcal{H}_0}ds$$

by Lemma 2.15. The sup here is bounded by $C(\bar{t})$ by direct computation and the second factor is arbitrarily small if $|t_2 - t_1|$ is sufficiently small. Next, the square of the $\mathcal{H}_1$ norm of the second term on the right in (2.71) is bounded by

$$C(\bar{t}) \sup_{\varepsilon\le\mu\le L} \int_0^{t_1} \mu^{-1}\big[e^{(s-t_2)/\mu} - e^{(s-t_1)/\mu}\big]^2 ds$$

which by the mean value theorem applied to the integrand is bounded by $C(\bar{t})|t_2 - t_1|^2/\varepsilon^3$. Finally by Lemma 2.15 the square of the $\mathcal{H}_1$ norm of the third term on the right in (2.71) is bounded by

$$C(\bar{t}) \int_0^t |\tilde{g}(s) - P_\varepsilon\tilde{g}(s)|^2_{\mathcal{H}_0}.$$

The integrand here goes to zero as $\varepsilon \to 0$ for almost all s and is bounded by $|\tilde{g}(s)|^2_{\mathcal{H}_0}$, which is integrable; this integral therefore goes to zero as $\varepsilon \to 0$. Combining these bounds, we conclude that the right side of (2.71) can be made arbitrarily small by choosing first ε small then requiring that t_2 and t_1 be sufficiently close depending on ε. This proves that $u : [0,\bar{t}] \to \mathcal{H}_1$ is (uniformly) continuous.

□

Exercises

2.1 Show that if u is the solution of (2.3) described in Theorem 2.1 and $w(t) = e^{-\lambda_0 t}u(t)$ for a constant λ_0, then w satisfies the corresponding problem with $F(t)$ replaced by $e^{-\lambda_0 t}F(t)$ and A replaced by $A+\lambda_0\langle\cdot,\cdot\rangle_{\mathcal{H}_0}$.

2.2 Assume that the hypotheses in (2.1) and (2.2) hold with A and F constant in t and $\lambda \leq 0$. Prove that there is an element $\bar{u} \in \mathcal{H}_1$ such that the solution u of Theorem 2.1 converges strongly in $\mathcal{H}_0$ to $\bar{u}$ as $t \to \infty$. Suggestion: look for a time-independent solution of (2.3).

2.3 The construction of the basis used in the proof of Theorem 2.1 depends on the theory of spectral measures and spectral integrals (see Theorem A.10 in the Appendix). An easier and more accessible construction is available for the case that $\mathcal{H}_0$ and $\mathcal{H}_1$ are separable Hilbert spaces with $\mathcal{H}_1$ dense and *compactly* contained in $\mathcal{H}_0$, as follows: Let $S \in \mathcal{L}(\mathcal{H}_0, \mathcal{H}_1)$ be the operator satisfying $\langle Sf, v\rangle_{\mathcal{H}_1} = \langle f, v\rangle_{\mathcal{H}_0}$ for all $f \in \mathcal{H}_0$ and $v \in \mathcal{H}_1$. Show that S is compact and self-adjoint as an element of $\mathcal{L}(\mathcal{H}_0)$. There is therefore an $\mathcal{H}_0$-orthonormal basis $\{\varphi_k\}_k$ for $\mathcal{H}_0$ whose elements are eigenvectors of S; that is, $S\varphi_k = \mu_k\varphi_k$ where the μ_k are positive and decrease to zero. Show that the set $\{\psi_k \equiv \mu_k^{1/2}\varphi_k\}_k$ is an $\mathcal{H}_1$-orthonormal basis for $\mathcal{H}_1$ and that the $\mathcal{H}_0$–projection onto the span of $\varphi_1, \ldots, \varphi_K$ restricted to $\mathcal{H}_1$ coincides with the $\mathcal{H}_1$-projection onto the span of $\psi_1, \ldots, \psi_K$. Then check the conclusions of Theorem A.10.

2.4 Let u be the solution of (2.3) described in Theorem 2.1(b) and let $\mathcal{H}^K$ be a finite dimensional subspace of $\mathcal{H}_1$. Define the so-called Galerkin approximation $u^K : [0,T) \to \mathcal{H}^K$ by the requirement that (2.6) holds for all $v^K \in \mathcal{H}^K$. Show that given $\bar{t} \in [0,T)$ there is a constant $C(\bar{t})$ depending on $\bar{t}$ and on A but independent of K, $\mathcal{H}^K$, ς_0 and F such that

$$\sup_{0\leq t\leq \bar{t}} |u(t) - u^K(t)|^2_{\mathcal{H}_0} + \int_0^{\bar{t}} |u(t) - u^K(t)|^2_{\mathcal{H}_1}\,dt$$

$$\leq C(\bar{t})\Big[\sup_{0\leq t\leq \bar{t}} |u(t) - \Pi^K u(t)|^2_{\mathcal{H}_0} + \int_0^{\bar{t}} |u(t) - \Pi^K u(t)|^2_{\mathcal{H}_1}\,dt\Big]$$

where $\Pi^K : \mathcal{H}_0 \to \mathcal{H}^K$ is the $\mathcal{H}_0$-orthogonal projection onto $\mathcal{H}^K$. Thus modulo the constant $C(\bar{t})$ and in the norm implied above, the Galerkin approximation u^K is as accurate as the $\mathcal{H}_0$-orthogonal projection $\Pi^K u$. Suggestion: first show that $u^K - \Pi^K u$ is orthogonal to $\mathcal{H}^K$ in the inner product $\langle\cdot,\cdot\rangle_{\mathcal{H}_0} + \int_0^t A(\cdot,\cdot)$.

2.5 While it is clear from the proof of Theorem 2.1 that each of the two summands on the left side of (2.4) is bounded by the expression on the right, it is less clear that their sum is so bounded. Prove this.

2.6 Let $\mathcal{H}$ be a separable Hilbert space and I a nontrivial interval in $\mathbf{R}$. Show that if $u, v \in wAC(I;\mathcal{H})$ then $\langle u, v\rangle_{\mathcal{H}}$ is absolutely continuous on I with derivative $\langle u', v\rangle_{\mathcal{H}} + \langle u, v'\rangle_{\mathcal{H}}$. Conclude that $|u|^2_{\mathcal{H}}$ is absolutely continuous with derivative $2\langle u, u'\rangle_{\mathcal{H}}$.

2.7 Let $F \in wAC(I; X^*)$ where I is a nontrivial interval and X is a separable Banach space.

(a) Fix a specific representative F' and show that for almost all $t \in I$ the difference quotients $(F(t+h) - F(t))/h$ are weak-$*$ convergent to $F'(t)$ as $h \to 0$. Suggestion: exploit the fact that almost every t is a Lebesgue point of $|F'(\cdot)|_{X^*}$.

(b) Assume in addition that $F(t) \in V$ for all $t \in I$ where V is a weak-$*$ sequentially closed subspace of X^*. Prove that $F'(t) \in V$ for almost all $t \in I$.

2.8 Let $\mathcal{H}$ be a separable Hilbert space, I a nontrivial interval in $\mathbf{R}$ and $u \in wAC(I;\mathcal{H})$. Show that given $\varepsilon > 0$ there is a $\delta > 0$ such that $\sum_i |u(b_i) - u(a_i)|_{\mathcal{H}} < \varepsilon$ for every finite collection of nonoverlapping intervals $\{[a_i, b_i]\}$ contained in I with $\sum_i |b_i - a_i| < \delta$.

2.9 Prove the assertion (2.20) made in the first paragraph of the proof of Theorem 2.5.

2.10 Prove the assertion in Remark 3 following the statement of Theorem 2.7 in section 2.3. Suggestion: given $F \in \mathcal{V}_j$ construct the solution z to the equation $A_\varepsilon(t)(z,\cdot) + \lambda\langle z,\cdot\rangle_{\mathcal{H}_0} = F$ by iteration in $\mathcal{H}_{j+2}$.

2.11 Suppose that the results of Theorem 2.7 hold for a particular solution u of (2.3) with $T = \infty$ and with the left side of (2.28) bounded independently of $\bar{t}$. Show that for $j = 0, \ldots, m-1$, $u^{(j)} \to 0$ strongly in $\mathcal{H}_{2m-2j-1}$ as $t \to \infty$. Suggestion: bound the total variation of $|u^{(j)}|^2_{\mathcal{H}_{2m-2j-1}}$ on $[0,\infty)$.

2.12 Show that the rate of regularization $|u(t)|_{\mathcal{H}_1} \le Ct^{-1/2}|\varsigma_0|_{\mathcal{H}_0}$ proved in Theorem 2.12 for the case that $g = 0$ and $\varsigma_0 \in \mathcal{H}_0$ is optimal in the sense that the bound $|u(t)|_{\mathcal{H}_1} \le Ct^{-p}|\varsigma_0|_{\mathcal{H}_0}$ cannot hold in general in the context of the theorem for any $p < 1/2$. Suggestion: consider the example in Remark 4 in section 2.6 with initial data $\hat{\varsigma}_0(\xi) = |\xi|^{-q}$ for $|\xi| \ge 1$ and for various $q > 0$; the constant C is of importance here.

2.13 Let $\mathcal{H}_0 = L^2(\mathbf{R}^n)$, $\mathcal{H}_1 = H^1(\mathbf{R}^n)$, $S \in \mathcal{L}(\mathcal{H}_0, \mathcal{H}_1)$ the operator defined by $\langle Sf, v\rangle_{\mathcal{H}_1} = \langle f, v\rangle_{\mathcal{H}_0}$ for $f \in \mathcal{H}_0$ and $v \in \mathcal{H}_1$, as in (2.59), and E the spectral measure of S regarded as a symmetric operator in $\mathcal{L}(\mathcal{H}_0)$. Identify $E(V)$ for Borel sets V in terms of the Fourier transform, apply the result to compute the spectral integral of a bounded measurable function, and then compute the first integral on the right side of (2.66) in terms of the standard heat kernel. Suggestion: Let $\varphi(\xi) = (1+|\xi|^2)^{-1}$ and show that for Borel sets V, $E(V) = \mathcal{F}^{-1}\mathcal{M}_{\chi_V \circ \varphi}\mathcal{F}$ where $\mathcal{F}$ is Fourier transform, χ_V is the characteristic function of V and $\mathcal{M}_f v = fv$ for $f \in L^\infty$ and $v \in L^2$.

2.14 Let $\mathcal{H}_0$ and $\mathcal{H}_1$ be separable Hilbert spaces with $\mathcal{H}_1$ dense and continuously contained in $\mathcal{H}_0$ and let $S \in \mathcal{L}(\mathcal{H}_0, \mathcal{H}_1)$ be the operator defined by $\langle Sf, v\rangle_{\mathcal{H}_1} = \langle f, v\rangle_{\mathcal{H}_0}$ for all $f \in \mathcal{H}_0$ and $v \in \mathcal{H}_1$. Let E be the spectral measure of S as a symmetric operator in $\mathcal{L}(\mathcal{H}_0)$ and for $s \geq 0$ define vector spaces

$$\tilde{\mathcal{H}}_s = \{v \in \mathcal{H}_0 : \int \mu^{-s} dE_{(v,v)} < \infty\}.$$

(Thus $\tilde{\mathcal{H}}_0 = \mathcal{H}_0$ and $\tilde{\mathcal{H}}_1 = \mathcal{H}_1$, by (2.63).) Show that $\tilde{\mathcal{H}}_s$ is a Hilbert space with inner product $\langle u, v\rangle_{\mathcal{H}_s} \equiv \int \mu^{-s} dE_{(u,v)}$. Now let $u(t)$ be as in (2.66) with λ, g and $\tilde{B}$ all zero. Show that if $\varsigma_0 \in \tilde{\mathcal{H}}_s$ with $s \in [0, 1]$ then $|u(t)|_{\mathcal{H}_1} \leq Ct^{(s-1)/2}|\varsigma_0|_{\mathcal{H}^s}$, and that if $\varsigma_0 \in \mathcal{H}_0$ then for all $s \geq 0$ $|u(t)|_{\tilde{\mathcal{H}}_s} \leq Ct^{-s/2}|\varsigma_0|_{\mathcal{H}_0}$.

2.15 Suppose that u satisfies the spectral representation (2.66) with g and B both zero and with the hypotheses and notations of Theorem 2.13 in force. Assume that there is an interval $(\mu_1, \mu_2]$ with $0 < \mu_1 < \mu_2 < L$ such that $E((\mu_1, \mu_2])\varsigma_0 = 0$. Let $P_1 = E((0, \mu_1])$ and $P_2 = E((\mu_2, L])$. Show that if $P_2\varsigma_0 \neq 0$ then the ratio $|P_1 u(t)|_{\mathcal{H}_0}/|u(t)|_{\mathcal{H}_0} \to 0$ as $t \to \infty$. This shows that u is well-approximated by $P_2 u$ for large time.

The requirement that $E((\mu_1, \mu_2])\varsigma_0 = 0$ may be a special property of ς_0 but also holds for all ς_0 if there is a "spectral gap," that is, an interval $(\mu_1, \mu_2]$ as above on which E is zero. For example, if $\mathcal{H}_1$ is compactly contained in $\mathcal{H}_0$, then E is zero outside a countable set whose only limit point is zero; in this case there are infinitely many such spectral gaps.

CHAPTER 3

Linear Parabolic Systems: Basic Theory

In this chapter we apply the abstract theory in sections 2.1, 2.6 and 2.7 to parabolic systems of partial differential equations. Specifically we let Ω be an open set in $\mathbf{R}^n$ and we take $\mathcal{H}_0$ and $\mathcal{H}_1$ to be closed subspaces of $[L^2(\Omega)]^N$ and $[H^1(\Omega)]^N$ respectively where $\mathcal{H}_1$ is dense in $\mathcal{H}_0$ and $N \geq 1$. Elements of these spaces will be (column) vector functions of $x \in \Omega$ with differentiation carried out component-wise. We will consider the following specific case of the abstract problem (2.3): find $u(t)$ for $t \in [0,T) \subseteq [0,\infty)$ such that for all $v \in \mathcal{H}_1$ and $t \in [0,T)$,

$$\text{(3.1)} \qquad \langle u(t), v\rangle_{L^2(\Omega)} = \langle \varsigma_0, v\rangle_{L^2(\Omega)} + \int_0^t \big[-A(s)(u(s),v) + F(s)\cdot v\big]ds$$

where $\varsigma_0 \in \mathcal{H}_0$,

$$\text{(3.2)} \quad A(u,v) = \int_\Omega \big[(a^{\eta,\omega}u_{x_\eta})\cdot v_{x_\omega} + (b^\eta u_{x_\eta})\cdot v + (c^\omega u)\cdot v_{x_\omega} + (du)\cdot v\big]dx + \int_{\partial\Omega} (fu)\cdot v\, d\sigma$$

(summation over repeated indices is understood) and

$$\text{(3.3)} \qquad F\cdot v = \int_\Omega (g\cdot v + G : \nabla_x v) + \int_{\partial\Omega} \Gamma\cdot v\, d\sigma.$$

Here $a^{\eta,\omega}$, b^η, c^ω and d are $N\times N$ matrix–valued functions of $(x,t) \in \Omega\times[0,T)$, f is an $N\times N$ matrix–valued function of $(x,t)\in\partial\Omega\times[0,T)$, and the dot denotes the usual inner product in $\mathbf{R}^N$; g and Γ are $\mathbf{R}^N$-valued, G is $N\times n$ matrix-valued, $\nabla_x v$ is the $N\times n$ x-derivative matrix, and $G:\nabla_x v$ is the scalar product $G^{ij}v^i_{x_j}$. Finally, σ denotes surface measure on $\partial\Omega$ (σ is constructed in section 3.2) and the boundary integrals involving f and Γ are omitted in the case that $\Omega = \mathbf{R}^n$.

The basic existence result is given in Theorem 3.6 below and may be paraphrased as follows: *if Ω is a uniformly Lipschitz domain (or satisfies the weaker cone condition if the boundary integrals are omitted), if the components of the coefficients of A and F are in $L^p_{loc}([0,T);L^s(\Omega))$ or $L^p_{loc}([0,T);L^s(\partial\Omega))$ for certain p and s, and if the ensemble of matrices $a^{\eta,\omega}$ satisfies a certain positive definiteness condition, then given $\varsigma_0\in\mathcal{H}_0$ there is a unique solution $u\in C([0,T);\mathcal{H}_0)$ of (3.1) whose Lebesgue equivalence class on $[0,T)$ is in $L^2_{loc}([0,T);\mathcal{H}_1)$ and which satisfies the energy equality (2.5) and the bound*

(2.4) *in term of* ς_0 *and* F. The precise statement and its proof are given in Theorem 3.6 in section 3.3 below and a parallel existence theory is given in section 3.4 for systems which are symmetric to leading order. Solutions of symmetric systems are somewhat more regular and can be represented in terms of an associated spectral measure if the $a^{\eta,\omega}$ are independent of t.

In section 3.1 we consider specific examples of such systems, paying particular attention to boundary conditions and the corresponding choices of subspaces $\mathcal{H}_0$ and $\mathcal{H}_1$. In section 3.2 we review basic material concerning boundary measures and normal vector fields, including the divergence and trace theorems for Sobolev functions (complete proofs of which are given in section A.4 in the Appendix) and we discuss an important question of analysis concerning measurability and differentiability on $\Omega \times [0,T)$ of elements of $L^2_{loc}([0,T); H^1(\Omega))$.

§3.1 Linear Parabolic Systems and Their Weak Forms

In this section we consider specific examples of initial-boundary value problems for systems of partial differential equations whose weak forms are included in (3.1)–(3.3). Our perspective will be that the weak form is the problem of interest and that the differential equation is a symbolic equivalent which may be useful in some situations. Differential equations will therefore be treated in a purely formal way in this section; recovery of differential equations from their weak forms is considered in Theorem 5.5 in Chapter 5.

Assume that Ω is a proper open subset of $\mathbf{R}^n$ with unit outer normal vector field ν and surface measure σ on $\partial\Omega$ and that the divergence theorem holds on Ω. We can then integrate by parts in (3.2) to obtain that

$$A(u,v) = -\int_\Omega (Lu)\cdot v\,dx + \int_{\partial\Omega} (L_\partial u)\cdot v\,d\sigma$$

where

$$Lu = (a^{\eta,\omega}u_{x_\eta})_{x_\omega} - b^\eta u_{x_\eta} + (c^\omega u)_{x_\omega} - du,$$

$$L_\partial u = \nu^\omega(a^{\eta,\omega}u_{x_\eta} + c^\omega u) + fu,$$

$$F\cdot v = \int_\Omega (g - \operatorname{Div} G)\cdot v\,dx + \int_{\partial\Omega} (G\nu + \Gamma)\cdot v\,d\sigma$$

and where $\operatorname{Div} G$ is the $\mathbf{R}^N$-valued function $(\operatorname{Div} G)^i = G^{ij}_{x_j}$. The strong form (2.13) of (3.1), that is, $\langle u_t, v\rangle_{\mathcal{H}_0} = -A(u,v) + F\cdot v$ for each t, may then be written

$$\int_\Omega \big(u_t - Lu - g + \operatorname{Div} G\big)\cdot v\,dx = \int_{\partial\Omega} \big(-L_\partial u + G\nu + \Gamma\big)\cdot v\,d\sigma. \tag{3.4}$$

Now consider specific examples in which u is a formal solution of (3.4) with initial value $u(\cdot,0) = \varsigma_0$:

Example 1. If $\mathcal{H}_0 = [L^2(\Omega)]^N$ and $\mathcal{H}_1 = [H_0^1(\Omega)]^N$, then the boundary integral in (3.4) is zero hence so is the integral on the left, and we conclude formally that u is a solution of the system

$$\begin{cases} u_t = Lu + g - \operatorname{Div} G, \ (x,t) \in \Omega \times (0,T), \\ u(x,t) = 0, \ (x,t) \in \partial\Omega \times [0,T), \\ u(x,0) = \varsigma_0(x), \ x \in \Omega. \end{cases}$$

The "essential" Dirichlet boundary condition that $u = 0$ on $\partial\Omega$ is a consequence of the requirement that $u(\cdot,t) \in \mathcal{H}_1$. Note that a nonhomogeneous boundary condition $u = \varphi$ on $\partial\Omega$ can be accommodated in this formalism by extending φ to a map on $\overline{\Omega}$ and replacing the unknown u by $u - \varphi$.

Example 2. If $\mathcal{H}_0 = [L^2(\Omega)]^N$ and $\mathcal{H}_1 = [H^1(\Omega)]^N$ then we can still apply (3.4) with test functions v which are zero on $\partial\Omega$ to obtain the same differential equation as in Example 1. The left side of (3.4) is therefore zero for arbitrary v, so that the term in parentheses on the right side is zero on $\partial\Omega$ (provided that the restrictions to $\partial\Omega$ of elements $v \in \mathcal{H}_1$ are suitably dense). That is,

$$\begin{cases} u_t = Lu + g - \operatorname{Div} G, \ (x,t) \in \Omega \times (0,T), \\ L_\partial u = G\nu + \Gamma, \ (x,t) \in \partial\Omega \times [0,T), \\ u(x,0) = \varsigma_0(x), \ x \in \Omega. \end{cases}$$

The "natural" boundary condition that $L_\partial u = G\nu + \Gamma$ is thus derived as a consequence of the weak form (3.1) rather than as a requirement that $u \in \mathcal{H}_1$. Notice that in the special case that $N = 1$ and L is the Laplace operator, $a^{\eta,\omega}$ is the Kronecker delta and $c^\omega = 0$ so that $L_\partial u = \nu^\omega u_{x_\omega} = \nabla_x u \cdot \nu$. The natural boundary condition for the Laplace operator is therefore the familiar Neumann condition $\dfrac{\partial u}{\partial n} \equiv \nabla_x u \cdot \nu = G\nu + \Gamma$ on $\partial\Omega$.

Example 3. Examples 1 and 2 can be generalized to cases in which different components of u satisfy different boundary conditions and these may be coupled and may differ on different components of $\partial\Omega$. Thus let $\{\mathcal{C}_l\}$ be the connected components of $\partial\Omega$ and for each l let Q_l be given mapping $\mathcal{C}_l$ into the set of $N_l \times N$ matrices, where either $1 \le N_l \le N$ and $Q_l(x)$ has full rank at each $x \in \mathcal{C}_l$ or $N_l = 0$ and Q_l is the $N \times N$ zero matrix on $\mathcal{C}_l$. Let $\mathcal{H}_0 = [L^2(\Omega)]^N$ and

$$\mathcal{H}_1 = \{v \in [H^1(\Omega)]^N \text{ such that } Q_l v = 0 \text{ on } \mathcal{C}_l \text{ for all } l\}.$$

As in Examples 1 and 2 we find that the quantity in the parentheses on the left side of (3.4) is zero, hence the integral on the right side is zero for arbitrary $v \in \mathcal{H}_1$. Thus at each $x \in \mathcal{C}_l$, $L_\partial u - G\nu - \Gamma \perp v$ for all v satisfying $Q_l v = 0$. Since the solution u itself is to be in $\mathcal{H}_1$, we conclude formally that

$$\begin{cases} u_t = Lu + g - \operatorname{Div} G, \ (x,t) \in \Omega \times (0,T), \\ Q_l u = 0, \ (x,t) \in \mathcal{C}_l \times [0,T), \\ L_\partial u - G\nu - \Gamma \in (\operatorname{nullspace} Q_l)^\perp, \ (x,t) \in \mathcal{C}_l \times [0,T), \\ u(x,0) = \varsigma_0(x), \ x \in \Omega \end{cases}$$

where $\perp$ denotes orthogonal complement in $\mathbf{R}^n$. Observe that the second and third conditions, which are to hold for all l, comprise respectively N_l and $N - N_l$ scalar conditions, as expected.

Example 4. In this example we let $n = N = 3$ and define

$$\mathcal{H}_1 = \{v \in [H_0^1(\Omega)]^3 \text{ such that } \operatorname{div} v = 0 \text{ in } \Omega\}$$

where div is the divergence with respect to x, and take $\mathcal{H}_0$ to be the L^2 closure of $\mathcal{H}_1$, whose elements are thus divergence–free in the sense of weak derivatives. Proceeding as in the previous examples, we find that the integral on the left in (3.4) is zero for vector fields v which are divergence–free. Therefore if Π denotes the L^2 projection onto $\mathcal{H}_0$, then u formally satisfies

$$\begin{cases} u_t = \Pi\big(Lu + g - \operatorname{Div} G\big), \ (x,t) \in \Omega \times (0,T), \\ \operatorname{div} u = 0, \ (x,t) \in \Omega \times (0,T), \\ u = 0, \ (x,t) \in \partial\Omega \times [0,T), \\ u(x,0) = \varsigma_0(x), \ x \in \Omega. \end{cases}$$

Example 5. In this example we reverse the logic and begin with an initial-boundary value problem, now with a nonstandard boundary condition, and illustrate its weak formulation in the format above in (3.1)–(3.3). Again let $n = N = 3$ and consider the initial-boundary value problem

$$\text{(3.5)} \qquad \begin{cases} u_t = (\mu u_{x_\eta})_{x_\eta} + \nabla_x(\lambda \operatorname{div} u) - \nabla_x P + g, \ (x,t) \in \Omega \times (0,T), \\ u = -\kappa[(\nabla_x u + \nabla_x^{tr} u)\nu]_{tan}, \ (x,t) \in \partial\Omega \times [0,T), \\ u(x,0) = \varsigma_0(x), \ x \in \Omega \end{cases}$$

where ^{tr} denotes transpose and *tan* denotes projection orthogonal to the normal vector on $\partial\Omega$. The differential equation here is a simplified, linearized model of Newton's law applied to a compressible fluid in which u, P and g represent the fluid velocity, pressure and external force, and μ and λ are positive functions describing certain material properties of the fluid (see Batchelor [**4**], for example, for a detailed discussion of the rational mechanics of fluids). The boundary condition, which is the point of interest here, expresses a proposal of Navier [**32**] that on the boundary of the region containing the fluid the velocity u is proportional to the tangential component of the "stress." The so-called stress tensor is the linear map taking the unit normal to a hypothetical hypersurface inside Ω, or to the boundary surface $\partial\Omega$, into the force exerted by the fluid on one side of the surface on the fluid on the other, or on the boundary. In standard compressible fluid mechanics the stress tensor is proportional to the matrix $\frac{1}{2}[(\nabla_x u + \nabla_x^{tr} u)]$; this accounts for the second equation in (3.5), which implies in particular that u is to be tangential to the boundary. We therefore define

$$\mathcal{H}_1 = \{v \in [H^1(\Omega)]^3 \text{ such that } v \cdot \nu = 0 \text{ on } \partial\Omega\}$$

and $\mathcal{H}_0 = [L^2(\Omega)]^3$. To derive the weak form we multiply the differential equation in (3.5) by $v \in \mathcal{H}_1$ and apply the divergence theorem; the result is that $\langle u_t, v\rangle_{\mathcal{H}_0} = -A(u,v) + F \cdot v$ where

$$A(u,v) = \int_\Omega \big(\mu \nabla_x u : \nabla_x v + \lambda \operatorname{div} u \operatorname{div} v\big) dx - \int_{\partial\Omega} (\mu \nabla_x u\, \nu) \cdot v \, d\sigma.$$

This appears to be inconsistent with (3.2), however, because the integrand in the boundary integral involves $\nabla_x u$, whose restriction to $\partial\Omega$ is not in general well-defined for $u(\cdot,t) \in [H^1(\Omega)]^3$ (see Theorem 3.2 below). We can reduce the order of u appearing here, however, by making use of the boundary condition in (3.5) as follows. If $v \in \mathcal{H}_1$ then v is tangential on the boundary, so that at points of $\partial\Omega$,

$$\begin{aligned}(\nabla_x u\, \nu) \cdot v &= [(\nabla_x u + \nabla_x^{tr} u)\nu]_{tan} \cdot v - [\nabla_x^{tr} u\, \nu]_{tan} \cdot v \\ &= -\kappa^{-1} u \cdot v - [\nabla_x^{tr} u\, \nu]_{tan} \cdot v.\end{aligned}$$

The order of the second term on the right here can be reduced by ordinary calculus: Let $\partial\Omega$ be given locally by $\{x : h(x) = 0\}$ where $h : \mathbf{R}^3 \to \mathbf{R}$ and $\nu = \nabla_x h/|\nabla_x h|$, and let $s \to x(s)$ be a curve taking values locally on $\partial\Omega$. Then differentiating the relation $u(x(s),t) \cdot \nabla_x h(x(s)) = 0$ with respect to s for fixed t we find that the tangential component of the vector $\nabla_x^{tr} u\, \nabla_x h + h'' u$ is zero on $\partial\Omega$ and therefore that $[\nabla_x u^{tr} \nu]_{tan} = -|\nabla_x h|^{-1}[h'' u]_{tan}$. Thus the boundary integral in the expression for A above may be written exactly as in (3.2) with $f = \mu\big(\kappa^{-1} I - |\nabla_x h|^{-1} h''\big)$.

Systems of partial differential equations such as those considered in the above examples are loosely described as "parabolic" if the associated bilinear forms A satisfy the fourth condition in (2.1). More precisely, the pair of operators (L, L_∂) is said to be *locally uniformly parabolic on* $[0,T)$ *with respect to the spaces* $\mathcal{H}_0$ *and* $\mathcal{H}_1$ if the four conditions in (2.1) are satisfied.

§3.2 Preliminaries

In this section we review various facts and notations concerning boundary measures, normal vectors and measurability on product spaces, preliminary to the application of Theorem 2.1 to the examples of section 3.1. Complete statements and proofs are given in sections A.4 and A.5 in the Appendix.

Boundary Measures and Trace and Divergence Theorems

The following theorem shows that if Ω is a Lipschitz domain then there is a Radon measure σ (see section A.1) on the Borel subsets of $\partial\Omega$ corresponding to surface area and there is a unit outer normal vector field ν defined σ-almost everywhere:

Theorem 3.1. *Let* Ω *be a Lipschitz domain in* $\mathbf{R}^n$ *as described in* Definition 1.7 *and let* $\mathcal{U}_i$, ψ_i *and* T_i *be as in that definition.*

(a) *Define a cover* $\{\mathcal{W}_i\}_i$ *of* $\partial\Omega$ *by* $\mathcal{W}_1 = \mathcal{U}_1$ *and* $\mathcal{W}_i = \mathcal{U}_i - \cup_{j=1}^{i-1} \mathcal{W}_i$ *for* $i \geq 2$; *and for* $f \in C_c(\partial\Omega)$ *define*

$$\mathcal{I}(f) = \sum_i \int_{T_i^{-1}(\partial\Omega\cap\mathcal{W}_i)} (f \circ T_i)(y)\sqrt{1 + |\nabla\psi_i(y)|^2}\, dy.$$

Then there is a unique Radon measure σ *on the Borel sets in* $\partial\Omega$ *such that*

$$\int_{\partial\Omega} f\, d\sigma = \mathcal{I}(f) \tag{3.6}$$

for all $f \in C_c(\partial\Omega)$.

(b) $L^p(\partial\Omega, \sigma)$ *is separable for* $p \in [1, \infty)$.

(c) *If* $f : \partial\Omega \to \mathbf{R}$ *is Borel measurable and nonnegative and* $E \subseteq \partial\Omega \cap \mathcal{U}_i$ *is a Borel set, then*

$$\int_E f\, d\sigma = \int_{T_i^{-1}(E)} (f \circ T_i)(y)\sqrt{1 + |\nabla\psi_i(y)|^2} dy \tag{3.7}$$

(*whose value could be* $+\infty$).

(d) *If* $f : \partial\Omega \to \mathbf{R}$ *is Borel measurable then* $f \in L^1(\partial\Omega)$ *if and only if* $\mathcal{I}(|f|) < \infty$ (*which is meaningful by* (c)), in which case

$$\int_{\partial\Omega} f d\sigma = \mathcal{I}(f^+) - \mathcal{I}(f^-)$$

where $f^+ = f \vee 0$ and $f^- = f^+ - f$.

(e) *There is a set* $E \subset \partial\Omega$ *of* σ*-measure zero and a measurable mapping* ν *from* $\partial\Omega - E$ *into the set of unit vectors in* $\mathbf{R}^n$ *such that for* $x_0 \in \partial\Omega - E$, $x_0 + r\nu(x_0)$ *is in* Ω *for small negative* r *and in* $(\overline{\Omega})^c$ *for small positive* r; *and if* $s \to x(s)$ *is a continuous mapping from a nontrivial interval into* $\partial\Omega$ *with* $x(0) = x_0$ *and* x *differentiable at* $s = 0$, *then* $\dot{x}(0) \cdot \nu(x_0) = 0$.

The *trace* of an element $f \in C_c(\mathbf{R}^n)$ with respect to a proper subset Ω of $\mathbf{R}^n$ is its restriction $f\big|_{\partial\Omega}$ to $\partial\Omega$. The trace operator is thus a linear map from $C_c(\mathbf{R}^n)$ to $C_c(\partial\Omega)$. The following theorem shows that the trace operator extends to an operator $\mathcal{T}$ on $W^{1,p}(\Omega)$ with values in $L^p(\partial\Omega)$:

Theorem 3.2. *Let* Ω *be a uniformly Lipschitz domain in* $\mathbf{R}^n$ *with parameters* ε_0 *and* L *as described in* Definition 1.7 *and let* $p \in [1, \infty)$. *Then there is an operator* $\mathcal{T} \in \mathcal{L}(W^{1,p}(\Omega), L^p(\partial\Omega))$ *such that for* $f \in C_c(\mathbf{R}^n)$, $\mathcal{T}(f\big|_{\overline{\Omega}}) = f\big|_{\partial\Omega}$. *The norm of* $\mathcal{T}$ *is determined by* n, L, ε_0 *and* p.

Finally, the divergence theorem holds for Sobolev functions on uniformly Lipschitz domains:

Theorem 3.3. *Let Ω be a uniformly Lipschitz domain in $\mathbf{R}^n$, ν the unit normal vector field on $\partial\Omega$ defined in* Theorem 3.1(e) *and $f \in W^{1,1}(\Omega)$. Then for $j = 1, \ldots, n$,*

$$\int_\Omega f_{x_j} dx = \int_{\partial\Omega} (\mathcal{T} f)\nu^j \, d\sigma. \tag{3.8}$$

The Spaces $L^{p,q}$

Next we consider an important issue concerning measurability on product spaces. The application of Theorem 2.1 to the problems discussed in section 3.1 will show that, under appropriate conditions, solutions are in the space $C([0,T); [L^2(\Omega)]^N)$ with equivalence classes in a subspace of $L^2_{loc}([0,T); [H^1(\Omega)]^N)$. It is natural to ask whether for such a solution u we can choose for each t a representative $w(\cdot,t)$ of $u(t)$, which is therefore a measurable function on Ω, in such a way that $w(\cdot,\cdot)$ is measurable on the product space $\Omega \times (0,T)$. If so, then w would be amenable to analyses involving the Fubini and Tonelli theorems, and its weak differentiability with respect to x on the $(n+1)$-dimensional domain $\Omega \times (0,T)$ could be considered.

This question is somewhat subtle. An example due to Sierpinski (see [**13**] pg. 69 or [**36**] pg. 152) shows that for each $t \in [0,1]$ there is (or may be—see the final sentence in this paragraph) a countable set $S_t \subseteq [0,1]$ such that

$$S \equiv \{(x,t) \in [0,1]^2 : t \in [0,1] \text{ and } x \in S_t\}$$

is not measurable with respect to Lebesque measure on $\mathbf{R}^2$. Its characteristic function is therefore not measurable on $[0,1]^2$ and no redefinition on a set of measure zero can make it so. On the other hand, if $u(t)$ is the Lebesgue equivalence class on $[0,1]$ of the characteristic function of S_t, then $u(t) = 0$ in $L^2([0,1])$ for all t, and therefore $u \in C([0,1]; L^2([0,1])$. The construction in the paragraph above, however, in which w is derived from u, could lead to a function w which is not in fact measurable on the product space, namely, the characteristic function of S. While rather striking and disconcerting, this example depends on the continuum hypothesis, and the continuum hypothesis is known to be undecidable.

Of course, we could take $w \equiv 0$ in the above example, thus obtaining a representative of u which *is* measurable on the product space. We will show that such a choice is always possible for the functions we will encounter. Here we give a minimal discussion, sufficient for the statement of the basic existence result in the next section; a complete and self-contained exposition is given in section A.5.

Fix a nontrivial interval $I \subset \mathbf{R}$, an open set Ω in $\mathbf{R}^n$ and $p \in [0,\infty]$ with Hölder conjugate p', and let $f \in L^p(I; L^2(\Omega))$. Then given $v \in L^2(\Omega)$ and $\psi \in L^{p'}(I)$, the integral $\int_I \left(\int_\Omega f(t)v \, dx \right) \psi(t) dt$ is well-defined, that is, is independent of representatives and is finite. Thus f defines a real-valued bilinear function on the vector space $L^2(\Omega) \times L^{p'}(I)$ in a natural way. We will see that this same action is induced by an equivalence class f_{m} whose members are measurable on the product space $\Omega \times I$. We begin with the following definition:

Definition 3.4. *Let I, Ω and p be as above. Then $L^{p,2}(\Omega \times I)$ is the set of equivalence classes $[\varphi]_{\Omega\times I}$ of functions φ on $\Omega \times I$ which are measurable with respect to the (Lebesgue) product measure on $\Omega \times I$ and for which the L^p norm on I of $|\varphi(\cdot,t)|_{L^2(\Omega)}$, denoted by $|\varphi|_{L^{p,2}(\Omega\times I)}$, is finite. The vector space $L^{p,2}_{(t)loc}(\Omega \times I)$ is the set of equivalence classes of functions which are measurable on $\Omega \times I$ and whose restrictions to $\Omega \times I'$ are in $L^{p,2}(\Omega \times I')$ for every compact interval $I' \subseteq I$. The cross product spaces $[L^{p,2}(\Omega \times I)]^N$ and $[L^{p,2}_{(t)loc}(\Omega \times I)]^N$ are then defined in the obvious ways.*

Notice that Tonelli's theorem underlies the definition: if φ is measurable on the product space, then $|\varphi(\cdot,t)|_{L^2(\Omega)}$ defines a measurable class on I. Moreover, the Fubini and Tonelli theorems show that, if $\varphi \in L^{p,2}(\Omega\times I)$, then given $v \in L^2(\Omega)$ and $\psi \in L^{p'}(I)$, the integral $\int_I \left(\int_\Omega \varphi(x,t)v(x)dx\right)\psi(t)dt$ is well-defined and is equal to the integral over $\Omega \times I$ with respect to the product measure.

The following results are extracted from the more extensive discussion in section A.5. We note first that if $f \in L^p(I; H^1(\Omega))$ and if φ is any representative of f, then for almost all t each weak derivative $\varphi_{x_i}(t)$ exists in the weak sense on Ω, thereby generating a weakly measurable map from I into $L^2(\Omega)$. Its equivalence class over I, which we denote by f_{x_j}, is then an element of $L^p(I; L^2(\Omega))$.

Theorem 3.5. *Let Ω be an open set in $\mathbf{R}^n$ and I a nontrivial interval in $\mathbf{R}$ and let $p \in [0,\infty]$ with Hölder conjugate p'.*

(a) *There is a one-to-one correspondence $f \longleftrightarrow f_{\mathrm{m}}$ between elements $f \in L^p(I; L^2(\Omega))$ and $f_{\mathrm{m}} \in L^{p,2}(\Omega \times I)$ such that for all $v \in L^2(\Omega)$ and $\psi \in L^{p'}(I)$,*

$$\int_I \left(\int_\Omega f(t)v\right)\psi(t)dt = \int_I \left(\int_\Omega f_{\mathrm{m}}(x,t)v(x)dx\right)\psi(t)dt.$$

(b) *If $f \in L^p(I; L^2(\Omega))$ and $\psi \in C_c^\infty(\Omega \times I^\circ)$ (I° is the interior of I) then*

$$\int_I \left(\int_\Omega f(t)\psi(\cdot,t)\right)dt = \int_I \left(\int_\Omega f_{\mathrm{m}}(x,t)\psi(x,t)dx\right)dt.$$

(c) *If $f \in L^p(I; H^1(\Omega))$ then f_{m} has partial derivatives $D_{x_j} f_{\mathrm{m}}$ in the weak sense on $\Omega \times I^\circ$ and $D_{x_j} f_{\mathrm{m}} = (f_{x_j})_{\mathrm{m}}$.*

(d) *If $f \in L^\infty(I; L^\infty(\Omega))$ and $v \in L^\infty(I; L^2(\Omega))$ then $fv \in L^\infty(I; L^2(\Omega))$ and $(fv)_{\mathrm{m}} = f_{\mathrm{m}} v_{\mathrm{m}}$.*

The first three of these results are special cases of Corollaries A.16 and A.17 and Theorem A.19 in the Appendix and the last is a special case of Lemma 5.6.

We remark that if $f \in C(I; L^2(\Omega))$ then its equivalence class $[f]_I$ is in $L^\infty_{(t)loc}(I; L^2(\Omega))$ and therefore its measurable counterpart is included in Definition 3.4. There will be no ambiguity in writing $f_{\mathfrak{m}}$ in place of the more complicated $([f]_{\mathcal{I}})_{\mathfrak{m}}$.

§3.3 The Basic Existence Theorem

In this section we apply Theorem 2.1(b) to obtain a basic result on the existence and uniqueness of solutions to the problems described in (3.1)–(3.3). We show that under fairly mild conditons on Ω, A, F and ς_0, solutions exist, are unique and satisfy the energy equality (2.5) and the bound (2.4). (The corresponding application of Theorem 2.1(a) is straightforward but more technical and will play no role in the subsequent development.)

Theorem 3.6. *Let $T \in (0, \infty]$ and assume the following:*

(a) Ω *is an open set in* $\mathbf{R}^n$, $n \geq 2$, *not necessarily bounded, satisfying one of the following: either* (I) Ω *is a uniformly Lipschitz domain strictly contained in* $\mathbf{R}^n$, *or* (II) Ω *satisfies the cone condition (this includes the case* $\Omega = \mathbf{R}^n$).

(b) $\mathcal{H}_0$ *and* $\mathcal{H}_1$ *are closed subspaces of* $[L^2(\Omega)]^N$ *and* $[H^1(\Omega)]^N, N \geq 1$, *respectively, with the inherited norms and with* $\mathcal{H}_1$ *dense in* $\mathcal{H}_0$;

(c) A *is as defined in* (3.2) *(the term* $\int_{\partial\Omega} (fu) \cdot v \, d\sigma$ *is omitted in case* (II)) *with coefficient matrices satisfying the following:*
- *the components of* $a^{\eta,\omega}$ *are elements of* $L^\infty_{loc}([0,T); L^\infty(\Omega))$
- *the components of* b^η *and* c^ω *are finite sums in which each summand is an element of* $L^\infty_{loc}([0,T); L^s(\Omega))$ *for some* $s \in (n, \infty]$
- *the components of* d *are finite sums in which each summand is in* $L^\infty_{loc}([0,T); L^r(\Omega))$ *for some* $r \in (n/2, \infty]$
- *in case* (I) *the components of* f *are finite sums in which each summand is in* $L^\infty_{loc}([0,T); L^s(\partial\Omega))$ *for some* $s \in (n, \infty]$

  (in each of the three items above the values of s *and* r *may depend on the component and the summand in question)*
- *given* $\bar{t} \in [0,T)$ *there is a positive number* $\theta(\bar{t})$ *such that for almost all* $(x,t) \in \Omega \times (0, \bar{t})$
$$(M^{tr} a^{\eta\omega}_{\mathfrak{m}}(x,t) M)^{\omega\eta} \geq 2\theta(\bar{t}) |M|^2$$
  for all $N \times n$ *matrices* M; *here* ^{tr} *denotes transpose, summation over* η *and* ω *is understood, any choice of matrix norm is allowed and* $a^{\eta\omega}_{\mathfrak{m}}$ *is any member of the equivalence class of matrix-valued*

functions on $\Omega \times [0,T)$ *whose elements correspond to those of* $a^{\eta,\omega}$ *as in* Theorem 3.5.

Then the following hold:

(d) *The conditions in* (2.1) *are satisfied with* $A \in L^\infty_{loc}([0,T); \mathcal{B}(\mathcal{H}_1 \times \mathcal{H}_1))$. *In particular, for* $\bar{t} \in [0,T)$ *and almost all* $t \in [0,\bar{t}]$,

$$\|A(t)\|_{\mathcal{B}(\mathcal{H}_1 \times \mathcal{H}_1)} \le C(\bar{t})$$

and

$$A(t)(v,v) \ge \theta(\bar{t})|v|^2_{H^1} - \lambda(\bar{t})|v|_{L^2} \tag{3.9}$$

for all $v \in \mathcal{H}_1$, *where* $\lambda \in L^\infty_{loc}([0,T))$ *and* $|\lambda|_{L^\infty([0,\bar{t}])}$ *and* $C(\bar{t})$ *depend on the cone parameters* δ *and* r *in* Definition 1.6(b) *and on upper bounds for* $\theta(\bar{t})^{-1}$, *for the norms of the coefficients of* A *on* $[0,\bar{t}]$ *and the numbers of summands described in* (c), *and for* $(s-n)^{-1}$ *and* $(r-n/2)^{-1}$ *for all values of* s *and* r *occurring in* (c).

(e) *Given*

- g *as in* (3.3) *with components which are finite sums of elements in* $L^2_{loc}([0,T); L^q(\Omega))$ *for some* $q \in [2n/(n+2), 2]$ *for* $n \ge 3$ *and in* $(1,2]$ *for* $n=2$
- G *as in* (3.3) *with components in* $L^2_{loc}([0,T); L^2(\Omega))$
- (*in case* (I)) Γ *as in* (3.3) *with components in* $L^2_{loc}([0,T); L^2(\partial\Omega))$
- $\varsigma_0 \in \mathcal{H}_0$,

there is a unique $u \in C([0,T); \mathcal{H}_0)$ *whose* $[0,T)$*-equivalence class is in* $L^2_{loc}([0,T); \mathcal{H}_1)$ *and which satisfies the conclusions of* Theorem 2.1:

- u *satisfies* (3.1) *for all* $v \in \mathcal{H}_1$
- *the energy equality*

$$\begin{aligned} &\tfrac{1}{2}|u(t)|^2_{[L^2(\Omega)]^N} + \int_0^t A(s)(u(s),u(s))ds \\ &\quad = \tfrac{1}{2}|\varsigma_0|^2_{[L^2(\Omega)]^N} + \int_0^t\int_\Omega [g\cdot u + G : \nabla_x u]\,dx + \int_0^t\int_{\partial\Omega} \Gamma \cdot u\, d\sigma \end{aligned} \tag{3.10}$$

  holds for all $t \in [0,T)$ (*integrals over* $\partial\Omega$ *are omitted in case* (II))
- *the bound*

$$\begin{aligned} &\sup_{0\le t\le \bar{t}} |u(t)|^2_{[L^2(\Omega)]^N} + \theta(\bar{t}) \int_0^{\bar{t}} e^{\int_t^{\bar{t}} 2\lambda(s)ds} |u(t)|^2_{[H^1(\Omega)]^N} dt \\ &\quad \le e^{\int_0^{\bar{t}} 2\lambda(t)dt} |\varsigma_0|^2_{[L^2(\Omega)]^N} + \theta(\bar{t})^{-1} \int_0^{\bar{t}} e^{\int_t^{\bar{t}} 2\lambda(s)ds} C_F(t)dt \end{aligned} \tag{3.11}$$

  holds for $t \in [0,T)$; *here*

$$C_F(t) = C_1 C_g(t) + C_G(t) + C_2 C_\Gamma(t)$$

where $C_g(t)$ is the sum of the squares of the $L^q(\Omega)$ norms in the first item in (e), *C_1 depends on n and in the case that $n = 2$ on an upper bound for $(q-2)^{-1}$ for all such q, $C_G(t)$ and $C_\Gamma(t)$ are the sums of the squares of the $L^2(\Omega)$ and $L^2(\partial\Omega)$ norms of the components of $G(t)$ and $\Gamma(t)$, and C_2 depends on an upper bound for the norm of the trace operator which maps $H^1(\Omega)$ to $L^2(\partial\Omega)$ (the term $C_2 C_\Gamma$ is omitted in case* (II))

- *$u_{\mathrm{m}} \in \left[L^{\infty,2}_{(t)loc}(\Omega \times [0,T))\right]^N$ and the weak derivatives $D_{x_\eta} u_{\mathrm{m}}$ exist on $\Omega \times (0,T)$ and are in $\left[L^{2,2}_{(t)loc}(\Omega \times [0,T))\right]^N$; and if $[C_c^\infty(\Omega)]^N \subset \mathcal{H}_1$ then the differential equation*

$$\frac{\partial}{\partial t} u_{\mathrm{m}} = \frac{\partial}{\partial x_\omega}\left(a^{\eta,\omega}_{\mathrm{m}} D_{x_\eta} u_{\mathrm{m}}\right) - b^\eta_{\mathrm{m}} D_{x_\eta} u_{\mathrm{m}} + \frac{\partial}{\partial x_\omega}\left(c^\omega_{\mathrm{m}} u_{\mathrm{m}}\right) - d_{\mathrm{m}} u_{\mathrm{m}} + g_{\mathrm{m}} - \operatorname{Div} G_{\mathrm{m}} \tag{3.12}$$

holds in the sense of distributions on $\Omega \times (0,T)$, where each term here is either locally integrable on $\Omega \times (0,T)$ or is a distribution derivative $\frac{\partial}{\partial t}, \frac{\partial}{\partial x_\omega}$ or Div *of a locally integrable function.*

Remarks:

(1) The conclusion in (d) that $\lambda \in L^\infty_{loc}([0,T))$ is stronger than the minimal requirement in (2.1) that $\lambda \in L^1_{loc}([0,T))$. This results from integrability hypotheses in (c) which are slightly stronger than is strictly necessary. Small improvements can be made at considerable technical expense involving more complicated index arithmetic, as will be evident from the proof.

(2) For $N = 1$ the coefficients $a^{\eta\omega}$ are scalar-valued and the last condition in (c) reduces to the requirement that, for almost all $(x,t) \in \Omega \times (0,\bar{t})$, $a^{\eta,\omega}_{\mathrm{m}}(x,t)\xi^\eta\xi^\omega \geq 2\theta(\bar{t})|\xi|^2$ for all $\xi \in \mathbf{R}^n$.

(3) See also Remarks (1), (2) and (4) following the statement of Theorem 2.1 for further comments related to the material of this section.

PROOF of Theorem 3.6. To prove (d) we need to check that the following hold for $z, w \in \mathcal{H}_1$: (*i*) the mapping $t \to A(t)(z,w)$ is measurable; (*ii*) $|A(t)(z,w)| \leq C(\bar{t})|z|_{\mathcal{H}_1}|w|_{\mathcal{H}_1}$ for $\bar{t} \in (0,T)$; and (*iii*) for $t \in [0,\bar{t}]$, $A(t)(z,z) \geq \theta(\bar{t})|z|^2_{\mathcal{H}_1} - \lambda(\bar{t})|z|^2_{\mathcal{H}_0}$ where $C(\bar{t})$ and $\lambda(\bar{t})$ are as described in (d).

Consider first the contribution to $A(z,w)$ from $(a^{\eta,\omega} z_{x_\eta}) \cdot w_{x_\omega}$. (We will suppress here the distinction between A and any specific weakly measurable representative on $[0,T)$.) Since $|z_{x_\eta}||w_{x_\omega}| \in L^1(\Omega)$ and $|a^{\eta,\omega}| \in$

$L^\infty_{loc}([0,T); L^\infty(\Omega))$, the mapping $t \to \int_\Omega \left(a^{\eta,\omega}(\cdot,t)z_{x_\eta}\right) \cdot w_{x_\omega} dx$ is a measurable function of t which is bounded in absolute value as required for (ii). Concerning (iii) we record that by the last hypothesis in (c),

$$\int_\Omega \left(a^{\eta,\omega}(\cdot,t)z_{x_\eta}\right) \cdot z_{x_\omega} dx = \int_\Omega \left[(\nabla_x z)^{tr} a^{\eta,\omega}(\cdot,t)\nabla_x z\right]^{\omega,\eta}$$
$$\geq 2\theta(\bar{t})|\nabla_x z|^2_{L^2} = 2\theta(\bar{t})\Big(|z|^2_{\mathcal{H}_1} - |z|^2_{\mathcal{H}_0}\Big).$$

Next, abusing notation slightly, we denote by $bz_x w$ any one of the summands in the expansion of the term $(b^\omega z_{x_\eta}) \cdot w$ in the expression (3.2) defining $A(z,w)$. Then for almost all t, $b(\cdot,t) \in L^s(\Omega)$ for a particular $s \in (n,\infty]$ so that the Hölder conjugate s' of s is in $[1, n/(n-1))$. Then $\gamma \equiv 2s'/(2-s')$ satisfies $1/\gamma + 1/2 = 1/s'$, and γ is in $[2,p^*)$ where $p^* = 2n/(n-2)$ for $n \geq 3$ and $p^* \in [2,\infty)$ is sufficiently large depending on s for $n=2$. It follows that $H^1(\Omega)$ is continuously contained in $L^p(\Omega)$ for $p \in [2,p^*]$ (see Theorem A.3; the cone condition is applied here) and therefore

$$|z_x w|_{L^{s'}} \leq C|z_x|_{L^2}|w|_{L^\gamma} \leq C|z|_{H^1}|w|_{H^1}. \tag{3.13}$$

Since $b \in L^\infty_{loc}([0,T); L^s(\Omega))$, the mapping $t \to \int_\Omega bz_x w\, dx$ is measurable in t with image bounded in absolute value by $C|b(\cdot,t)|_{L^s(\Omega)}|w|_{H^1}|z|_{H^1}$ for almost all t, and is therefore in $L^\infty_{loc}([0,T))$ as required for (i) and (ii). For (iii) we recall that $\gamma \in [2,p^*)$, so that, applying (3.13) with $w = z$ and interpolating the L^γ norm between and 2 and p^* we obtain that for some $\tau \in (0,1]$ and arbitrary $\delta > 0$,

$$\left|\int_\Omega bz_x z\, dx\right| \leq C|b|_{L^s}|z_x|_{L^2}|z|^\tau_{L^2}|z|^{1-\tau}_{L^{p^*}}$$
$$\leq C|b|_{L^s}|z|^{2-\tau}_{H^1}|z|^\tau_{L^2}$$
$$\leq \delta|z|^2_{H^1} + C(\delta)|z|^2_{L^2}.$$

Bounds for the terms $(c^\omega z)\cdot w_{x_\omega}$ and $(dz)\cdot w$ in (3.2) are obtained in a similar way. For the boundary integral in case (I) we denote by fwz any one of the summands in the expression for $(fw)\cdot z$. Then $f \in L^\infty_{loc}([0,T); L^s(\partial\Omega))$ where s and its conjugate s' are in the same ranges as in (3.13), and therefore by Theorem 3.2,

$$|zw|_{L^{s'}(\partial\Omega)} \leq C|zw|_{W^{1,s'}(\Omega)}$$
$$\leq C(|zw|_{L^{s'}(\Omega)} + |zw_x|_{L^{s'}(\Omega)} + |z_x w|_{L^{s'}(\Omega)}).$$

The argument is now the same as for the term $(bz_{x_\eta})\cdot w$, proceeding from (3.13) above.

These observations prove that $A \in L^\infty_{loc}([0,T); \mathcal{B}(\mathcal{H}_1 \times \mathcal{H}_1))$ with the required bound and that, given $\bar{t} \in (0,T)$ and $\delta > 0$, there is a constant C depending on $\bar{t}$ and δ and on the same quantities listed in (d) for λ such that for almost all $t \in [0, \bar{t}]$ and for all $z \in [H^1(\Omega)]^N \supseteq \mathcal{H}_1$,

$$A(t)(z, z) \geq 2\theta(\bar{t})|z|^2_{H^1(\Omega)} - \delta|z|^2_{H^1(\Omega)} - C(\delta)|z|^2_{L^2(\Omega)}.$$

Thus (iii) holds for appropriate choices of constants δ and $\lambda(\bar{t})$. This completes the proof of (d).

To prove (e) we have to check that the mapping F defined in (3.3) is in $L^2_{loc}([0, T); \mathcal{H}_1^*)$. The required measurability and boundedness are obvious for the contributions from the terms involving G and Γ (in case (I)). For the term involving a particular summand of a particular component of g, still denoted $g \in L^2_{loc}([0, T); L^q(\Omega))$, we have that, by the hypothesis on q, the Hölder conjugate q' of q is in the range $[2, p^*]$, where p^* is as above preceding (3.13). Therefore for $v \in H^1$, $\left| \int_\Omega g(\cdot, t) \cdot v \, dx \right| \leq C|g(\cdot, t)|_{L^q(\Omega)}|v|_{H^1(\Omega)}$. The contribution of the term involving g to $|F(t)|^2_{\mathcal{H}_1^*}$ is therefore bounded by $C_1 C_g(t)$, which is in $L^1_{loc}([0, T))$, as required.

Theorem 2.1(b) now applies to show that (3.1)–(3.3) has a unique solution u as described in (e) and which satisfies (3.10) and (3.11). The statements that $u_{\mathrm{m}} \in \left[L^{\infty,2}_{(t)loc}\right]^N$ and $D_{x_j} u_{\mathrm{m}} \in \left[L^{2,2}_{(t)loc}\right]^N$ then follow from Theorem 3.5(c).

To prove (3.12) we fix $\varphi \in C_c^\infty(\Omega \times (0, T))$ so that by Theorem 3.5(b),

$$\int_{\Omega \times (0,T)} u_{\mathrm{m}} \varphi_t = \lim_{h \to 0} h^{-1} \int_0^T \int_\Omega u(t)[\varphi(\cdot, t) - \varphi(\cdot, t - h)] dx \, dt.$$

Moving the difference quotient onto u and applying (3.2) and the fact that $\varphi(\cdot, t) \in \mathcal{H}_1$, we then obtain

$$\begin{aligned} \int_{\Omega \times (0,T)} u_{\mathrm{m}} \varphi_t = \lim_{h \to 0} h^{-1} \int_0^T \int_t^{t+h} & \Big[A(s)(u(s), \varphi(\cdot, t)) \\ & - \int_\Omega \big(g(s) \cdot \varphi(\cdot, t) + G : \nabla \varphi(\cdot, t) \big) dx \Big] ds \, dt. \end{aligned}$$

For fixed h we can reverse the s and t integrals above (to justify this approximate φ by a function which is piecewise constant in t and which when substituted into the brackets on the right yields a measurable function of (s, t)) then take the limit as $h \to 0$ to obtain

$$\begin{aligned} \int_{\Omega \times (0,T)} u_{\mathrm{m}} \varphi_t &= \int_0^T \int_\Omega \left[(a^{\eta,\omega} u_{x_\eta}) \cdot \varphi_{x_\omega} + \ldots - g \cdot \varphi - G : \nabla \varphi \right] dx ds \\ &= \int_{\Omega \times (0,T)} \left[(a^{\eta,\omega}_{\mathrm{m}} u_{\mathrm{m} x_\eta}) \cdot \varphi_{x_\omega} + \ldots + \big(- g_{\mathrm{m}} + \operatorname{Div} G_{\mathrm{m}} \big) \cdot \varphi \right], \end{aligned}$$

the last step here being justified by Theorem 3.5(d).

□

§3.4 Parabolic Systems with Symmetry

In this section we consider parabolic systems satisfying the symmetry conditions of Theorems 2.11 and 2.12. Specifically, we examine solutions of the problem (3.1)–(3.3) with the terms involving c^j, G and Γ absent, with $d = d_1 + d_2$ where d_1 and f are symmetric, and under the assumption that $(a^{\eta,\omega})^{tr} = a^{(\omega,\eta)}$ for all η and ω. The problem is therefore to find $u \in C([0,T);\mathcal{H}_0^{weak})$ with equivalence class in $L^2([0,T);\mathcal{H}_1)$ such that

$$\langle u(t), v\rangle_{L^2(\Omega)} = \langle \varsigma_0, v\rangle_{L^2(\Omega)} + \int_0^t \big[- A_{sym}(s)(u(s),v) - B(s)(u(s),v) + \langle g(s), v\rangle_{L^2(\Omega)}\big] ds \tag{3.14}$$

for all $v \in \mathcal{H}_1$, where

$$A_{sym}(z,w) = \int_\Omega \big[(a^{\eta,\omega} z_{x_\eta}) \cdot w_{x_\omega} + (d_1 z)\cdot w\big] dx + \int_{\partial\Omega} (fz)\cdot w \, d\sigma \tag{3.15}$$

and

$$B(z,w) = \int_\Omega \big[(b^\eta z_{x_\eta}) \cdot w + (d_2 z)\cdot w\big] dx \tag{3.16}$$

for $z, w \in \mathcal{H}_1$. Again, $a^{\eta,\omega}$, b^η, d_1 and d_2 are $N\times N$ matrix-valued and g is an $\mathbf{R}^N$-valued function of $(x,t) \in \Omega\times[0,T)$; and f is an $N \times N$ matrix-valued function of $(x,t) \in \partial\Omega \times [0,T)$ (the term involving f is omitted if $\Omega = \mathbf{R}^n$).

The following theorem gives the application of Theorems 2.11 and 2.12 to the above problem, part (a) for initial data $\varsigma_0 \in \mathcal{H}_1$ and part (b) for $\varsigma_0 \in \mathcal{H}_0$:

Theorem 3.7. *Let T, Ω, $\mathcal{H}_0$ and $\mathcal{H}_1$ satisfy hypotheses* (a) *and* (b) *of* Theorem 3.6 *and let $A = A_{sym} + B$ be as above in* (3.15)–(3.16) *with coefficients satisfying the following:*

- *The components of each $a^{\eta,\omega}$ are in $wAC^2([0,T);L^\infty(\Omega))$, those of d_1 are finite sums of elements of $wAC^2([0,T);L^{r_1}(\Omega))$ for values of r_1 in $(n/2,\infty]$, and in case* (I) *of* Theorem 3.6 *those of f are finite sums of elements of $wAC^2([0,T);L^s(\partial\Omega))$ for values of s in $(n,\infty]$.*

- *The components of each b^η are in $L^2_{loc}([0,T);L^\infty(\Omega))$ and those of d_2 are finite sums in which each summand is in $L^2_{loc}([0,T);L^{r_2}(\Omega))$ for some $r_2 \in [n,\infty]$ if $n \geq 3$ and in $(2,\infty]$ if $n = 2$.*

- *For each t and for all η and ω, $(a^{\eta,\omega})^{tr} = a^{(\omega,\eta)}$ a.e. in Ω, and at each t, d_1 is symmetric a.e. in Ω and in case* (I) *of* Theorem 3.6 *f is symmetric a.e. in $\partial\Omega$.*

- *The matrix-valued functions $a_{\mathrm{m}}^{\eta,\omega}$ satisfy the coerciveness condition in the last item in* Theorem 3.6(c).

(Different values of r_1, r_2 and s are allowed for different components and summands of d_1, d_2 and f.)

(a) *Then given $\varsigma_0 \in \mathcal{H}_1$ and $g \in L^2_{loc}([0,T);\mathcal{H}_0)$,*

- *there is a unique $u \in wAC^2_{loc}([0,T);\mathcal{H}_0) \cap C([0,T);\mathcal{H}_1^{weak})$ which satisfies (3.14) for all $v \in \mathcal{H}_1$ and $t \in [0,T)$;*
- *and given specific representatives $t \mapsto u'(t)$, $A(t)$ and (in case (I)) $f(t)$ into $\mathcal{H}_0$, $\mathcal{B}(\mathcal{H}_1 \times \mathcal{H}_1)$ and $L^2(\partial\Omega)$ respectively, there is a set $E \subset [0,T)$ of measure zero such that for $t \notin E$ and for all $v \in \mathcal{H}_1$,*

$$\begin{aligned}\int_\Omega u'(t)\cdot v\,dx = -\int_\Omega \big[(a^{\eta,\omega}u_{x_\eta})\cdot v_{x_\omega} + (b^\eta u_{x_\eta})\cdot v \\ + ((d_1+d_2)u)\cdot v\big]dx \\ -\int_{\partial\Omega}(fu)\cdot v\,d\sigma + \int_\Omega g\cdot v\,dx;\end{aligned} \tag{3.17}$$

- *the energy equality*

$$\begin{aligned}\tfrac12|u(t)|^2_{L^2(\Omega)} + \int_0^t\int_\Omega \big[(a^{\eta,\omega}u_{x_\eta})\cdot u_{x_\omega} + (b^\eta u_{x_\eta})\cdot u \\ +((d_1+d_2)u)\cdot u\big]dx + \int_0^t\int_{\partial\Omega}(fu)\cdot u\,d\sigma \\ = \tfrac12|\varsigma_0|^2_{L^2(\Omega)} + \int_0^t\int_\Omega g\cdot u\,dx\end{aligned} \tag{3.18}$$

  holds for all $t \in [0,T)$;
- *and if $\bar t \in (0,T)$ then there is a constant $C(\bar t)$ depending on $\bar t$, $n, \mathcal{H}_0, \mathcal{H}_1$, on the values of r_1, r_2 and s and the numbers of summands referred to in the hypotheses and on the norms on $[0,\bar t]$ of the coefficients $a^{\eta,\omega}$, b^η, d_1, d_2 and f in the hypotheses and on $\theta(\bar t)$, such that*

$$\begin{aligned}\sup_{0\le t\le \bar t}|u(t)|^2_{\mathcal{H}_1} + \int_0^{\bar t}|u'(t)|^2_{\mathcal{H}_0}dt \\ \le C(\bar t)\Big[|\varsigma_0|^2_{\mathcal{H}_1} + \int_0^{\bar t}|g(t)|^2_{L^2(\Omega)}dt\Big].\end{aligned} \tag{3.19}$$

(b) *In place of the hypotheses of* (a) *assume that g is as in* (3.14) *with components which are finite sums of elements in $L^2_{loc}([0,T);L^q(\Omega))$ for values of $q \in [2n/(n+2),2]$ for $n \ge 3$ and in $(1,2]$ for $n = 2$ and with $t^{\delta/2}g(t)$ in $L^2_{loc}([0,T);L^2_{loc}(\Omega))$ for some $\delta \ge 1$, and that $\varsigma_0 \in \mathcal{H}_0$.* Then

- *there is a unique*

$$u \in C([0,T);\mathcal{H}_0) \cap wAC^2((0,T);\mathcal{H}_0) \cap C((0,T);\mathcal{H}_1^{weak})$$

with equivalence class in $L^2_{loc}([0,T);\mathcal{H}_1)$ *which satisfies* (3.14) *for all* $v \in \mathcal{H}_1$ *and* $t \in [0,T)$;

- *the strong form* (3.17) *of* (3.14) *holds for almost all* t *and the energy equality* (3.18) *holds for all* $t \in [0,T)$;

- *given* $\bar{t} \in [0,T)$ *there is a constant* $C(\bar{t})$ *depending on the same quantities as in* (3.19) *and on* δ *and the values of* q *occurring in the hypotheses above such that*

$$\sup_{0<t\leq\bar{t}} t^\delta |u(t)|^2_{\mathcal{H}_1} + \int_0^{\bar{t}} t^\delta |u'(t)|^2_{\mathcal{H}_0} dt$$
$$\leq C(\bar{t})\Big[|\varsigma_0|^2_{\mathcal{H}_0} + \int_0^{\bar{t}} \big(C_g(t) + t^\delta |g(t)|^2_{L^2(\Omega)}\big) dt\Big]$$

where $C_g(t)$ *is the sum of the squares of the* $L^q(\Omega)$ *norms referred to above.*

Proof. The hypotheses of Theorem 2.11 are easily checked, most of the required estimates having already been given in the proof of Theorem 3.6. The conclusions therefore follow directly from the conclusions of Theorems 2.11 and 2.12.

□

We conclude this section with two results for the case that A_{sym} is independent of t. These show that the spectral representations in (2.66) and (2.67) hold and that the weak continuity $t \mapsto u(\cdot,t) \in \mathcal{H}_1^{weak}$ in Theorem 3.7 can be improved to strong continuity.

Theorem 3.8. *Let* T, Ω, $\mathcal{H}_0$ *and* $\mathcal{H}_1$ *satisfy hypotheses* (a) *and* (b) *of* Theorem 3.6 *and let* $A = A_{sym} + B$ *be as above in* (3.15)–(3.16) *with coefficients satisfying the following:*

- $a^{\eta,\omega} \in L^\infty(\Omega)$, *the components of* d_1 *are finite sums of elements of* $L^{r_1}(\Omega)$ *for values of* $r_1 \in (n/2,\infty]$ *and in case* (I) *of* Theorem 3.6 *the components of* f *are finite sums of elements* $L^s(\Omega)$ *for values of* s *in* $(n,\infty]$

- *for all* η *and* ω, $(a^{\eta,\omega})^{tr} = a^{(\omega,\eta)}$ *a.e. in* Ω, d_1 *is symmetric a.e. in* Ω *and in case* (I) f *is symmetric a.e. in* $\partial\Omega$

- *there is a positive number* θ *such that for almost all* $x \in \Omega$,

$$(M^{tr} a^{\eta,\omega}(x) M)^{\omega,\eta} \geq 2\theta |M|^2$$

for all $N \times n$ *matrices* M

- $b^\eta \in L^2_{loc}([0,T); L^\infty(\Omega))$ *and the components of* d_2 *are finite sums of elements of* $L^2_{loc}([0,T); L^{r_2}(\Omega))$ *for values of* $r_2 \in [n, \infty]$ *when* $n \geq 3$ *and in* $(2, \infty]$ *when* $n = 2$.

(Different values of r_1, r_2 *and* s *are allowed for different components and summands of* d_1, d_2 *and* f*).*

Let $\varsigma_0 \in \mathcal{H}_0$ *and* $g \in L^2_{loc}([0,T); L^2(\Omega))$ *and let* u *be the unique solution of* (3.14) *described in* Theorem 3.7(b) (*or* Theorem 3.7(a) *if* $\varsigma_0 \in \mathcal{H}_1$). *Then*

$$u \in C((0,T); \mathcal{H}_1) \text{ and if } \varsigma_0 \in \mathcal{H}_1 \text{ then } u \in C([0,T); \mathcal{H}_1).$$

In addition, there is a λ *such that if* $A_0 \equiv A_{sym} + \lambda \langle \cdot, \cdot \rangle_{[L^2(\Omega)]^N}$ *then* A_0 *is an* $\mathcal{H}_1$*-equivalent inner product and there is a self-adjoint operator* $S \in \mathcal{L}(\mathcal{H}_0, \mathcal{H}_1)$ *satisfying* (2.59). *The solution* u *then satisfies the representation* (2.66) *in which* E *is the spectral measure of* S. *In the case that* Ω *is bounded,* S *is a compact operator and the representation* (2.66) *takes the discrete form* (2.67) *in which* $\mu_j \to 0^+$ *are the eigenvalues of* S *and* $\{\varphi_j\}_j$ *is the corresponding* $\mathcal{H}_0$*-orthonormal system of eigenfunctions.*

PROOF. The conclusions are immediate from Theorem 2.13 and Corollary 2.14. Observe that the last statement of the theorem holds because Ω is bounded, which implies that $H^1(\Omega)$ is compactly contained in $L^2(\Omega)$ (Theorem A.5) and consequently S is a compact operator in $\mathcal{L}(\mathcal{H}_0)$. □

See Remarks (1)–(4) following the statement of Theorem 2.12 for comments related to the material of this section.

Exercises

(See also Exercises 2.1–2.4 and 2.12–2.15.)

3.1 Give the details of the application of Theorem 2.1(b) to the problem (3.1)–(3.3) for the case that $n = 1$ and Ω is an open interval in $\mathbf{R}$.

3.2 Give the details of the application of Theorem 2.1(b) to the problem (3.1)–(3.3) in which $\Omega = \mathbf{R}^n$ and the products $g(t) \cdot v$ and $G(t) : \nabla_x v$ are replaced by $\chi\,(g^i(t) * v^i)$ and $\chi\,(G^{ij}(t) * v^i_{x_j})$ where $*$ denotes spatial convolution in $\mathbf{R}^n$ and χ is the characteristic function of a ball in $\mathbf{R}^n$. Suggestion: apply Young's inequality (see [**13**] pg. 241 for example): if $p^{-1} + q^{-1} = r^{-1} + 1$, where $1 \leq p, q, r \leq \infty$, then for $\varphi \in L^p$ and $\psi \in L^q$ (on $\mathbf{R}^n$), $\varphi * \psi \in L^r$ with norm bounded by $C|\varphi|_{L^p}|\psi|_{L^q}$ where C depends on n, p and q.

3.3 Give the details of the application of Theorem 2.1(b) to the system in Exercise 1.5.

3.4 Assume that the hypotheses of Theorem 3.6 hold with $T = \infty$, Ω bounded, $\mathcal{H}_1 = H^1(\Omega)$, b, c, f, G and Γ all zero and d a constant $N \times N$

matrix. Show that the average value $\overline{u}(t)$ of $u(\cdot,t)$ over Ω satisfies an ordinary differential equation which can be solved explicitly in terms of the matrix exponential $\exp(td)$. Show also that if d is positive definite, say $\xi \cdot (d\xi) \geq \varepsilon|\xi|^2$ for all $\xi \in \mathbf{R}^N$ and some $\varepsilon > 0$, and if the components of g are in $L^2([0,\infty); L^2(\Omega))$, then $|u(\cdot,t) - \overline{u}(t)|_{L^2(\Omega)} \to 0$ as $t \to \infty$. Thus the solution is asymptotially spatially homogeneous with its average coalescing with an orbit of an associated ordinary differential equation.

CHAPTER 4

Elliptic Systems: Higher Order Regularity

In this chapter we prove two regularity results for weak solutions of elliptic systems of second order. The first is independent of boundary conditions and shows that solutions have weak derivatives which are locally bounded in L^2 of the spatial domain (*interior* regularity). The second applies to a broad class of boundary-value problems in domains with smooth boundaries and shows that these derivatives are in L^2 of the entire domain (*global* regularity). In both cases the underlying partial differential equations hold in the sense of weak derivatives, and if the order of weak differentiability is sufficiently high, then in the sense of continuous derivatives as well. These local and global results are stated in Theorems 4.1 and 4.2 respectively and are proved in subsequent sections. Both are fundamental in the theory of partial differential equations and Theorem 4.2 will be an essential ingredient in the regularity theory for parabolic systems in Chapters 5 and 6.

Specifically, we again let Ω be an open set in $\mathbf{R}^n$, $n \geq 2$, not necessarily bounded, and $\mathcal{H}_1$ a closed subspace of $H^1(\Omega)^N$, $N \geq 1$, and consider the bilinear form $A \in \mathcal{B}(\mathcal{H}_1 \times \mathcal{H}_1)$ of Chapter 3, now with coefficients independent of t:

$$
\begin{aligned}
A(u,v) = \int_\Omega \big[(a^{\eta,\omega} u_{x_\eta}) \cdot v_{x_\omega} + (b^\eta u_{x_\eta}) \cdot v \\
+ (c^\omega u) \cdot v_{x_\omega} + (du) \cdot v\big]\, dx + \int_{\partial\Omega} (fu) \cdot v \, d\sigma
\end{aligned}
\tag{4.1}
$$

where $a^{\eta,\omega}$, b^η, c^ω and d are $N \times N$–matrix valued functions of $x \in \Omega$, f is an $N \times N$–matrix valued function of $x \in \partial\Omega$ and the dot denotes the usual inner product in $\mathbf{R}^N$. We impose fairly general conditions on these coefficient functions and consider the regularity of an element $u \in \mathcal{H}_1$ which is assumed to satisfy

$$A(u,v) = F \cdot v \tag{4.2}$$

for all $v \in \mathcal{H}_1$, where $F \in \mathcal{H}_1^*$ is given by

$$F \cdot v = \int_\Omega (g \cdot v + G : \nabla_x v) + \int_{\partial\Omega} \Gamma \cdot v \, d\sigma. \tag{4.3}$$

Here g and Γ are $\mathbf{R}^N$-valued, G is $N \times n$–matrix-valued, $\nabla_x v$ is the $N \times n$ x–derivative matrix and $G : \nabla_x v$ is the scalar product $G^{ij} v^i_{x_j}$. (Boundary integrals are omitted in both (4.1) and (4.3) if $\Omega = \mathbf{R}^n$.)

In Theorem 4.1 we consider the case that (4.2) holds for all $v \in [H_0^1(\Omega)]^N$, so that the boundary integrals are absent and no regularity conditions on $\partial\Omega$ are imposed. The components of u will then be in $H^{k+2}(\Omega_0)$ for all open subsets Ω_0 of Ω a positive distance from $\partial\Omega$ provided that those of g are in $H^k(\Omega)$ and those of G are in $H^{k+1}(\Omega)$ (in the special case that $\Omega = \mathbf{R}^n$, u will be in $H^{k+2}(\mathbf{R}^n)$). In Theorem 4.2 we strengthen the hypotheses by requiring that (4.2) holds for all $v \in \mathcal{H}_1$, which will be the subspace of $[H^1(\Omega)]^N$ discussed in Example 3 of section 3.1 corresponding to a variety of possible essential, natural and mixed boundary conditions, and requiring further that Ω and Γ satisfy certain C^{k+2} and H^{k+1} regularity conditions respectively.

The existence and uniqueness of a "solution" u of (4.2) can be proved when A is coercive in the sense that $A(u,u) \geq C|u|^2_{\mathcal{H}_1}$ for a positive constant C and for all $u \in \mathcal{H}_1$. However, by allowing fairly general lower order terms in (4.1) we preclude coerciveness in general and therefore can make no statement about existence or uniqueness. In fact, solutions may fail to exist for some g, G and Γ, and there may be multiple solutions for others. See the discussion in [**12**], pg. 321, for example. In particular there may be nonzero u for which (4.2) holds with g, G and Γ all zero. Consequently, in the generality considered here, solutions cannot be bounded in a given Sobolev norm, say, solely by a sum of bounds for g, G, and Γ.

§4.1 Statements of Results

Throughout this chapter Ω will be an open set in $\mathbf{R}^n$, $n \geq 2$. For the global regularity result of Theorem 4.2, Ω will be a C^{k+2}_{bdd} domain, not necessarily bounded (Definition 1.7(c)) and the integer $k \geq 0$ will be fixed throughout.

Hypotheses on Coefficients

First we make measurability and positivity conditions similar to those in Theorem 3.6:

(4.4)

The components of the $N \times N$ matrices $a^{\eta,\omega}$, b^{η}, c^{ω} and d are Lebesgue equivalence classes of measurable functions on Ω, those of f are σ-equivalence classes of measurable functions on $\partial\Omega$ (σ is the boundary measure described in Theorem 3.1*; this condition is omitted if $\Omega = \mathbf{R}^n$), and there is a positive number θ such that for any given representatives $a^{\eta,\omega}$ and for a.a. $x \in \Omega$,*

$$(M^{tr} a^{\eta,\omega}(x) M)^{\omega,\eta} \geq 2\theta |M|^2$$

for all $N \times n$ matrices M. Here ^{tr} denotes transpose, summation over η and ω is understood and any matrix norm is allowed.

Before proceeding we need to introduce a measure of differentiability for functions defined on the boundary of Ω. Let $\mathcal{B}_i$ and T_i be as in Definition 1.7 and for $\varphi : \partial\Omega \to \mathbf{R}$, $q \in [1, \infty]$ and a multi-index α with $|\alpha| \leq k + 2$ define

$$\langle\varphi\rangle^{\alpha}_{L^q,\partial\Omega} \equiv \begin{cases} \left(\sum_i |D^{\alpha}_{\tilde{y}}(\varphi \circ T_i)|^q_{L^q(\mathcal{B}_i)}\right)^{1/q}, & q \in [1, \infty), \\ \sup_i |D^{\alpha}_{\tilde{y}}(\varphi \circ T_i)|_{L^\infty(\mathcal{B}_i)}, & q = \infty, \end{cases} \tag{4.5}$$

provided that the derivatives $D^{\alpha}_{\tilde{y}}(\varphi \circ T_i)$ exist in the weak sense on each $\mathcal{B}_i$.

The following conditions on the coefficients of A in (4.1) will suffice for most applications but are somewhat stronger than necessary:

$$\begin{cases} \textit{the components of each } a^{\eta,\omega} \textit{ and each } c^{\omega} \textit{ are in } W^{k+1,\infty}(\Omega) \\ \textit{the components of each } b^{\eta} \textit{ and } d \textit{ are in } W^{k,\infty}(\Omega) \\ \langle f\rangle^{\alpha}_{L^\infty,\partial\Omega} < \infty \textit{ for all multi-indices } \alpha \textit{ with } |\alpha| \leq k+1 \\ \quad (\textit{for } \text{Theorem 4.2 } \textit{only}). \end{cases} \tag{4.6}$$

Weaker but more technical conditions can be given in terms of index sets $\mathcal{I}_l \subseteq [2, \infty]$ defined as follows for nonnegative integers l:

$$\mathcal{I}_l = \begin{cases} [n/l, \infty], & 0 \leq l < n/2, \\ (2, \infty], & l = n/2, \\ [2, \infty], & l > n/2. \end{cases} \tag{4.7}$$

Observe that I_l increases with l, that $\infty \in I_l$ for every l and that $I_0 = \{\infty\}$. (The dependence of $\mathcal{I}_l$ on n is suppressed.) We can now describe weaker sufficient conditions on coefficients and in doing so we will abuse notation somewhat and make the simple statement "$D^\alpha d \in L^{\mathcal{I}_{|\alpha|+1}}$", for example, in place of the precise but more complicated statement "each component $d^{\mu\nu}$ of the matrix-valued function d has a weak derivative $D^\alpha d^{\mu\nu}$ which is a finite sum $\sum_i d^{\mu\nu}_{\alpha,i}$ where each $d^{\mu\nu}_{\alpha,i}$ is in $L^{s^{\mu\nu}_{\alpha,i}}(\Omega)$ for some $s^{\mu\nu}_{\alpha,i} \in \mathcal{I}_{|\alpha|+1}$." The most general hypotheses on the coefficients of A that the theory accommodates are then as follows:

$$\begin{cases} D^\alpha a \in L^{\mathcal{I}_{(|\alpha|-1)\vee 0}} \textit{ and } D^\alpha c \in L^{\mathcal{I}_{|\alpha|}} \textit{ for } 0 \leq |\alpha| \leq k+1 \\ D^\alpha b \in L^{\mathcal{I}_{|\alpha|}} \textit{ and } D^\alpha d \in L^{\mathcal{I}_{|\alpha|+1}} \textit{ for } 0 \leq |\alpha| \leq k \\ \langle f\rangle^{\alpha}_{L^{s_\alpha},\partial\Omega} < \infty \textit{ for } 0 \leq |\alpha| \leq k+1 \textit{ and some } s_\alpha \in \mathcal{I}_{|\alpha|} \\ \quad (\textit{for } \text{Theorem 4.2 } \textit{only}). \end{cases} \tag{4.8}$$

Observe that the conditions in (4.6) imply those in (4.8) because $\infty \in \mathcal{I}_l$ for every l.

We denote by C_A an upper bound for θ^{-1} and for the norms, seminorms and numbers of summands in the descriptions above in (4.8). If n is even and if any summand of any derivative of any component of any of the coefficients in (4.8) is required to be in $L^s(\Omega)$ or $L^s(\mathcal{B}_i)$ for some s in $\mathcal{I}_{n/2}$, then C_A must also be an upper bound for $(s-2)^{-1}$.

Observe that (4.4) and (4.8) are stronger than the hypotheses on the coefficients of A in Theorem 3.6. Consequently $A \in \mathcal{B}(\mathcal{H}_1 \times \mathcal{H}_1)$ and as in (3.9),

$$A(v,v) \geq \theta |v|^2_{[H^1(\Omega)]^N} - C|v|^2_{[L^2(\Omega)]^N} \tag{4.9}$$

for all $v \in [H^1(\Omega)]^N$, where the positive constant C depends on upper bounds for θ^{-1} and C_A.

We can now state a local regularity result in which boundary considerations enter neither the hypotheses nor the conclusions and solutions are shown to be in $[H^{k+2}]^N$ on interior sets:

Theorem 4.1. *Let k be a nonnegative integer and assume the following:*

- *Ω is an open set in $\mathbf{R}^n$, $n \geq 2$, not necessarily bounded;*
- *A is the bilinear form defined in (4.1) with the boundary integral omitted and with coefficients $a^{\eta,\omega}$, b^{η}, c^{ω}, and d satisfying the positive definiteness condition in (4.4) and either of the sets of regularity conditions (4.6) or (4.8) with references to f omitted and with the constant C_A as defined above.*

Then the following hold:

(a) *If $\Omega = \mathbf{R}^n$: there is a constant C depending on n, N and on upper bounds for k and C_A such that if $u \in [H^1(\mathbf{R}^n)]^N$ satisfies*

$$A(u,v) = \int_{\mathbf{R}^n} \big(g \cdot v + G : \nabla_x v\big) dx$$

for all $v \in [C_c^\infty(\mathbf{R}^n)]^N$ where the components of g are in $H^k(\mathbf{R}^n)$ and those of G are in $H^{k+1}(\mathbf{R}^n)$, then $u \in H^{k+2}(\mathbf{R}^n)$ and

$$|u|_{[H^{k+2}(\mathbf{R}^n)]^N} \leq C\big(|u|_{[L^2(\mathbf{R}^n)]^N} + |g|_{[H^k(\mathbf{R}^n)]^N} + |G|_{[H^{k+1}(\mathbf{R}^n)]^{N\times n}}\big).$$

(b) *If Ω is a proper open subset of $\mathbf{R}^n$: given a proper open subset Ω_0 of Ω with $\gamma \equiv \operatorname{dist}(\Omega_0, \partial\Omega) > 0$, there is a constant C depending on n, N and on upper bounds for k, C_A and γ^{-1} such that if $u \in [H^1(\Omega)]^N$ satisfies*

$$A(u,v) = \int_{\Omega} \big(g \cdot v + G : \nabla_x v\big) dx$$

for all $v \in [C_c^\infty(\Omega)]^N$, hence for all $v \in [H_0^1(\Omega)]^N$, where the components of g are in $H^k(\Omega)$ and those of G are in $H^{k+1}(\Omega)$, then $u \in [H^{k+2}(\Omega_0)]^N$ and

$$|u|_{[H^{k+2}(\Omega_0)]^N} \leq C\big(|u|_{[L^2(\Omega)]^N} + |g|_{[H^k(\Omega)]^N} + |G|_{[H^{k+1}(\Omega)]^{N\times n}}\big).$$

(c) *If* Ω *and* Ω_0 *are as in* (b), *then the components of* $a^{\eta,\omega}u_{x_\eta}$ *and* $c^\omega u$ *are in* $H^{k+1}(\Omega_0) \subseteq H^1(\Omega_0)$ *and those of* $b^\eta u_{x_\eta}$ *and* du *are in* $H^k(\Omega_0) \subseteq L^2(\Omega_0)$, *and the equation*

$$-(a^{\eta,\omega}u_{x_\eta})_{x_\omega} + b^\eta u_{x_\eta} - (c^\omega u)_{x_\omega} + du = g - \operatorname{Div} G, \tag{4.10}$$

in which subscripts denote weak derivatives in Ω *and* $\operatorname{Div} G$ *is the vector field whose ith component is* $G^{ij}_{x_j}$, *holds in* $[H^k(\Omega_0)]^N$. *If* $\Omega = \mathbf{R}^n$ *as in* (a) *then these same statements hold with* Ω_0 *replaced by* $\mathbf{R}^n$.

(d) *Let* $k_0 = k_0(n)$ *be the positive integer defined by*

$$k_0 - 1 \leq n/2 < k_0 \tag{4.11}$$

and let $\lambda = k_0 - n/2$ *if* $k_0 - n/2 < 1$ *and choose* λ *arbitrarily in* $(0,1)$ *if* $k_0 - n/2 = 1$. *Assume that* $k = k_0 + l$ *where* l *is a nonnegative integer. Then in either of the cases* (a) *or* (b) *there are specific representatives* $a^{\eta,\omega}$, b^η, c^ω, d, g, G *and* u *such that, for every ball* B *whose closure is contained in* Ω *the components of* b^η, d *and* g *are in* $C^{l,\lambda}_{\mathrm{bdd}}(B)$ (*see* section 1.2), *those of* $a^{\eta,\omega}$, c^ω *and* G *are in* $C^{l+1,\lambda}_{\mathrm{bdd}}(B)$, *and those of* u *are in* $C^{l+2,\lambda}_{\mathrm{bdd}}(B)$; *and* (4.10) *holds pointwise in* Ω *with subscripts denoting continuous derivatives in the sense of calculus.*

Theorem 4.1 is proved in section 4.2.

Boundary Conditions and the space $\mathcal{H}_1$

The second result of this chapter concerns global regularity of solutions of (4.1)–(4.3) in which Ω is a proper subset of $\mathbf{R}^n$ and both solutions and test functions satisfy certain essential boundary conditions. Briefly, for each connected component $\mathcal{C}_l$ of $\partial\Omega$ we allow N_l linearly independent zero-conditions on $u|_{\mathcal{C}_l}$, where $N_l \in \{0, \dots, N\}$. Thus if $N_l = 0$ then no condition is imposed and if $N_l = N$ then elements v of $\mathcal{H}_1$ will be required to satisfy $v = 0$ on $\mathcal{C}_l$. For intermediate values of N_l the boundary condition will take the form $Q_l v = 0$ on $\mathcal{C}_l$, where Q_l maps $\mathcal{C}_l$ into the set of $N_l \times N$ matrices and has full rank at each x. The detailed hypotheses are as follows:

Ω *is a* C^{k+2}_{bdd} *domain in* $\mathbf{R}^n$, *not necessarily bounded, where* $n \geq 2$ *and* $k \geq 0$ (Definition 1.7(c)), *and there is a positive constant* C_{BC}, *and for each connected component* $\mathcal{C}_l$ *of* $\partial\Omega$ *there is an integer* $N_l \in \{0, \dots, N\}$ *such that for each* l *for which* $N_l \geq 1$ (*if any*) *the following hold: there are continuous mappings* Q_l *and* S_l *from* $\mathcal{C}_l$ *into the set of* $N_l \times N$ *matrices and* $N \times N$ *matrices respectively such that*

$$Q_l S_l^{-1} = \begin{bmatrix} I_{N_l \times N_l} & 0 \end{bmatrix} \text{ on } \mathcal{C}_l \tag{4.12}$$

and such that, in the notation of Definition 1.7, *for all* i *for which* $\mathcal{C}_l \cap \mathcal{U}_i \neq \phi$, *the functions* $Q_l \circ T_i$ *and* $S_l^{\pm 1} \circ T_i$ *have continuous, bounded derivatives up to order* $k+2$ *on* $T_i^{-1}(\mathcal{C}_l \cap \mathcal{U}_i) \subset \mathcal{B}_i$, *and for* $\tilde{y} \in T_i^{-1}(\mathcal{C}_l \cap \mathcal{U}_i)$ *and* $|\alpha| \leq k+2$,

$$|D^\alpha_{\tilde{y}}(Q_l \circ T_i)(\tilde{y})|,\ |D^\alpha_{\tilde{y}}(S_l^{\pm 1} \circ T_i)(\tilde{y})| \leq C_{BC}. \tag{4.13}$$

(Observe that if $\mathcal{C}_l \cap \mathcal{U}_i$ is nonempty, then it coincides with $\partial\Omega \cap \mathcal{U}_i$, whose inverse image under T_i is an open subset of $\mathcal{B}_i$.) We take $C_{BC} = 1$ if $N_l = 0$ for all l. The space $\mathcal{H}_1$ will then be the closed subspace of elements $v \in [H^1(\Omega)]^N$ such that $Q_l v = 0$ on $\mathcal{C}_l$ in the sense of the trace theorem (Theorem 3.2) *for all l for which $N_l \geq 1$.*

We can now state the global regularity result in which solutions are shown to be in $H^{k+2}(\Omega)$:

Theorem 4.2. *Let k be a nonnegative integer and assume the following:*

- *Ω is a C^{k+2}_{bdd} domain in $\mathbf{R}^n$ where $n \geq 2$ (Ω need not be bounded)*
- *the hypotheses and notations regarding $\mathcal{H}_1$ and the N_l, Q_l and S_l in the paragraph above (in which (4.13) occurs) are in effect*
- *the coefficients $a^{\eta,\omega}$, b^η, c^ω, d and f of the bilinear form A in (4.1) satisfy the positive definiteness condition (4.4) and either of the sets of regularity conditions (4.6) or (4.8).*

Then there is a positive constant C determined by n, N and by upper bounds for k, M^{k+2}_Ω (defined in Definition 1.7*), C_A and C_{BC} (defined in the subsections above) such that if $u \in \mathcal{H}_1$ satisfies*

$$(4.14)\qquad A(u,v) = \int_\Omega (g \cdot v + G : \nabla_x v) + \int_{\partial\Omega} \Gamma \cdot v \, d\sigma$$

for all $v \in \mathcal{H}_1$ and for some $g \in [H^k(\Omega)]^N$, $G \in [H^{k+1}(\Omega)]^{N\times n}$ and $\Gamma \in [L^2(\partial\Omega)]^N$ satisfying

$$(4.15)\qquad C_\Gamma \equiv \Big(\sum_i |\Gamma \circ T_i|^2_{[H^{k+1}(\mathcal{B}_i)]^N} \Big)^{1/2} < \infty$$

in the notation of Definition 1.7*, then the following hold:*

(a) *$u \in [H^{k+2}(\Omega)]^N$ and*

$$(4.16)\quad |u|_{[H^{k+2}(\Omega)]^N} \leq C\big(|u|_{[L^2(\Omega)]^N} + |g|_{[H^k(\Omega)]^N} + |G|_{[H^{k+1}(\Omega)]^{N\times n}} + C_\Gamma \big).$$

(b) *The components of $a^{\eta,\omega} u_{x_\eta}$ and $c^\omega u$ are in $H^{k+1}(\Omega) \subseteq H^1(\Omega)$ and those of $b^\eta u_{x_\eta}$ and du are in $H^k(\Omega) \subseteq L^2(\Omega)$; and (4.10) holds with each term in $H^k(\Omega)$ and all derivatives taken in the weak sense in Ω.*

The components of $a^{\eta,\omega} u_{x_\eta}$, $c^\omega u$, fu, G and Γ are in $L^2(\partial\Omega)$ and the following boundary conditions hold: if $N_l = N$ (see the paragraph which includes (4.12)) then $u = 0$ on $\mathcal{C}_l$; if $N_l = 0$ then

$$(4.17)\qquad \nu^\omega (a^{\eta,\omega} u_{x_\eta} + c^\omega u) + fu = G\nu + \Gamma$$

on $\mathcal{C}_l$, where ν is the outer normal vector (see Theorem 3.1*); and if $N_l \in \{1, \ldots, N-1\}$ then $Q_l u = 0$ on $\mathcal{C}_l$, where Q_l is as in (4.13), and*

$$(4.18)\qquad \nu^\omega (a^{\eta,\omega} u_{x_\eta} + c^\omega u) + fu - G\nu - \Gamma \in (\text{null space}\, Q_l)^\perp$$

where $^\perp$ *denotes orthogonal complement in* $\mathbf{R}^N$, σ*-almost everywhere on* $\mathcal{C}_l$ *for any given representatives of these functions.*

(c) *Let* k_0 *and* λ *be as in* Theorem 4.1(d) *and assume that* $k = k_0 + l$ *where* l *is a nonnegative integer. Then there are specific representatives* $a^{\eta,\omega}$, b^η, c^ω, d, g, G *and* u *such that for every ball* B *in* $\mathbf{R}^n$ *the components of* b^η, d, *and* g *are in* $C^{l,\lambda}_{\mathrm{bdd}}(\Omega \cap B)$, *those of* $a^{\eta,\omega}$, c^ω *and* G *are in* $C^{l+1,\lambda}_{\mathrm{bdd}}(\Omega \cap B)$ *and those of* u *are in* $C^{l+2,\lambda}_{\mathrm{bdd}}(\Omega \cap B)$, *these functions and the corresponding derivatives therefore having continuous extensions to* $\overline{\Omega \cap B}$*; and the equation* (4.10) *and the boundary conditions described in* (b) *hold pointwise on* $\overline{\Omega}$ *and on* $\partial\Omega$ *respectively with subscripts denoting continuous derivatives in the sense of calculus or the extensions of these derivatives to* $\overline{\Omega}$.

The proofs of Theorems 4.1 and 4.2 are connected in a fairly lengthy analysis given in the following two sections, beginning with an analysis in Proposition 4.3 below of regularity of solutions of a certain model problem. The model problem has the same form as that of (4.1)–(4.3) but with boundary integrals omitted, Ω replaced by a simpler model domain, the support of the solution assumed to intersect the boundary in a subset of a hyperplane, and the corresponding matrices Q_l in (4.13) constant and with a very simple form. This analysis will then carry over immediately to give a proof of the interior regularity result Theorem 4.1. Then in section 4.3 we prove the global regularity result Theorem 4.2 by showing first that the boundary integrals in (4.8) can be absorbed into terms involving integrals over Ω and second by applying a partition of unity to combine the regularity results of Theorem 4.1 in the interior with the model problem analysis near the boundary.

§4.2 Interior Regularity: Proof of Theorem 4.1

In this section we prove regularity of weak solutions of the model problem outlined above and then apply the analysis in its proof to verify Theorem 4.1.

Proposition 4.3. *Let integers* $n \geq 2$, $N \geq 1$, $N' \in \{0, \dots, N\}$ *and* $k \geq 0$ *be given, together with positive numbers* ε *and* r. *Assume the following:*

- $\Omega' \equiv \{z = (\tilde{z}, z_n) \in \mathbf{R}^n : \tilde{z} \in B_r \text{ and } -\psi(\tilde{z}) < z_n < 0\}$ *where* B_r *is a ball of radius* r *in* $\mathbf{R}^{n-1}$ *and* $\psi : B_r \to (\varepsilon, \infty)$ *has continuous, bounded derivatives up to order* $k+2$;
- $\mathcal{H}'_1$ *is the subspace of elements* v *in* $[H^1(\Omega')]^N$ *such that* $v(\tilde{z}, z_n) = 0$ *for* $\tilde{z} \in \partial B_r$ *or* $z_n = -\psi(\tilde{z})$, *and such that if* $N' \geq 1$ *then* $v^i(\tilde{z}, 0) = 0$ *for* $i = 1, \dots, N'$ *(this condition is vacuous if* $N' = 0$*);*

- *$A' \in \mathcal{B}(\mathcal{H}_1' \times \mathcal{H}_1')$ is a bilinear form as in* (4.1) *with the boundary integral omitted and with coefficients $a', \ldots, d'$ satisfying* (4.4) *and either* (4.6) *or* (4.8) *with corresponding constant $C_{A'}$ as described immediately before the statement of* Theorem 4.1.

For $\sigma \in (0, \varepsilon \wedge r)$ define

$$\Omega'_\sigma \equiv \{z = (\tilde{z}, z_n) \in \Omega' : \tilde{z} \in B_{r-\sigma} \text{ and } -\psi(\tilde{z}) + \sigma < z_n < 0\}. \tag{4.19}$$

Then given such σ there is a constant C depending on n, N and on upper bounds for k, ε^{-1}, r^{-1}, bounds for the derivatives of ψ up to order $k+2$, and on $C_{A'}$ and σ^{-1} such that, if $u' \in \mathcal{H}_1'$ has support in $\overline{\Omega'_\sigma}$ and satisfies

$$A'(u', v) = \int_{\Omega'} (g' \cdot v + G' : \nabla_z v) dz \tag{4.20}$$

for all $v \in \mathcal{H}_1'$ and for some g' and G' whose components are in $H^k(\Omega')$ and $H^{k+1}(\Omega')$ respectively, then $u' \in [H^{k+2}(\Omega')]^N$ and

$$|u'|_{[H^{k+2}(\Omega')]^N} \leq C\big(|u'|_{[L^2(\Omega')]^N} + |g'|_{[H^k(\Omega')]^N} + |G'|_{[H^{k+1}(\Omega')]^{N\times n}}\big). \tag{4.21}$$

Observe that Ω' is a uniformly Lipschitz but not C^{k+2}_{bdd} domain. No difficulty results, however, because the support of u' is assumed to be disjoint from the nonsmooth part of $\partial\Omega'$.

The overall proof of Proposition 4.3 is by induction on k together with two essential ideas. The first is the observation that a difference quotient $\delta u'$ of u' in a direction transverse to the hyperplane $\{z_n = 0\}$ is in $\mathcal{H}_1'$ (the special form of the essential boundary conditions is required here) and satisfies a problem with the same bilinear form A' but with different g' and G'. Thus if the lemma holds with k replaced by $i - 1 < k$, then the induction hypothesis will apply to show that $\delta u' \in H^{i+1}$ with norm bounded independently of the spatial increment. Subsequences of difference quotients therefore converge weakly in H^{i+1}, and their limits are necessarily the weak derivative of u' in the transverse direction. This shows that all the weak derivatives of u' of order less than or equal to $i+2$ exist and are in L^2, with the possible exception of the highest derivative in the z_n-direction. The second essential idea is the observation that this latter weak derivative can be expressed in terms of the other highest order terms in (4.20) by application of the positivity hypothesis (4.4).

Two preliminary discussions are required for the proof of Proposition 4.3. The first concerns multipliers, which are functions ζ defined on an open set $\mathcal{W} \in \mathbf{R}^n$ which for given integers $0 \leq i \leq l$ satisfy conditions insuring that $\zeta v \in H^i(\mathcal{W})$ for all $v \in H^l(\mathcal{W})$; the second concerns difference quotients and their relation to weak derivatives and multipliers.

Multipliers

Definition 4.4. *Let $\mathcal{W}$ be an open set in $\mathbf{R}^n$, $n \geq 2$, and let $l \geq i \geq 0$ be integers. Then $\mathcal{M}^{l,i}(\mathcal{W})$ is the set of elements $\zeta \in L^1_{loc}(\mathcal{W})$ such that for all multi-indices α of length less than or equal to i, the weak derivative $D^\alpha\zeta$ exists and is a finite sum $\sum_m \zeta_{\alpha,m}$ where each $\zeta_{\alpha,m}$ is an element of $L^{s_{\alpha,m}}(\mathcal{W})$ for some $s_{\alpha,m} \in \mathcal{I}_{l-i+|\alpha|}$ (see (4.7)).*

The subset $\tilde{\mathcal{M}}^{l,i}$ is the set of elements $\zeta \in \mathcal{M}^{l,i}$ for which each $s_{\alpha,m} \in \tilde{\mathcal{I}}_{l-i+|\alpha|}$, where $\tilde{\mathcal{I}}_j$ is the set obtained by removing the left–hand endpoint of $\mathcal{I}_j$, with the exception that $\tilde{\mathcal{I}}_0 = \mathcal{I}_0 = \{\infty\}$.

The following lemma shows that members of $\mathcal{M}^{l,i}$ define multiplication operators from H^l to H^i. Part (a) will be applied in the present chapter and part (b) will be needed later in chapter 5.

Lemma 4.5. *Let $\mathcal{W}$ be an open set in $\mathbf{R}^n$, $n \geq 2$, satisfying the cone condition* Definition 1.6(b) *and let $l \geq i \geq 0$ be integers.*

(a) *If $\zeta \in \mathcal{M}^{l,i}(\mathcal{W})$, then the operator L_ζ defined by $L_\zeta v = \zeta v$ is a bounded linear transformation from $H^l(\mathcal{W})$ into $H^i(\mathcal{W})$ and there is a constant C such that*

$$|\zeta|_{\mathcal{M}^{l,i}(\mathcal{W})} \equiv |L_\zeta|_{\mathcal{L}(H^l,H^i)} \leq C \sum_{|\alpha|\leq i} \sum_m |\zeta_{\alpha,m}|_{L^{s_{\alpha,m}}} \tag{4.22}$$

where the $\zeta_{\alpha,m}$ and $s_{\alpha,m}$ are as in Definition 4.4. *The constant C depends on n, l and i and on upper bounds for the inverses of the two cone parameters in* Definition 1.6(b)*; in the case that n is even and $n/2 \in \{l-i,\dots,l\}$, C depends also on an upper bound for $(s_{\alpha,m}-2)^{-1}$ for α satisfying $l-i+|\alpha| = n/2$.*

(b) *If $\zeta \in \tilde{\mathcal{M}}^{l,i}(\mathcal{W})$ then there is a number $\delta \in (0,1)$ depending on n, l, i and the specific values of $s_{\alpha,m} \in \tilde{I}_{l-i+|\alpha|}$ such that, for $v \in H^i(\mathcal{W})$,*

$$|\zeta v|_{H^i(\mathcal{W})} \leq C\Big(\sum_{|\alpha|\leq i} \sum_m |\zeta_{\alpha,m}|_{L^{s_{\alpha,m}}}\Big)|v|^{1-\delta}_{H^l(\mathcal{W})}|v|^{\delta}_{H^i(\mathcal{W})} \tag{4.23}$$

where C is as in (a).

Proof. We will apply the fact that if $\mathcal{W}$ is an open set in $\mathbf{R}^n$ satisfying the cone condition and if m is a nonnegative integer, then $H^m(\mathcal{W})$ is continuously included in $L^r(\mathcal{W})$ for all r in the set $\mathcal{I}'_m$ defined by

$$\mathcal{I}'_m = \begin{cases} [2, 2n/(n-2m)]\,, & 0 \leq m < n/2 \\ [2,\infty)\,, & m = n/2 \\ [2,\infty]\,, & m > n/2 \end{cases} \tag{4.24}$$

(see Theorem A.3(a)). The norm of the inclusion is evidently independent of r (by interpolation) unless $m = n/2$, in which case it depends also on a finite upper bound for r. By checking cases we find that if $r^{-1} + s^{-1} = \frac{1}{2}$ then $s \in \mathcal{I}_m$ (see (4.7)) if and only if $r \in \mathcal{I}'_m$.

To prove (a) we let v be an element of $H^l(\mathcal{W})$, α a multi-index with $|\alpha| \leq i$ and $\alpha = \beta + \gamma$. Without loss of generality we can take $D_z^\beta \zeta \in L^s$ for some $s \in \mathcal{I}_{l-i+|\beta|}$. Define r by $r^{-1} + s^{-1} = \frac{1}{2}$ so that $r \in \mathcal{I}'_{l-i+|\beta|}$. Then since $l - i + |\beta| = l - i + |\alpha| - |\gamma| \leq l - |\gamma|$ we have that $D^\gamma v \in H^{l-|\gamma|} \subset H^{l-i+|\beta|}$, which is continuously included in L^r. Thus

$$D^\beta \zeta \, D^\gamma v \in L^s \cdot L^r \subset L^2. \tag{4.25}$$

This proves that $D^\alpha(\zeta v) \in L^2$ for all $|\alpha| \leq i$, as required. Observe that the inclusion constant in (4.25) is independent of r, hence of s, unless $l - i + |\beta| = n/2$, in which case it depends on an upper bound for r, hence on an upper bound for $(s-2)^{-1}$.

To prove (b) we derive a bound for the L^2 norm of the term on the left in (4.25) assuming now that $s \in \tilde{\mathcal{I}}_{l-i+|\beta|}$. In this case there is a $\delta \in (0,1)$ such that $\tilde{s} \equiv (1-\delta)s \in \mathcal{I}_{l-i+|\beta|}$, so that if $\tilde{r}$ is defined by $\tilde{r}^{-1} + \tilde{s}^{-1} = \frac{1}{2}$, then $\tilde{r} \in \mathcal{I}'_{l-i+|\beta|}$. Consequently since $l - \gamma = l - i + |\beta|$,

$$|D^\gamma v|_{L^{\tilde{r}}} \leq C |D^\gamma v|_{H^{l-i+|\beta|}} \leq C |v|_{H^l}.$$

Applying this together with the fact that $\tilde{s}/2(1-\delta)$, $\tilde{r}/2(1-\delta)$ and $1/\delta$ are conjugate indices, we obtain for the case $\tilde{s} < \infty$ that

$$\begin{aligned}
|D^\beta \zeta \, D^\gamma v|_{L^2}^2 &= \int |D^\beta \zeta|^2 |D^\gamma v|^{2(1-\delta)} |D^\gamma v|^{2\delta} \\
&\leq C \Big(\int |D^\beta \zeta|^{\tilde{s}/(1-\delta)} \Big)^{2(1-\delta)/\tilde{s}} \Big(\int |D^\gamma v|^{\tilde{r}} \Big)^{2(1-\delta)/\tilde{r}} \Big(\int |D^\gamma v|^2 \Big)^\delta \\
&= C |D^\beta \zeta|_{L^s}^2 |v|_{H^l}^{2(1-\delta)} |v|_{H^i}^{2\delta};
\end{aligned}$$

a similar bound holds for the case that $\tilde{s} = \infty$. To complete the proof of (b) we take the minimum over all such δ occurring in the above argument noting that if $0 < a < b$, then $a^\delta b^{1-\delta}$ is a decreasing function of δ.

□

Difference Quotients

Now fix $\mu \in \{1, \dots, n-1\}$, let $e_\mu \in \mathbf{R}^n$ be the corresponding standard basis vector, and let Ω' and Ω'_σ be as in (4.19) for a particular $\sigma \in (0, \varepsilon \wedge r)$. Then if $h > 0$ is sufficiently small depending on σ, $z \pm h e_\mu$ will be in Ω' if $z \in \Omega'_\sigma$, and if $\zeta : \Omega \to \mathbf{R}$ we can therefore define

$$\begin{cases} (\tau_h^\pm \zeta)(z) = \zeta(z \pm h e_\mu) \\ (\delta_h^\pm \zeta)(z) = \pm h^{-1}\big[\zeta(z \pm h e_\mu) - \zeta(z)\big]. \end{cases} \tag{4.26}$$

It is easy to check that these definitions extend to Lebesgue equivalence classes, that the product rule

$$\delta_h^\pm(\zeta\eta) = \zeta(\delta_h^\pm \eta) + (\delta_h^\pm \zeta)\,(\tau_h^\pm \eta) \tag{4.27}$$

holds where defined, and that if $\zeta, \eta \in L^2(\Omega)$ with at least one having support in Ω'_σ, then

$$\int_\Omega (\delta_h^\pm \zeta)\,\eta = -\int_\Omega \zeta\,(\delta_h^\mp \eta) \tag{4.28}$$

for h sufficiently small depending on σ. Further properties of difference quotients are given in the following lemma:

Lemma 4.6. *Let Ω' be as in the statement of* Proposition 4.3 *and let $\sigma \in (0, \varepsilon \wedge r)$ and $q \in (1, \infty]$ with Hölder conjugate q'. Then the following hold for h sufficiently small depending on σ:*

(a) *For integers $l \geq 0$ the maps $\varphi \in C_c^\infty(\Omega') \to \overline{\varphi}_h^\pm$ defined by*

$$\overline{\varphi}_h^\pm(z) = \int_0^1 \varphi(z \pm hse_\mu)ds, \ z \in \Omega'_\sigma,$$

extend to bounded linear transformations from $W^{l,q'}(\Omega')$ to $W^{l,q'}(\Omega'_\sigma)$ with norm less than or equal to one.

(b) *If ζ and its weak derivative ζ_{z_μ} are in $L^q(\Omega')$, then $\delta_h^\pm \zeta \in L^q(\Omega'_\sigma)$, $|\delta_h^\pm \zeta|_{L^q(\Omega'_\sigma)} \leq |\zeta_{z_\mu}||_{L^q(\Omega')}$, and for all $v \in L^{q'}(\Omega')$ with support in Ω'_σ,*

$$\int_{\Omega'} (\delta_h^\pm \zeta)\, v\, dz = \int_{\Omega'} (\zeta_{z_\mu})\, \overline{v}_h^\mp\, dz. \tag{4.29}$$

(c) *If $\zeta \in L^1_{loc}(\Omega')$ and if its weak derivative ζ_{z_μ} is in $W^{l,q}(\Omega')$ for some nonnegative integer l, then $\delta_h^\pm \zeta \in W^{l,q}(\Omega'_\sigma)$ and*

$$|\delta_h^\pm \zeta|_{W^{l,q}(\Omega'_\sigma)} \leq C |\zeta_{z_\mu}|_{W^{l,q}(\Omega')}$$

where C depends on n, l, i and on upper bounds for ε^{-1}, r^{-1} (defined in Proposition 4.3*), σ^{-1} and the Lipschitz constant for ψ in B_r.*

(d) *If $\zeta \in L^1_{loc}(\Omega')$ and if its weak derivative ζ_{z_μ} is in $\mathcal{M}^{l,i}(\Omega')$ for some nonnegative integers $i \leq l$, then $\delta_h^\pm \zeta \in \mathcal{M}^{l,i}(\Omega'_\sigma)$ and*

$$|\delta_h^\pm \zeta|_{\mathcal{M}^{l,i}(\Omega'_\sigma)} \leq C \sum_{|\alpha| \leq i} \sum_m |(\zeta_{z_\mu})_{\alpha,m}|_{L^{s_{\alpha,m}}(\Omega')}$$

where $(\zeta_{z_\mu})_{\alpha,m}$ *and* $s_{\alpha,m}$ *are as in* Definition 4.4 (*with* ζ *replaced by* ζ_{z_μ}) *and* C *depends on* n, l, i *and on upper bounds for* ε^{-1}, r^{-1} (*defined in* Proposition 4.3), *the Lipschitz constant for* ψ *in* B_r *and in the case that* n *is even and* $n/2 \in \{l-i,\ldots,l\}$, *on an upper bound for* $(s_{\alpha,m}-2)^{-1}$ *for all* α *satisfying* $l-i+|\alpha| = n/2$.

PROOF. Part (a) follows immediately from the bound $|\overline{\varphi}_h^{\pm}|_{W^{l,q'}(\Omega'_\sigma)} \le |\varphi|_{W^{l,q'}(\Omega')}$ and the fact that C_c^∞ is dense in $W^{l,q'}(\Omega')$. To prove (b) we first take $v = \varphi \in C_c^\infty(\Omega'_\sigma)$ so that by (4.28) and the assumption on the support of φ,

$$\begin{aligned}\int_{\Omega'} (\delta_h^{\pm}\zeta)\,\varphi\,dz &= -\int_{\Omega'} \zeta\,(\delta_h^{\mp}\varphi)dz \\ &= -\int_{\Omega'} \zeta(z) D_{z_\mu}\Big(\int_0^1 \varphi(z \mp hse_\mu ds\Big)dz = \int_{\Omega'} \zeta_{z_\mu}(z)\overline{\varphi}_h^{\mp}(z)dz.\end{aligned}$$

The bound stated in (b) now follows and (c) and (d) are immediate from (b) and the fact that the operators $\delta_h^{\pm}$ commute with weak differentiation.

□

PROOF of Proposition 4.3.

We will prove the following statement by induction on $i = -1,\ldots,k$:

$\mathcal{P}_{(i)}$: *If* $u' \in \mathcal{H}'_1$ *has support in* Ω'_σ *for some* $\sigma \in (0, \varepsilon \wedge r)$ *and satisfies*

$$A'(u',v) = \int_{\Omega'} (g'\cdot v + G' : \nabla_z v)dz \tag{4.30}$$

for all $v \in \mathcal{H}'_1$, *for* some *bilinear form* A' *satisfying the hypotheses of* Proposition 4.3 *and for some* g' *and* G' *whose components are in* $[H^{i\vee 0}(\Omega')]^N$ *and* $[H^{i+1}(\Omega')]^{N\times n}$ *respectively, then* $u' \in [H^{i+2}(\Omega')]^N$ *and*

$$\begin{aligned}|u'|_{[H^{i+2}(\Omega')]^N} \le C\big(&|u'|_{[L^2(\Omega')]^N} + |g'|_{[H^{i\vee 0}(\Omega')]^N} \\ &+ |G'|_{[H^{i+1}(\Omega')]^{N\times n}}\big)\end{aligned} \tag{4.31}$$

where C *is as described in* Proposition 4.3.

The statement $\mathcal{P}_{(-1)}$ is true by (4.9), so we may assume that $i \ge 0$ and prove $\mathcal{P}_{(i)}$ assuming $\mathcal{P}_{(i-1)}$. There is no loss of generality in taking $i = k$. We therefore assume that $\mathcal{P}_{(k-1)}$ has been proved and fix A' and u' as in the hypotheses in $\mathcal{P}_{(k)}$. In particular $u' \in \mathcal{H}'_1$ has support in $\overline{\Omega}'_\sigma$ and satisfies (4.30) with the given A' and where the components of g' and G' are in $H^k(\Omega')$ and $H^{k+1}(\Omega')$ respectively. Abbreviating $[H^k(\Omega')]^N$ by $H^k(\Omega')$, etc., we therefore have from $\mathcal{P}_{(k-1)}$ that

$$|u'|_{H^{k+1}(\Omega')} \le C\big(|u'|_{L^2(\Omega')} + |g'|_{H^k(\Omega')} + |G'|_{H^{k+1}(\Omega')}\big) \equiv CC_{\mathcal{G}'}. \tag{4.32}$$

A convenient reduction is now immediate. Since

$$(b')^\eta \in \mathcal{M}^{k,k},\ (c')^\omega \in \mathcal{M}^{k+1,k+1},\ d \in \mathcal{M}^{k+1,k} \tag{4.33}$$

by (4.8) and Lemma 4.5, we have that $(b')^\eta u'_{x_\eta}$, $d'u' \in H^k(\Omega')$ and $c'u' \in H^{k+1}(\Omega')$ with respective norms bounded by $C\mathcal{C}_{\mathcal{G}'}$. These terms can therefore be absorbed into g' and G', affecting only a change in the constant C in (4.32). We therefore proceed to show that $u' \in H^{k+2}(\Omega')$ and $|u'|_{H^{k+2}(\Omega')} \le C\mathcal{C}_{\mathcal{G}'}$ assuming that A' is fixed as in the statement of Proposition 4.3 but with the terms $(b')^\eta u'_{x_\eta}$, $c'u'$ and $d'u'$ absent. The argument will be divided into three steps, as follows.

Step 1: An integration by parts. Let $\tau_h^\pm$ and $\delta_h^\pm$ be the operators defined in (4.26) for a given $\mu \in \{1, \ldots, n-1\}$ (μ will be fixed through Step 2). Choose a function $\chi \in C_c^\infty(\mathbf{R}^n)$ satisfying $\chi \equiv 1$ on $\overline{\Omega}'_{\sigma/2}$ and $(\operatorname{supp}\chi \cap \overline{\Omega}') \subset \overline{\Omega}'_{\sigma/4}$ and whose derivatives up to order $k+2$ are bounded by C. We claim that, for h sufficiently small depending on σ, there are functions g''_h and G''_h with components in $H^{(k-1)\vee 0}$ and H^k respectively satisfying

$$|g''|_{H^{(k-1)\vee 0}(\Omega')},\ |G''_h|_{H^k(\Omega')} \le C\mathcal{C}_{\mathcal{G}'} \tag{4.34}$$

and such that

$$A'(\delta_h^+ u', v) + A'(u', \delta_h^-(\chi v)) = \int_{\Omega'} (g''_h \cdot v + G''_h : \nabla_z v) dz \tag{4.35}$$

for all $v \in \mathcal{H}'_1(\Omega')$.

To prove this we first note that since u' has support in $\overline{\Omega}'_\sigma$, the integrals on the left may be taken over $(\Omega'_\sigma + B_h) \cap \Omega' \subset \Omega'_{\sigma/2}$, on which $\chi = 1$. Applying (4.27) and (4.28) to each summand in the obvious way, denoting weak derivatives with subscripts and suppressing the prime on a, we obtain

$$\begin{aligned}\int_{\Omega'} &\left[(a^{\eta,\omega}\delta_h^+ u'_{z_\eta}) \cdot v_{z_\omega} + (a^{\eta,\omega} u'_{z_\eta}) \cdot (\delta_h^-(\chi v)_{z_\omega})\right] dz \\ &= \int_{\Omega'} \left[(a^{\eta,\omega}\delta_h^+ u'_{z_\eta}) \cdot v_{z_\omega} - \delta_h^+(a^{\eta,\omega} u'_{z_\eta}) \cdot v_{z_\omega}\right] dz \\ &= -\int_{\Omega'} \left[((\delta_h^+ a^{\eta,\omega})(\tau_h^+ u'_{z_\eta})) \cdot v_{z_\omega}\right] dz.\end{aligned}$$

This has the form of the terms on the right in (4.35) with $g''_h = 0$ and the nonzero components of G''_h being components of $(\delta_h^+ a^{\eta,\omega})(\tau_h^+ u')_{z_\eta}$. The H^k norm of the latter is bounded by

$$|(\delta_h^+ a^{\eta,\omega})(\tau_h^+ u')_{z_\eta})|_{H^k(\Omega'_{\sigma/2})} \le |\delta_h^+ a^{\eta,\omega}|_{\mathcal{M}^{k,k}(\Omega'_{\sigma/2})} |u'|_{H^{k+1}(\Omega')} \le C\mathcal{C}_{\mathcal{G}'}$$

by (4.32), Lemma 4.5, Lemma 4.6(d) and by the first hypothesis in (4.8), which includes the requirement that $D^\alpha a^{\eta,\omega}_{z_\mu} \in L^{\mathcal{I}_{|\alpha|}}$ (in the shorthand of (4.8)) for $|\alpha| \le k$. This proves the claim (4.34)–(4.35).

Step 2: H^{k+1} bound for transverse derivatives. We first claim that, for h sufficiently small depending on σ, there are new functions g''_h and G''_h, still satisfying (4.34), such that

$$A'(\delta_h^+ u', v) = \int_{\Omega'} (g''_h \cdot v + G''_h : \nabla_z v)\, dz \tag{4.36}$$

for all $v \in \mathcal{H}'_1$. To prove this we first choose a function $\tilde{\chi} \in C_c^\infty$ which is identically one on $\operatorname{supp}\chi + B_h$ and such that $(\operatorname{supp}\tilde{\chi}) \cap \overline{\Omega}' \subset \overline{\Omega}'_{\sigma/8}$. Applying (4.28)–(4.30), we compute the second term on the left side of (4.35):

$$\begin{aligned}
&\int_{\Omega'} \left[\tilde{\chi} g' \cdot \delta_h^-(\chi v) + G' : \nabla_z \delta_h^-(\chi v)\right] dz \\
&\quad = \int_{\Omega'} \left[(\overline{\tilde{\chi} g'})_h^+ \cdot (\chi v)_{z_\mu} - (\delta_h^+ G') : \nabla_z(\chi v)\right] dz \\
&\quad = \int_{\Omega'} \Big[\big(\chi_{z_\mu}(\overline{\tilde{\chi} g'})_h^+ - (\delta_h^+ G')\nabla_z \chi\big) \cdot v + \chi\big((\overline{\tilde{\chi} g'})_h^+ e_\mu^{tr} - \delta_h^+ G'\big) : \nabla_z v\Big]\, dz.
\end{aligned}$$

This has the form of the integral on the right side of (4.36) in which the $H^{(k-1)\vee 0}(\Omega')$ norm of the contribution to g''_h is bounded by

$$|\chi_{z_\mu}(\overline{\tilde{\chi} g'})_h^+) - (\delta_h^+ G')\nabla_z \chi|_{H^{(k-1)\vee 0}(\Omega'_{\sigma/8})} \le C\big(|g'|_{H^{(k-1)\vee 0}} + |G'|_{H^k}\big) \le C C_{\mathcal{G}'}$$

by Lemma 4.6; and the $H^k(\Omega')$ norm of the contribution to G''_h is bounded by

$$|\chi\big((\overline{g'})_h^+ e_\mu^{tr} - \delta_h^+ G'\big)|_{H^k(\Omega'_{\sigma/8})} \le C\big(|g'|_{H^k} + |G'|_{H^{k+1}}\big) \le C C_{\mathcal{G}'}.$$

This proves the claim (4.36).

Observe that $\delta u' \in \mathcal{H}'_1$ (the specific form of the essential boundary condition in the definition of $\mathcal{H}'_1$ is important here). Therefore $\mathcal{P}_{(k-1)}$ applies to (4.36) to show that, for h small depending on σ, $\delta_h^+ u' \in [H^{k+1}(\Omega')]^N$ with $[H^{k+1}(\Omega')]^N$-norm bounded by $C C_{\mathcal{G}'}$. Since this bound is independent of h, there is a sequence $h \to 0$ such that $\delta_h^+ u'$ converges weakly in $[H^{k+1}(\Omega')]^N$, say to w. Then for $\varphi \in [C_c^\infty(\Omega')]^N$,

$$\begin{aligned}
\int_{\Omega'} u' \cdot \varphi_{z_\mu} dz &= \lim_{h\to 0} \int_{\Omega'} u' \cdot \delta_h^- \varphi\, dz \\
&= -\lim_{h\to 0} \int_{\Omega'} \delta_h^+ u' \cdot \varphi\, dz = -\int_{\Omega'} w \cdot \varphi\, dz.
\end{aligned}$$

This proves that w is the weak D_{z_μ}-derivative of u'. Since $\mu \in \{1, \ldots, n-1\}$ was arbitrary, we conclude that for all multi-indices α of length less than or equal to $k+2$, with the possible exception of $(0, \ldots, k+2)$, the weak derivative $D_z^\alpha u'$ exists, is an element of $[L^2(\Omega')]^N$ and satisfies the bound $|D_z^\alpha u'|_{L^2(\Omega')} \le C C_{\mathcal{G}'}$, where C and $C_{\mathcal{G}'}$ are as described in Proposition 4.3 and (4.32) respectively.

Step 3: H^{k+1} bound for the normal derivative. The required bound for the remaining derivative of order $k+2$ can now be deduced from (4.30) and the bounds obtained in Step 2. In this argument we mix D^α notation with subscripts to denote weak derivatives in Ω' and we take all integrals over Ω'.

Let $\varphi \in [C_c^\infty(\Omega')]^N$ and let α be a multi-index with $|\alpha| \le k$. Applying (4.30) with $v = D^\alpha \varphi \in \mathcal{H}_1'$ and integrating by parts we obtain

$$\left|\int (a^{\eta,\omega} u'_{z_\eta}) \cdot D^\alpha \varphi_{z_\omega} dz\right| = \left|\int [(D^\alpha g' - D^\alpha \mathrm{Div}\, G') \cdot \varphi] dz\right| \le C C_{\mathcal{G}'} |\varphi|_{L^2} \tag{4.37}$$

by the assumptions on g' and G'. Next we show that, with the possible exception of the the term for which $\eta = \omega = n$, the integral of each of the summands on the left in (4.37) is bounded in absolute value by $C C_{\mathcal{G}'}|\varphi|_{L^2}$. First note that, by (4.8) and Lemma 4.5, $a^{\eta,\omega} \in \mathcal{M}^{k+1,k+1} \subset \mathcal{M}^{k+1,k}$ and $a^{\eta,\omega}$, $a^{\eta,\omega}_{z_\omega} \in \mathcal{M}^{k,k}$. Thus if $\eta \ne n$ is fixed then the term in question is

$$\left|\int D^\alpha (a^{\eta,\omega} u'_{z_\eta})_{z_\omega} \cdot \varphi\, dz\right| \le \sup_\omega |a^{\eta,\omega} u'_{z_\eta}|_{H^{k+1}} |\varphi|_{L^2} \le C C_{\mathcal{G}'} |\varphi|_{L^2},$$

and if $\omega \ne n$ is fixed the term is

$$\begin{aligned}\left|\int D^\alpha (a^{\eta,\omega} u'_{z_\eta z_\omega} + a^{\eta,\omega}_{z_\omega} u'_{z_\eta}) \cdot \varphi\, dz\right| &\le (|a^{\eta,\omega} u'_{z_\eta z_\omega}|_{H^k} + |a^{\eta,\omega}_{z_\omega} u'_{z_\eta}|_{H^k}) |\varphi|_{L^2} \\ &\le \left(\sup_\eta \left[|a^{\eta,\omega}|_{\mathcal{M}^{k,k}} |u'_{z_\omega}|_{H^{k+1}}\right] + |a^{\eta,\omega}_{z_\omega}|_{\mathcal{M}^{k,k}} |u'_{z_\eta}|_{H^k}\right) \varphi|_{L^2}.\end{aligned}$$

(sum over η in the second term on the right). We therefore conclude from (4.37) that

$$\left|\int (a^{n,n} u'_{z_n}) \cdot D^\alpha \varphi_{z_n} dz\right| \le C C_{\mathcal{G}'} |\varphi|_{L^2}. \tag{4.38}$$

Now replace the $\mathbf{R}^N$-valued function φ by $\varphi\xi$ where $\varphi \in C_c^\infty(\Omega')$ is scalar-valued and $\xi \in \mathbf{R}^N$ is a (constant) unit vector, and let $\zeta(x) \equiv (a^{n,n}(x))^{tr}\xi$ and $\alpha = (0, \ldots, k)$. If $k \ge 1$ the integral on the left in (4.38) is then

$$\begin{aligned}\int (\zeta \cdot u'_{z_n}) D^k_{z_n} \varphi_{z_n} dz &= (-1)^k \int D^k_{z_n} (\zeta \cdot u'_{z_n}) \varphi_{z_n} dz \\ &= (-1)^k \int D^{k-1}_{z_n} (\zeta \cdot u'_{z_n z_n} + \zeta_{z_n} \cdot u'_{z_n}) \varphi_{z_n} dz.\end{aligned}$$

The second term on the right here is bounded by

$$|\zeta_{z_n} \cdot u'_{z_n}|_{H^k} |\varphi|_{L^2} \le |a_{z_n}|_{\mathcal{M}^{k,k}} |u'|_{H^{k+1}} |\varphi|_{L^2} \le C C_{\mathcal{G}'} |\varphi|_{L^2}$$

and if $k \ge 2$ the first term can be written

$$(-1)^k \int D^{k-2}_{z_n} (\zeta \cdot u'_{z_n z_n z_n} + \zeta_{z_n} u'_{z_n z_n}) \varphi_{z_n} dz.$$

Now repeat the argument: the second term here is bounded by $C C_{\mathcal{G}'}|\varphi|_{L^2}$, exactly as in the previous step. Continuing in this way we thus find that

$$\left|\int [((a^{n,n})^{tr}\xi) \cdot D^{k+1}_{z_n} u'] \varphi_{z_n} dz\right| \le C C_{\mathcal{G}'} |\varphi|_{L^2}.$$

This proves that the weak derivative $D_{z_n}\left[\xi \cdot \left(a^{n,n} D_{z_n}^{k+1} u'\right)\right]$ exists and is bounded in L^2 by $CC_{\mathcal{G}'}$. Letting ξ range over standard basis vectors we conclude that the same is true for the components of the vector $a^{n,n} D_{z_n}^{k+1} u'$. To complete the argument we observe that the positivity hypothesis in (4.4) shows that the moduli of the (possibly complex) eigenvalues of $a^{n,n}$ are bounded below away from zero on Ω', hence so is its determinant. Then since the components of $a^{n,n}_{z_n}$ are in $W^{1,\infty}$ by (4.8), so too are those of $(a^{n,n})^{-1}$ and its transpose. Combining this with the previous observation we conclude that $D_{z_n} D_{z_n}^{k+1} u'$ exists and is bounded in $L^2(\Omega')$ by $CC_{\mathcal{G}'}$. This completes the proof of Step 3 and together with the conclusion above in Step 2 shows that u' is in $H^{k+2}(\Omega')$ and satisfies the bound in (4.21). This in turn proves $\mathcal{P}_{(k)}$ and completes the proof of Proposition 4.3.

□

PROOF of Theorem 4.1.

The proof of Theorem 4.1(a) is essentially contained in that of Proposition 4.3 but is substantially simpler. First for the case that Ω is a proper subset of $\mathbf{R}^n$ we proceed by induction, as described at the beginning of the previous subsection, assuming that $u \in H^{k+1}(\Omega_1)$ where Ω_1 is open, $\Omega_0 \subset \Omega_1 \subset \Omega$, $\operatorname{dist}(\Omega_0, \partial\Omega_1) \geq \gamma/2$, $\operatorname{dist}(\Omega_1, \partial\Omega) \geq C^{-1}$ and Ω_1 satisfies the cone condition with parameters bounded below by C^{-1} (so that Lemma 4.5 applies). Then for $h > 0$ sufficiently small depending on γ and now for arbitrary $\mu \in \{1, \ldots, n\}$, $x \pm h e_\mu$ will be in Ω_1 if $x \in \Omega_0$. We can then define difference operators $\delta_h^{\pm}$ and a cutoff function $\chi \in C_c^\infty(\Omega_1)$ with $\chi \equiv 1$ on Ω_0 (in place of the cutoff function χ in Step 1 of the proof of Proposition 4.3) and then proceed as in Steps 1 and 2 to prove the claim in (4.36). This leads immediately to the conclusion that $u_{x_\mu} \in H^{k+1}(\Omega_0)$ with the required bound. The argument in Step 3 is avoided because μ is arbitrary. The proof is even simpler for part (b) in which $\Omega = \mathbf{R}^n$: the cutoff function χ is avoided altogether and the conclusion of the induction is that $u_{x_\mu} \in H^{k+1}(\mathbf{R}^n)$.

To prove (c) we check from (4.8) that a and c are in $\mathcal{M}^{k+1,k+1}(\Omega)$ and that b and d are in $\mathcal{M}^{k+1,k}(\Omega)$. The components of $a^{\eta,\omega} u_{x_\eta}$ and $c^\omega u$ are therefore in $H^{k+1}(\Omega_0) \subset H^1(\Omega_0)$ and those of $b^\eta u_{x_\eta}$ and du are in $H^k(\Omega_0) \subset L^2(\Omega_0)$. We can therefore take $v \in H_0^1(\Omega_0)$ in (4.2) and apply the divergence theorem to obtain that

$$\int_{\Omega_0} \left[(-a^{\eta,\omega} u_{x_\eta})_{x_\omega} + b^\eta u_{x_\eta} - (c^\omega u)_{x_\omega} + du - g + \operatorname{Div} G\right] \cdot v \, dx = 0.$$

It follows that the term in brackets is zero in $L^2(\Omega_0)$ hence in $H^k(\Omega_0)$.

To prove (d) we note that $\mathcal{I}_l \subseteq [2, \infty]$ for every l (see (4.7)) so that by (4.8) the components of $a^{\eta,\omega}$ and c^ω are in $H^{k+1}(B)$ and those of b^η and d are in $H^k(B)$ for every ball B whose closure is contained in Ω. The conclusions of (d) therefore follow directly from Theorem A.6 together with the inclusions $u \in H^{k+2}(B)$, $G \in H^{k+1}(B)$ and $g \in H^k(B)$.

□

§4.3 Global Regularity: Proof of Theorem 4.2

In this section we prove Theorem 4.2 by applying Theorem 4.1 away from the boundary of Ω and a reduction to the model problem of Proposition 4.3 near the boundary.

PROOF of Theorem 4.2. First, Theorem A.26 shows that the integrals over $\partial\Omega$ in (4.1)–(4.3) involving f and Γ can be expressed as integrals over Ω of terms having the same forms as those involving b, c, d, g and G. It is easily checked from (4.8) and (4.15) that if the former are absorbed into the latter, then the augmented coefficients b, c and d and the augmented inhomogeneous terms g and G continue to satisfy the hypotheses of Theorem 4.2. We may therefore assume that the boundary integrals in (4.1)–(4.3) are absent.

We prove the theorem by induction on $k \geq -1$ (for *all* A satisfying the given hypotheses), as in the proof of Proposition 4.3. The case $k = -1$ again follows from (4.9) and for the induction step we assume that the theorem holds with k replaced by $k-1$, that the hypotheses of the theorem hold exactly as stated and therefore that

$$|u|_{H^{k+1}(\Omega)} \leq C\mathcal{C}_{\mathcal{G}} \equiv C\big(|u|_{L^2(\Omega)} + |g|_{H^k(\Omega)} + |G|_{H^{k+1}(\Omega)}\big) \tag{4.39}$$

(we again abbreviate $[H^k(\Omega)]^N$ by $H^k(\Omega)$, etc.) where here and throughout the proof C will denote a generic positive constant as described in the statement of the theorem. We have to show that

$$|u|_{H^{k+2}(\Omega)} \leq C\mathcal{C}_{\mathcal{G}}. \tag{4.40}$$

We can eliminate the terms involving b, c and d by observing that, by (4.8) and Lemma 4.5,

$$b^\eta \in \mathcal{M}^{k,k},\ c^\omega \in \mathcal{M}^{k+1,k+1} \text{ and } d \in \mathcal{M}^{k+1,k}, \tag{4.41}$$

so that $b^\eta u_{x_\eta}$, $d \in H^k(\Omega)$ and $cu \in H^{k+1}(\Omega)$ with respective norms bounded by $C\mathcal{C}_{\mathcal{G}}$. These terms can therefore be absorbed into g and G, and we may proceed assuming that the terms involving b, c and d in the definition of A are absent.

We begin by applying Theorem 1.8 with $X = \partial\Omega$ and $\delta > 0$ determined by $\underline{r}$, $\underline{s}$, and ε_0 (see Definition 1.7(c)) to obtain a partition of unity $\{\varphi_i\}$ of $\partial\Omega$ subordinate to the cover $\{\mathcal{U}_i\}$ and satisfying the conclusions of Theorem 1.8. We will obtain the required H^{k+2} bound for u restricted to $\partial\Omega \cap \mathcal{U}_i$, where i is now fixed, as follows. Let $\partial\Omega\cap\mathcal{U}_i$ be contained in the connected component $\mathcal{C}_l$ of $\partial\Omega$ and let $S = S_l$ be as in (4.13) if $1 \leq N_l \leq N-1$ and $S \equiv I$ otherwise. Then define the following in terms of the notations in Definition 1.7:

- $\Omega' \equiv \{z = (\tilde{z}, z_n) \in \mathbf{R}^n : \tilde{z} \in \mathcal{B}_i \text{ and } -\psi_i(\tilde{z}) < z_n < 0\}$
- $R(\tilde{z}, z_n) \equiv R_i(\tilde{z}, z_n + \psi_i(\tilde{z}))$.

Thus R is a C^{k+2} diffeomorphism of Ω' onto $\Omega \cap \mathcal{U}_i$ with derivatives up to order $k+2$ bounded pointwise by C. Continuing,

- let $\mathcal{H}_1'$ be the set of elements $v' \in [H^1(\Omega')]^N$ such that $v'(z) = 0$ if $\tilde{z} \in \partial\mathcal{B}_i$ or if $z_n = -\psi_i(\tilde{z})$, and if $N_l \geq 1$ then $(v')^j(\tilde{z}, 0) = 0$ for $\tilde{z} \in \mathcal{B}_i$ and $j = 1, \ldots, N_l$
- let $u' \equiv (\varphi_i \circ R)(S \circ R \circ \Pi)(u \circ R)$ where $\Pi(\tilde{z}, z_n) \equiv (\tilde{z}, 0)$ (the parentheses here indicate the multiplication of a scalar times a matrix times a vector)
- define $\mathcal{T}$ mapping functions on Ω' to functions on Ω by

$$(\mathcal{T}v')(x) = \begin{cases} \left[(S^{-1} \circ R \circ \Pi \circ R^{-1})(x)\right](v' \circ R^{-1})(x), \ x \in \Omega \cap \mathcal{U}_i, \\ 0, \ x \in (\Omega \cap \mathcal{U}_i^c) \end{cases}$$

  (S^{-1} denotes a matrix inverse and R^{-1} a functional inverse)
- define $A'(w', v') \equiv A(\mathcal{T}w', \mathcal{T}v')$.

The following are then easily checked:

- $u' \in \mathcal{H}_1' \cap H^{k+1}(\Omega')$ with $|u'|_{H^{k+1}(\Omega')} \leq C|\varphi_i u|_{H^{k+1}(\Omega)}$
- $(\mathcal{T}u')(x) = \begin{cases} (\varphi_i u)(x), \ x \in \Omega \cap \mathcal{U}_i \\ 0, \ x \in (\Omega \cap \mathcal{U}_i^c) \end{cases}$
- $\mathcal{T}$ is a bounded linear transformation from $\mathcal{H}_1'$ to $\mathcal{H}_1$ with norm bounded by C ($\mathcal{H}_1$ is defined in the paragraph in which (4.13) occurs); and if $v' \in \mathcal{H}_1' \cap H^m(\Omega')$ for some $m \in \{1, \ldots, k+2\}$ then $\mathcal{T}v' \in H^m(\Omega \cap \mathcal{U}_i)$ and $|\mathcal{T}v'|_{H^m(\Omega \cap \mathcal{U}_i)} \leq C|v'|_{H^m(\Omega')}$
- $A' \in \mathcal{B}(\mathcal{H}_1' \times \mathcal{H}_1')$ with norm bounded by C, and for all $v' \in \mathcal{H}_1'$, $A'(v', v') \geq C^{-1}|v'|^2_{H^1(\Omega')} - C|v'|^2_{L^2(\Omega')}$.

We also claim the following:

- A' has the same form as A in (4.1) but with the boundary integral omitted, Ω replaced by Ω', and $a^{\eta,\omega}$, b^η, c^ω and d replaced by $(a')^{\eta,\omega}$, $(b')^\eta$, $(c')^\omega$ and d', these coefficients satisfying the positivity and regularity conditions in (4.4) and (4.8) with Ω replaced by Ω' and C_A replaced by CC_A. In particular, A' satisfies the third itemized hypothesis of Proposition 4.3.

To prove this we let $w', v' \in \mathcal{H}_1'$ and $w = \mathcal{T}w'$ and $v = \mathcal{T}v'$. Then by the chain rule applied in the definition of $\mathcal{T}$ we find that to highest order, $v_{x_\eta} \circ R = X^{q\eta}\tilde{S}v'_{z_q} + \ldots$ (sum over q) where $\tilde{S} = S^{-1} \circ R \circ \Pi$ is C^{k+2} and

$X^{q\eta} = (R^{-1})^q_{x_\eta} \circ R$ is C^{k+1} with derivatives bounded in Ω'. Computing the highest–order terms in A' we then have that

$$\begin{aligned} A'(w', v') = A(w, v) &= \int_{\Omega \cap \mathcal{U}_i} (a^{\eta\omega} w_{x_\eta}) \cdot v_{x_\omega} dx \\ &= \int_{\Omega'} \Big(\big[(a^{\eta\omega} \circ R)(w_{x_\eta} \circ R)\big] \cdot (v_{x_\omega} \circ R) \Big) dz \\ &= \int_{\Omega'} \Big(X^{p\eta} X^{q\omega} \big[(a^{\eta\omega} \circ R)(\tilde{S} w'_{z_p})\big] \cdot (\tilde{S} v'_{z_q}) + \dots \Big) dz. \end{aligned}$$

This shows that $(a')^{pq} = X^{p\eta} X^{q\omega} \tilde{S}^{tr} (a^{\eta,\omega} \circ R)\tilde{S}$ (sum over η and ω; note that $a^{\eta\omega}$ and $(a')^{pq}$ are matrices and $X^{p\eta}$ and $X^{q\omega}$ are scalars). Thus for $|\alpha| \le k+1$, each component of $D^\alpha_z (a')^{pq}$ is a finite sum of derivatives $D^\beta_x a^{\eta\omega}$ of order $|\beta| \le |\alpha|$, modulo multiplication by bounded functions and composition with R. Since the $a^{\eta\omega}$ satisfy (4.8) so too do the $(a')^{pq}$ because these conditions are stronger for lower order derivatives and the sets I_l in (4.7) are increasing in l. Similar arguments apply to terms involving $(b')^\eta, (c')^\omega$ and d'. This proves that the coefficients of A' satisfy the required regularity conditions (4.8). To check the positivity condition in (4.4) we let M be an $N \times n$ matrix and compute

$$\begin{aligned} \big(M^{tr} (a')^{pq} M\big)^{qp} =& X^{p\eta} X^{q\omega} \big(M^{tr} \tilde{S}^{tr} (a^{\eta,\omega} \circ R) \tilde{S} M\big)^{qp} \\ =& \big((\tilde{S} M X)^{tr} (a^{\eta,\omega} \circ R)(\tilde{S} M X)\big)^{\omega\eta} \\ \ge& C^{-1} |\tilde{S} M X|^2 \ge C^{-1} |M|^2 \end{aligned}$$

by the positivity condition satisfied by the $a^{\eta\omega}$. This proves the above claim.

The last hypothesis of Proposition 4.3 to check is the following:

- there are functions g' and G' in $H^k(\Omega')$ and $H^{k+1}(\Omega')$ respectively such that

$$A'(u', v') = \int_{\Omega'} (g' \cdot v' + G' : \nabla_z v')\, dz \tag{4.42}$$

  for all $v' \in \Omega'$ and with respective norms bounded by $CC_{\mathcal{G},i}$, where

$$C_{\mathcal{G},i} = |u|_{H^{k+1}(\Omega \cap \mathcal{U}_i)} + |g|_{H^k(\Omega \cap \mathcal{U}_i)} + |G|_{H^{k+1}(\Omega \cap \mathcal{U}_i)}. \tag{4.43}$$

To prove this we let $v' \in \mathcal{H}'_1$ be given, $v = \mathcal{T} v'$ and X and $\tilde{S}$ be as above. Then

$$\begin{aligned} A'(u', v') &= A(u, \varphi_i v) + \big[A(\varphi_i u, v) - A(u, \varphi_i v)\big] \\ &= \int_{\Omega \cap \mathcal{U}_i} \big[g \cdot (\varphi_i v) + G : \nabla(\varphi_i v)\big] dx \\ &\quad + \int_{\Omega \cap \mathcal{U}_i} \big[(a^{\eta\omega} (\varphi_i u)_{x_\eta}) \cdot v_{x_\omega} - (a^{\eta,\omega} u_{x_\eta}) \cdot (\varphi_i v)_{x_\omega}\big] dx. \end{aligned}$$

Our hypotheses on g and G show that the contributions to g' and G' from the first integral on the right are bounded in H^k and H^{k+1} respectively by $CC_{\mathcal{G},i}$. The second integral may be written

$$\int_{\Omega'}\Big[(\varphi_{i,x_\eta}\circ R)\big(a^{\eta\omega}\circ R)(u\circ R)\big)\cdot(X^{q\omega}\tilde{S}v'_{z_q})$$
$$-(\varphi_{i,x_\omega}\circ R)\big(a^{\eta,\omega}\circ R)\big)(u_{x_\eta}\circ R)\cdot(\tilde{S}v')\Big]dz.$$

Recalling that X and $\tilde{S}$ have bounded derivatives on Ω' up to orders $k+1$ and $k+2$ respectively, we thus have that the $H^k(\Omega')$ norm of the contribution of the above integrand to the g' term in (4.42) is bounded by

$$C\sup_\omega |a^{\eta\omega}u_{x_\eta}|_{H^k(\Omega\cap\mathcal{U}_i)}\le C\sup_\omega|a^{\eta,\omega}|_{\mathcal{M}^{k,k}}|u|_{H^{k+1}(\Omega\cap\mathcal{U}_i)}\le CC_{\mathcal{G},i},$$

and the $H^{k+1}(\Omega')$ norm of the contribution to the G' term in (4.42) is bounded by

$$C\sup_{\eta,\omega}|a^{\eta,\omega}u|_{H^{k+1}(\Omega\cap\mathcal{U}_i)}\le C\sup_{\eta,\omega}|a^{\eta,\omega}|_{\mathcal{M}^{k+1,k+1}}|u|_{H^{k+1}(\Omega\cap\mathcal{U}_i)}\le CC_{\mathcal{G},i},$$

as required.

We have now checked that the hypotheses of Proposition 4.3 apply to (4.42) and therefore that $u'\in H^{k+2}(\Omega')$ with $|u'|_{H^{k+2}(\Omega')}\le CC_{\mathcal{G},i}$. It follows that $\varphi_i u\in H^{k+2}(\Omega)$ with

$$|\varphi_i u|_{H^{k+2}(\Omega)}\le CC_{\mathcal{G},i}. \tag{4.44}$$

Now recall from Theorem 1.8 that $\sum_i\varphi_i\equiv 1$ in a neighborhood $\mathcal{V}$ of $\partial\Omega$ for which $\operatorname{dist}(\partial\mathcal{V},\partial\Omega)\ge C^{-1}$. We can therefore choose $\varphi_0\in C_c^\infty(\mathbf{R}^n)$ with φ_0 identically one on $\Omega_0\equiv\Omega-\overline{\mathcal{V}}$, with support in Ω a positive distance from $\partial\Omega$ bounded below by C^{-1} and with derivatives of order up to $k+2$ bounded by C. Thus

$$u=\varphi_0u+(1-\varphi_0)\sum_i\varphi_iu. \tag{4.45}$$

Theorem 4.1 applies to show that the first term on the right here is in $H^{k+2}(\Omega)$ with norm bounded by $CC_{\mathcal{G}}$ ($C_{\mathcal{G}}$ is defined in (4.39)). Concerning the second term, we have that for finite sums over i,

$$\Big|\sum_i\varphi_iu\Big|^2_{H^{k+2}(\Omega)}=\sum_i\sum_{|\alpha|\le k+2}\int_\Omega|D^\alpha(\varphi_iu)|^2$$
$$=\sum_i|\varphi_iu|^2_{H^{k+2}(\Omega\cap\mathcal{U}_i)}\le C\sum_i C^2_{\mathcal{G},i}$$

by (4.44). However, the last condition in Definition 1.7 applied to each of the integrals in the definition (4.43) of $C_{\mathcal{G},i}$ together with the induction hypothesis (4.39) shows that

$$\sum_i C^2_{\mathcal{G},i}\le C\big(|u|^2_{H^{k+1}(\Omega)}+|g|^2_{H^k(\Omega)}+|G|^2_{H^{k+1}(\Omega)}\big)=CC^2_{\mathcal{G}}.$$

The sum in (4.45) therefore converges strongly in $H^{k+2}(\Omega)$ with limit bounded by $CC_{\mathcal{G}}$, so that $u \in H^{k+2}(\Omega)$ and $|u|_{H^{k+2}(\Omega)} \le CC_{\mathcal{G}}$. This completes the proof of Theorem 4.2(a).

Next, the assertion in (b) that (4.10) holds has already been proved in Theorem 4.1(c), and to prove that the boundary conditions hold as stated in (b) we fix a connected component $\mathcal{C}_l$ of $\partial\Omega$ and let N_l, Q_l and S_l be as in (4.12) and (4.13), taking S_l to be the identity matrix on $\mathcal{C}_l$ if $N_l = 0$. The essential boundary condition $Q_l u = 0$ on $\mathcal{C}_l$ holds (if $N_l > 0$) because $u \in \mathcal{H}_1$. Concerning the natural boundary condition we assume that $N_l \in \{0, \dots, N-1\}$ and let $x_0 \in \partial\Omega \cap \mathcal{U}_i \subseteq \mathcal{C}_l$ be a σ-Lebesgue point of representatives of the functions appearing in (4.18). Let ξ_0 be an arbitrary N-vector in the null space of $Q_l(x_0)$ if $N_l > 0$ and an arbitrary vector in $\mathbf{R}^n$ if $N_l = 0$. We will construct corresponding C^{k+2} test functions $v_h \in \mathcal{H}_1$ as follows. Define $\xi(x) = S_l^{-1}(x)S_l(x_0)\xi_0$ on $\mathcal{C}_l$, so that by (4.12) and (4.13), $\xi \circ T_i \in C^{k+2}_{bdd}(\mathcal{B}_i)$ and $Q_l(x)\xi(x) = 0$ on $\mathcal{C}_l$. Next extend the domain of ξ by choosing a nonnegative function $\chi_h \in C_c^\infty(\mathbf{R}^n)$ having support in the ball of radius h centered at $R_i^{-1}(x_0)$ and whose integral over $\partial\Omega$ equals one; then define $v_h = (\chi_h \circ R_i^{-1})(\xi \circ R_i \circ \Pi \circ R_i^{-1})$ on $\Omega \cap \mathcal{U}_i$, where $\Pi(\tilde{y}, y_n) = (\tilde{y}, \psi_i(\tilde{y}))$, and $v_h = 0$ elsewhere in Ω. Then $v_h \in \mathcal{H}_1$ if h is sufficiently small, and we can therefore apply (4.1), integrate by parts, and apply (4.10) to obtain

$$\int_{\partial\Omega} \left[\nu^\omega(a^{\eta,\omega}u_{x_\eta} + c^\omega u) + fu - G\nu - \Gamma\right] \cdot v_h \, d\sigma = 0.$$

Letting $h \to 0$, we conclude that, at x_0,

$$\left[\nu^\omega(a^{\eta,\omega}u_{x_\eta} + c^\omega u) + fu - G\nu - \Gamma\right] \cdot \xi_0 = 0.$$

Thus if $N_l = 0$, ξ is arbitrary and the term in brackets is zero; otherwise the term in brackets is orthogonal to every ξ_0 in the null space of Q_l, as asserted in (b).

The proofs of the regularity statements in (c) are identical to those of Theorem 4.1(d) with B now an arbitrary ball in $\mathbf{R}^n$, and the remaining statements then follow immediately from these and the conclusions in (b).

□

Exercises

4.1 Let $\tau_h^\pm$ and $\delta_h^\pm$ be as in (4.26) for a given $\mu \in \{1, \dots, n\}$, let $p \in [1, \infty)$ and abbreviate $L^p(\mathbf{R}^n) = L^p$.

- Show that if $u \in L^p$ then $\tau_h^+ u \to u$ in L^p as $h \to 0$ but τ_h^+ does not converge to the identity in the norm of $\mathcal{L}(L^p)$.
- Let $u \in L^p$ and show that $u \in W^{1,p}(\mathbf{R}^n)$ if and only if there is a positive number h_0 and a constant C such that $|\delta_h^+ u|_{L^p} \le C$ for all h with $|h| \le h_0$ and all μ.

These results hold as well with τ_h^+ and δ_h^+ replaced by τ_h^- and δ_h^-.

4.2 Suppose that $\zeta \in L^1_{loc}(\mathbf{R}^n)$ and that there is a constant C such that for all $v \in H^1(\mathbf{R}^n)$ the product ζv is in $H^1(\mathbf{R}^n)$ with norm bounded by $C|v|_{H^1(\mathbf{R}^n)}$. Show that ζ satisfies the condition in (4.22) for $\alpha = 0$ but need not satisfy this condition for $|\alpha| = 1$. Suggestion: for the latter let Ω_j be disjoint balls in $\mathbf{R}^2$ and choose $p_j > q_j > 2$ with $p_j, q_j \to 2$ as $j \to \infty$. Construct $\zeta_j \in C^1_c(\Omega_j)$ such that $\nabla_x \zeta_j \in (L^{q_j} - L^{p_j})(\Omega_j)$.

4.3 Suppose that the hypotheses and notations in Theorem 3.7 are in force and that u is the solution of (3.14)–(3.16) described in that theorem. Assume also that the coefficients $a^{\eta,\omega}$, b^η, d_1, d_2 and f satisfy the conditions in (4.4) and (4.8) with $k = 0$, uniformly on compact intervals in $[0, T)$, and that $g \in L^2_{loc}([0, T); L^2(\Omega))$. Show that $u \in L^2_{loc}([0, T); H^2(\Omega))$.

4.4 Complete the details of the following example in which the hypotheses of Theorem 4.2 hold with $k = 0$ with the exception that Ω is a uniformly Lipschitz domain but not a C^2_{bdd} domain. Let $x = (r\cos\theta, r\sin\theta)$ be the usual polar coordinate representation in $\mathbf{R}^2 - \{x : x_2 \le 0\}$ in which $r(x) > 0$ and $|\theta(x)| < \pi$, and let Ω be the set of points x for which $r(x) \in (0, 1)$ and $|\theta(x)| < 2\pi/3$. The function $w(x) = r(x)^{3/4}\cos(3\theta(x)/4)$ is then the real part of the analytic complex-valued function $(x + iy)^{3/4}$ and is therefore harmonic in Ω. Define $u(x) = \varphi(r(x))w(x)$ where φ is a nonincreasing C^∞ function on $\mathbf{R}$ with $\varphi(r) = 1$ or 0 according as $r \le 1/3$ or $r \ge 2/3$. Then u is in $H^1_0(\Omega)$ and satisfies the weak form of the equation $\Delta u = -g$ for a particular $g \in L^2(\Omega)$; that is,

$$\int_\Omega \nabla_x u \cdot \nabla_x v \, dx = \int_\Omega g \, v \, dx$$

for all $v \in H^1_0(\Omega)$. Check that u is not in $H^2(\Omega)$. (The fact that Ω is not convex also plays a role in this example; see Grisvard [**20**] for a thorough discussion.)

4.5 Assume that the hypotheses and notations of Theorem 4.2 are in force for a given $k \ge 0$ and in addition that A is coercive on $\mathcal{H}_1 \times \mathcal{H}_1$; that is, there is a positive number λ such that $A(u, u) \ge \lambda |u|^2_{\mathcal{H}_1}$ for all $u \in \mathcal{H}_1$. There is then an operator $S \in \mathcal{L}([L^2(\Omega)]^N, \mathcal{H}_1)$ such that $A(Sf, v) = \langle f, v \rangle_{L^2}$ for all $v \in \mathcal{H}_1(\Omega)$. Show that any eigenfunction of S must be in $[H^{k+2}(\Omega)]^N$; consequently if these hypotheses hold for all $k \ge 0$ then any such eigenfunction is in $C^\infty(\Omega)$ and all its derivatives have continuous extensions to $\overline{\Omega}$.

CHAPTER 5

Parabolic Systems: Higher Order Regularity

In this chapter we derive higher order regularity results for solutions of the parabolic systems considered in Chapter 3, now corresponding to a wide range of essential, natural and mixed boundary conditions. The general problem is restated below in section 5.1 and the main result is given in Theorem 5.5: under appropriate regularity conditions on coefficients, inhomogeneities and spatial domains and with compatible initial data, solutions are m times differentiable with respect to $t \in (0, T)$ and $2m + 1$ times differentiable with respect to $x \in \Omega$, where $m \geq 1$ will be fixed throughout. Differentiability is interpreted in three distinct ways in three parts of the theorem: (a) weak absolute continuity in time of the Sobolev space–valued solution u and Bochner space differentiability of its Lebesgue equivalence class on $[0, T)$ (Definition A.18), (b) differentiability in the weak sense on $\Omega \times (0, T)$ of the measurable counterpart $u_{\mathfrak{m}}$, and (c) continuous differentiability of a representative $u_{\mathfrak{c}}$ of $u_{\mathfrak{m}}$ in the usual calculus sense when m is sufficiently large. Initial layer regularization and higher order regularity away from $t = 0$ for solutions of problems with incompatible initial data are described in Theorem 5.8.

§5.1 Boundary Conditions, Coefficients and Inhomogeneities

The formal problem is as in Chapter 3: let Ω be an open set in $\mathbf{R}^n$, $n \geq 2$, $T \in (0, \infty]$ and $N \geq 1$, and let $\mathcal{H}_1$ be a closed subspace of $[H^1(\Omega)]^N$ which is dense in $[L^2(\Omega)]^N$ (specific choices for $\mathcal{H}_1$ are described below). Then given $\varsigma_0 \in [L^2(\Omega)]^N$ the problem is to find $u : [0, T) \to [L^2(\Omega)]^N$ whose $(0, T)$-equivalence class is a weakly measurable map into $\mathcal{H}_1$ and which satisfies

$$(5.1)\quad \langle u(t), v\rangle_{[L^2(\Omega)]^N} = \langle \varsigma_0, v\rangle_{[L^2(\Omega)]^N} + \int_0^t \big[-A(s)(u(s), v) + F(s) \cdot v\big] ds$$

for all $v \in \mathcal{H}_1$ where $A : [0, T) \to \mathcal{B}(\mathcal{H}_1 \times \mathcal{H}_1)$ and $F : [0, T) \to \mathcal{H}_1^*$ are given by

$$(5.2)\quad \begin{aligned} A(u, v) = \int_\Omega \big[(a^{\eta,\omega} u_{x_\eta}) \cdot v_{x_\omega} + (b^\eta u_{x_\eta}) \cdot v + (c^\omega u) \cdot v_{x_\omega} + (du) \cdot v\big] dx \\ + \int_{\partial\Omega} (fu) \cdot v \, d\sigma \end{aligned}$$

(summation over repeated indices η and ω is understood) and

$$F \cdot v = \int_\Omega (g \cdot v + G : \nabla_x v) + \int_{\partial\Omega} \Gamma \cdot v \, d\sigma. \tag{5.3}$$

Here $a^{\eta,\omega}$, b^η, c^ω and d are $N \times N$ matrix-valued functions of $(x,t) \in \Omega \times [0,T)$, f is an $N \times N$ matrix–valued function of $(x,t) \in \partial\Omega \times [0,T)$ and the dot denotes the usual inner product in $\mathbf{R}^N$; g and Γ are $\mathbf{R}^N$-valued, G is $N \times n$ matrix-valued, $\nabla_x v$ is the $N \times n$ x-derivative matrix, $G : \nabla_x v$ is the scalar product $G^{ij} v^i_{x_j}$ and σ is the boundary measure described in Theorem 3.1 (the boundary integrals involving f and Γ are omitted in the case that $\Omega = \mathbf{R}^n$). We saw in Theorem 3.6 that under fairly mild regularity hypotheses on A and F this problem has a unique solution $u \in C([0,T); [L^2(\Omega)]^N)$ with $[0,T)$-equivalence class in $L^2_{loc}([0,T); \mathcal{H}_1)$.

Higher order regularity properties of u will be derived from the application of Theorems 2.7 and 2.10. Three preliminary discussions will be required: description of the Hilbert subspaces $\mathcal{H}_j$ of $[H^1(\Omega)]^N$ and subspaces $\mathcal{V}_j$ of $\mathcal{H}_1^*$ and verification of the hypotheses preceding Theorem 2.7(a); formulation of the assumptions on the coefficients of A, construction of the bilinear forms $A^{(j)}$ and verification of the hypotheses in Theorem 2.7(b) and (c); and formulation of the assumptions on the inhomogeneities g, G and Γ and verification of the hypotheses (2.26)–(2.27) in Theorem 2.7.

Spatial Domain and Boundary Conditions. A specific realization of the general Hilbert space structure of Theorem 2.7 will now be described and will be fixed for this and the following chapter. This structure corresponds to the essential boundary conditions in Example 3 of section 3.1 and (4.13) in Chapter 4 with k now replaced by $2m - 1 \geq 1$. Definitions, notations and hypotheses are as follows:

Definition 5.1. *Fix an ensemble*

$$N,\ n,\ m,\ \Omega,\ \{N_l, Q_l, S_l\},\ C_{BC},\ \{\mathcal{H}_j\}_{j=0}^{2m+1},\ \{\mathcal{V}_j\}_{j=-1}^{2m-1} \tag{5.4}$$

in which N, n, m are integers with $N \geq 1$, $n \geq 2$ and $m \geq 1$; Ω is either $\mathbf{R}^n$, in which case N_l, Q_l, S_l and C_{BC} are not defined and play no role, or Ω is a C^{2m+1}_{bdd} domain in $\mathbf{R}^n$ (Definition 1.7), *not necessarily bounded, and N_l, Q_l and S_l are defined for each connected component $\mathcal{C}_l$ of $\partial\Omega$ as follows (as in the italicized paragraph preceding the statement of* Theorem 4.2 *but with k replaced by $2m-1$): $N_l \in \{0, \ldots, N\}$ and Q_l and S_l map $\mathcal{C}_l$ into the set of $N_l \times N$ matrices and $N \times N$ matrices respectively; if $0 < N_l < N$ then*

$$Q_l S_l^{-1} = \begin{bmatrix} I_{N_l \times N_l} & 0 \end{bmatrix}$$

on $\mathcal{C}_l$, where $I_{N_l \times N_l}$ is the $N_l \times N_l$ identity; if $N_l = 0$ then $Q_l = 0$ and if $N_l = N$ then $Q_l = I_{N \times N}$; and in either of the latter two cases $S_l = I_{N \times N}$. Also, in the notation of Definition 1.7, *for all i for which $\mathcal{U}_i$ intersects $\mathcal{C}_l$, $Q_l \circ T_i$ and $S_l^{\pm} \circ T_i$ are in $C^{2m+1}_{bdd}(\mathcal{B}_i)$ with derivatives bounded pointwise in $\mathcal{B}_i$ by C_{BC}.*

In all cases $\mathcal{H}_0$ *will be* $[L^2(\Omega)]^N$ *and if* $\Omega = \mathbf{R}^n$ *then* $\mathcal{H}_j$ *will be* $[H^j(\mathbf{R}^n)]^N$ *for* $j \geq 1$. *If* Ω *is a proper subset of* $\mathbf{R}^n$ *then* $\mathcal{H}_1$ *will be the set of elements* $v \in [H^1(\Omega)]^N$ *such that* $Q_l \mathcal{T} v = 0$ ($\mathcal{T}$ *is the trace operator of* Theorem 3.3) *on* $\mathcal{C}_l$ *for each* l (*this condition is vacuous if* $N_l = 0$) *and* $\mathcal{H}_j = \mathcal{H}_1 \cap [H^j(\Omega)]^N$ *for* $j \geq 1$.

Finally, $\mathcal{V}_{-1} \equiv \mathcal{H}_1^*$ *and for* $j = 0, \ldots, 2m-1$, $\mathcal{V}_j$ *is the set of elements* $F \in \mathcal{H}_1^*$ *for which there is a representation* (5.3) *in which the components of* g *are in* $H^j(\Omega)$, *those of* G *are in* $H^{j+1}(\Omega)$, *and either* $\Omega = \mathbf{R}^n$ *and the boundary integral involving* Γ *is absent, or in the notation of* Definition 1.7, $\sum_i |\Gamma \circ T_i|^2_{H^{j+1}(\mathcal{B}_i)} < \infty$.

The above structure is consistent with the requirements of Theorem 2.7:

Lemma 5.2. *The ensemble* (5.4) *in* Definition 5.1 *satisfies the hypotheses in the statement preceding part* (a) *of* Theorem 2.7.

PROOF. We need only identify a Hilbert space structure on $\mathcal{V}_j$. Observe that if $w \in \mathcal{H}_{j+2}$ then the mapping $v \in \mathcal{H}_1 \mapsto \langle w, v \rangle_{[H^1(\Omega)]^N}$ is in $\mathcal{V}_j$, and conversely given $F \in \mathcal{V}_j$ there is a unique $w_F \in \mathcal{H}_{j+2}$ for which $F \cdot v = \langle w_F, v \rangle_{[H^1(\Omega)]^N}$ for all $v \in \mathcal{H}_1$, by Theorem 4.2. Thus $\mathcal{V}_j$ can be identified with the Hilbert space $\mathcal{H}_{j+2}$ and so is a Hilbert space itself, as required.

□

As the proof suggests, we could have taken

$$\mathcal{V}_j = \{\langle z, \cdot \rangle_{\mathcal{H}_1} \text{ such that } z \in \mathcal{H}_{j+2}\}$$

and wrtten $F = \langle z, \cdot \rangle_{[H^1(\Omega)]^N}$ in place of the more complicated expression in (5.3). Applications typically present g, G and Γ, however, but not z, and this has determined our choice.

Hypotheses on Coefficients. These can be given both in terms of $W^{k,q}(\Omega)$–valued mappings on $[0, T)$ as well as in terms of weak derivatives of equivalence classes of functions which are measurable on $\Omega \times (0, T)$ (the same is true for the conclusions of the major theorems of this chapter). Both formulations are important, and to clarify the connection we extract several results from section A.5 in the Appendix where complete, detailed statements and proofs are given. These generalize Definition 3.4 and Theorem 3.5 and are summarized as follows, in which Ω is an arbitrary open set in $\mathbf{R}^n$, I a nontrivial interval in $\mathbf{R}$, $p \in [1, \infty]$, $q \in (0, \infty]$ and prime denotes Hölder conjugate:

- (Definition A.13) The space $L^{p,q}(\Omega \times I)$ is the Lebesgue equivalence class of real-valued functions w on $\Omega \times I$ which are measurable with respect to the product measure on $\Omega \times I$ and for which $\big| |w(\cdot, t)|_{L^q(\Omega)} \big|_{L^p(I)}$ (which is meaningful by Tonelli's theorem) is finite; and $L^{p,q}_{(t)loc}(\Omega \times I)$ is the set of such equivalence classes which are in $L^{p,q}(\Omega \times I')$ for every compact subinterval I' of I.

- (Corollary A.16) There is a one-to-one correspondence $w \leftrightarrow w_{\mathfrak{m}}$ between elements $w \in L^p(I; L^q(\Omega))$ and elements $w_{\mathfrak{m}} \in L^{p,q}(\Omega \times I)$ in which w and $w_{\mathfrak{m}}$ define the same action on $L^{p'}(I) \times L^{q'}(\Omega)$ by iterated integration.
- (Theorem A.23) If $w \in L^p(I; L^q(\Omega))$ then $w \in L^p(I; W^{k,q}(\Omega))$ (whose definition requires a brief discussion–see Definition A.22) if and only if for every multi-index α with $|\alpha| \leq k$, the derivative $D_x^\alpha w_{\mathfrak{m}}$ with respect to $x \in \Omega$ exists in the weak sense on $\Omega \times I^\circ$ and is in $L^{p,q}(\Omega \times I)$.
- (Theorem A.24) If $w \in L^1_{loc}(I; L^q(\Omega))$ then the equivalence class w contains an element w_{AC} which is in $wAC(I; W^{k,q}(\Omega))$ if and only if for every multi-index α with $|\alpha| \leq k$, the derivatives $D_x^\alpha w_{\mathfrak{m}}$ and $D_t D_x^\alpha w_{\mathfrak{m}}$ with respect to $x \in \Omega$ and $t \in I$ exist in the weak sense on $\Omega \times I^\circ$ and are in $L^{1,q}_{(t)loc}(\Omega \times I)$.

The correspondence $w \leftrightarrow w_{\mathfrak{m}}$ extends componentwise to vector–valued and matrix–valued functions in the obvious way. Also, if $w \in C(I; L^q(\Omega))$, then its equivalence class $[w]_I$ is in $L^\infty_{loc}(I; L^q(\Omega))$, and no ambiguity will result from our writing $w_{\mathfrak{m}}$ in place of $([w]_I)_{\mathfrak{m}}$.

We now describe hypotheses on the coefficients of A required for the higher order regularity result of the next section. The following positive definiteness assumption is equivalent to the last hypothesis in the basic existence result Theorem 3.6(c):

For each η and ω in $\{1, \ldots, n\}$, $a^{\eta,\omega}$ is a $[0,T)$–equivalence class of mappings from $[0,T)$ into the set of $N \times N$ matrices, each component of which is in $L^\infty_{loc}([0,T); L^\infty(\Omega))$; and given $\bar{t} \in [0,T)$ there is a positive number $\theta(\bar{t})$ such that for almost all $(x,t) \in \Omega \times (0,\bar{t})$,

$$(M^{tr} a_{\mathfrak{m}}^{\eta,\omega}(x,t) M)^{\omega,\eta} \geq 2\theta(\bar{t})|M|^2 \tag{5.5}$$

for all $N \times n$ matrices M; (here ^{tr} denotes transpose, summation over η and ω is understood and any choice of matrix norm is allowed).

Next, in order to formulate regularity conditions for functions defined on $\partial\Omega$ we recall that if $\psi : \partial\Omega \to \mathbf{R}$ and α is a multi-index, then in the notation of Definition 1.7,

$$\langle\psi\rangle^\alpha_{L^q,\partial\Omega} \equiv \begin{cases} \left(\sum_i |D_{\tilde{y}}^\alpha(\psi \circ T_i)|^q_{L^q(\mathcal{B}_i)}\right)^{1/q}, & q \in (1,\infty), \\ \sup_i |D_{\tilde{y}}^\alpha(\psi \circ T_i)|_{L^\infty(\mathcal{B}_i)}, & q = \infty \end{cases} \tag{5.6}$$

provided the above derivatives exist in the weak sense on each $\mathcal{B}_i$. Also, if $I \subset \mathbf{R}$ is a nontrivial interval, $j \geq 0$ and $\psi \in L^1_{loc}(I; L^q(\partial\Omega))$, we define

$$\langle\psi\rangle^{(\alpha,j)}_{L^q,\partial\Omega} \equiv \begin{cases} \left(\sum_i |D_t^j D_{\tilde{y}}^\alpha[(\psi \circ T_i)_{\mathfrak{m}}]|^q_{L^q(\mathcal{B}_i)}\right)^{1/q}, & q \in (1,\infty), \\ \sup_i |D_t^j D_{\tilde{y}}^\alpha[(\psi \circ T_i)_{\mathfrak{m}}]|_{L^\infty(\mathcal{B}_i)}, & q = \infty, \end{cases} \tag{5.7}$$

provided the above derivatives exist in the weak sense on each $\mathcal{B}_i \times I^\circ$. If so, then $\langle\psi\rangle^{(\alpha,j)}_{L^q,\partial\Omega}$ is a measurable function of t.

We now describe three sets of regularity conditions for the coefficients of A, each more general but also more complicated than the previous. The first set is unnecessarily strong but is easy to apply and suffices for many applications:

(5.8)

- *Each component of each of* $a^{\eta,\omega}$, b^η, c^ω *and* d *is an element* h *in* $L^\infty_{loc}([0,T); L^\infty(\Omega))$ *satisfying the following: for* $j = 0,\ldots,m-1$,

$$D_t^j D_x^\alpha h_{\mathrm{m}} \in L^{\infty,\infty}_{(t)loc}(\Omega \times [0,T))$$
$$D_t^{j+1} D_x^\alpha h_{\mathrm{m}} \in L^{2,\infty}_{(t)loc}(\Omega \times [0,T))$$

for $|\alpha| \le 2m-2j$ *if* h *represents a component of* $a^{\eta,\omega}$ *or* c^ω *and for* $|\alpha| \le 2m-2j-1$ *if* h *represents a component of* b^η *or* d.

- *(If* $\Omega \ne \mathbf{R}^n$*) each component* h *of* f *is in* $L^\infty_{loc}([0,T); L^\infty(\partial\Omega))$ *and for* $j = 0,\ldots,m-1$ *and* $|\alpha| \le 2m-2j$,

$$\langle h\rangle^{(\alpha,j)}_{L^\infty,\partial\Omega} \in L^\infty_{loc}([0,T))$$
$$\langle h\rangle^{(\alpha,j+1)}_{L^\infty,\partial\Omega} \in L^2_{loc}([0,T)).$$

(Readers for whose purposes the above conditions suffice may choose to proceed directly to Lemma 5.3 below.)

More general regularity hypotheses can be formulated in which less stringent conditions are imposed on higher order derivatives (which is to be expected, since a term such as $D_t^j D_x^\alpha(a^{\eta,\omega} u_{x_\eta})$, for example, involves products of higher derivatives of one factor and lower derivatives of the other). An intermediate set of hypotheses is available offering a broader range of allowable Lebesgue spaces for higher order t–derivatives of the coefficients, although not for higher order x–derivatives. These are stated in terms of index sets $\tilde{\mathcal{I}}_k$, which are the sets $\mathcal{I}_k$ defined in (A.1) with the left-hand endpoints removed, with the exception that $\tilde{\mathcal{I}}_0 = \{\infty\}$. Thus

$$\tilde{\mathcal{I}}_k = \begin{cases} \{\infty\}, & k = 0, \\ (n/k, \infty], & 0 < k \le n/2, \\ (2,\infty], & k > n/2. \end{cases} \tag{5.9}$$

The following intermediate regularity conditions will suffice for most applications:

(5.10)

- *Each component of each* $a^{\eta,\omega}$ *is an element* h *in* $L^\infty_{loc}([0,T);W^{1,\infty}(\Omega))$; *and given* $j \in \{0,\ldots,m-1\}$ *and* α *with* $|\alpha| \le 2m-2j$, *there is an* $s \in \tilde{\mathcal{I}}_{2j}$, *which may depend on* η,ω, *the particular component of* $a^{\eta,\omega}$, j *and* α, *such that*
$$D_t^j D_x^\alpha h_{\mathrm{m}} \in L^{\infty,s}_{(t)loc}(\Omega \times [0,T))$$
$$D_t^{j+1} D_x^\alpha h_{\mathrm{m}} \in L^{2,s}_{(t)loc}(\Omega \times [0,T)).$$
- *The above conditions hold as well for the components of* b^η, c^ω *and* d *with values of* s *in* $\tilde{\mathcal{I}}_{2j+1}$, $\tilde{\mathcal{I}}_{2j+1}$ *and* $\tilde{\mathcal{I}}_{2j+2}$ *respectively and with the upper bound for* $|\alpha|$ *replaced by* $2m-2j-1$, $2m-2j$ *and* $2m-2j-1$ *respectively.*
- *(If* $\Omega \neq \mathbf{R}^n$*) each component* h *of* f *is in* $L^\infty_{loc}([0,T);L^\infty(\partial\Omega))$ *and for each* $j = 0,\ldots,m-1$ *there is an* $s \in \tilde{\mathcal{I}}_{2j+1}$, *which may depend on the particular component of* f *and on* j, *such that for* $|\alpha| \le 2m-2j$,
$$\langle h\rangle^{(\alpha,j)}_{L^s,\partial\Omega} \in L^\infty_{loc}([0,T))$$
$$\langle h\rangle^{(\alpha,j+1)}_{L^s,\partial\Omega} \in L^2_{loc}([0,T)).$$

Observe that the conditions in (5.8) imply those in (5.10) with $s=\infty$ in each case. (Again, readers for whose purposes the above conditions suffice may choose to proceed directly to Lemma 5.3 below.)

Finally we describe the most general regularity hypotheses on coefficients which the theory can accommodate. These are stated in terms of index sets $\mathcal{I}_k$ and $\mathcal{K}_k$, defined in (A.1) and (A.2), and $\tilde{\mathcal{I}}_k$ defined above in (5.9). First, the following conditions guarantee that the bilinear form A in (5.2) satisfies the absolute continuity hypotheses (*ci*–*ciii*) of Theorem 2.7:

(5.11)

- *Each component of each* $a^{\eta,\omega}$ *is an element* h *in* $L^\infty_{loc}([0,T);L^q(\Omega))$ *with* $q=\infty$; *and given* $j \in \{0,\ldots,m-1\}$ *there is an* $s \in \mathcal{I}_{2j+1}$, *which may depend on* η,ω, *the particular component* h *of* $a^{\eta,\omega}$ *and* j, *such that*
$$D_t^j h_{\mathrm{m}} \in L^{\infty,s}_{(t)loc}(\Omega \times [0,T))$$
$$D_t^{j+1} h_{\mathrm{m}} \in L^{2,s}_{(t)loc}(\Omega \times [0,T)).$$
- *The above conditions hold as well for the components of* b^η, c^ω *and* d *with values of* q *in* $\{\infty\}$, $\{\infty\}$ *and* $\mathcal{I}_1$ *respectively and* $\mathcal{I}_{2j+1}$ *replaced by* $\mathcal{K}_{2j+1}$, $\mathcal{I}_{2j+2}$ *and* $\mathcal{K}_{2j+2}$ *respectively.*

The following hypotheses will be applied to verify the ellipticity condition (b) of Theorem 2.7:

(5.12)

- *For all η and ω, for each component h of $a^{\eta,\omega}$ and for $|\alpha| \leq 2m$, there is an $s \in \mathcal{I}_{(|\alpha|-1)\vee 0}$ such that $D_x^\alpha h_{\mathrm{m}} \in L^{\infty,s}_{(t)loc}(\Omega \times [0,T))$.*
- *The above condition holds as well for the components of b^η, c^ω and d with values of s in $\mathcal{I}_{|\alpha|}$, $\mathcal{I}_{|\alpha|}$ and $\mathcal{I}_{|\alpha|+1}$ respectively and with the upper bound for $|\alpha|$ replaced by $2m-1$, $2m$ and $2m-1$ respectively.*

The following hypotheses will be applied to verify condition (civ) of Theorem 2.7:

(5.13)

- *If $m \geq 2$ then for all η and ω, for each component h of $a^{\eta,\omega}$, and for $j = 1, \ldots, m-1$ and $|\alpha| \leq 2m-2j$, there is an $s \in \tilde{\mathcal{I}}_{2j+|\alpha|}$ such that $D_t^j D_x^\alpha h_{\mathrm{m}} \in L^{\infty,s}_{(t)loc}(\Omega \times [0,T))$.*
- *The above condition holds as well for the components of b^η, c^ω and d with $\tilde{\mathcal{I}}_{2j+|\alpha|}$ replaced by $\tilde{\mathcal{I}}_{2j+|\alpha|+1}$, $\tilde{\mathcal{I}}_{2j+|\alpha|+1}$ and $\tilde{\mathcal{I}}_{2j+|\alpha|+2}$ respectively and the upper bound for $|\alpha|$ replaced by $2m-2j-1$, $2m-2j$ and $2m-2j-1$ respectively.*

Finally, for the case that $\Omega \neq \mathbf{R}^n$, we require that $f \in L^\infty_{loc}([0,T); L^\infty(\partial\Omega))$ and that each component h of f satisfies the following:

(5.14)

- *for each α with $|\alpha| \leq 2m$ there is an $s \in \mathcal{I}_{|\alpha|}$ such that*

$$\langle h \rangle^{(\alpha,0)}_{L^s,\partial\Omega} \in L^\infty_{loc}([0,T))$$

- *for all $j = 0, \ldots, m-1$ there is an $s \in \mathcal{I}_{2j+2}$ such that*

$$\langle h \rangle^{(0,j)}_{L^s,\partial\Omega} \in L^\infty_{loc}([0,T))$$
$$\langle h \rangle^{(0,j+1)}_{L^s,\partial\Omega} \in L^2_{loc}([0,T))$$

- *and if $m \geq 2$ then for $j = 1, \ldots, m-1$ and $|\alpha| \leq 2m-2j$ there is an $s \in \tilde{\mathcal{I}}_{2j+|\alpha|+1}$ such that*

$$\langle h \rangle^{(\alpha,j)}_{L^s,\partial\Omega} \in L^\infty_{loc}([0,T)).$$

Observe that all the above conditions are satisfied automatically in the important case that the coefficient matrices are independent of x and t.

We now construct bilinear forms $A^{(j)}$ satisfying the hypotheses of Theorem 2.7:

Lemma 5.3. *Assume that the hypotheses and notations in* Definition 5.1 *are in force, let* $T \in (0, \infty]$ *and let* $\mathcal{I}_k$ *and* $\mathcal{K}_k$ *be the index sets defined in* (A.1) *and* (A.2). *Assume that* $N \times N$ *matrix-valued* $[0, T)$*–equivalence classes* $a^{\eta,\omega}$, b^η, c^ω, d *and* f *are given satisfying the descriptions and regularity conditions in one of* (5.8), (5.10) *or* (5.11)–(5.14) *and that the positive definiteness condition* (5.5) *holds. Then:*

(a) *For* $j = 0, \dots, m$ *there are mappings* $(a^{(j)})^{\eta,\omega}$ *such that for* $j < m$ *each component of each* $(a^{(j)})^{\eta,\omega}$ *is in* $wAC^2_{loc}([0, T); L^s(\Omega))$ *for some* $s \in \mathcal{I}_{2j+1}$ *depending on* j, η, ω *and the particular component; the derivative of each such component is the equivalence class in* $L^2_{loc}([0, T); L^s(\Omega))$ *of the corresponding component of* $(a^{(j+1)})^{\eta,\omega}$ *if* $j < m - 1$ *and the corresponding component of* $(a^{(m)})^{\eta,\omega} \in L^2_{loc}([0, T); L^s(\Omega))$ *if* $j = m$; $(a^{(0)})^{\eta,\omega}$ *is an element of the equivalence class* $a^{\eta,\omega}$ *as a mapping into the set of measurable mappings on* Ω, *and more generally,*

$$[(a^{(j)})^{\eta,\omega}]_{\mathrm{m}} = D^j_t a^{\eta,\omega}_{\mathrm{m}} \tag{5.15}$$

for $j = 0, \dots, m$.

The same statements hold as well for $(b^{(j)})^\eta$, $(c^{(j)})^\omega$ *and* $d^{(j)}$ *with* $\mathcal{I}_{2j+1}$ *replaced by* $\mathcal{K}_{2j+1}$, $\mathcal{I}_{2j+2}$, *and* $\mathcal{K}_{2j+2}$ *respectively and in the case that* $\Omega \neq \mathbf{R}^n$, *for* $f^{(j)}$ *with* Ω *replaced by* $\partial\Omega$ *and* $\mathcal{I}_{2j+1}$ *by* $\mathcal{I}_{2j+2}$.

(b) *For* $j = 0, \dots, m - 1$, $|\alpha| \le 2m - 2j$ *and any particular component* h *of any* $a^{\eta,\omega}$, *the Bochner space derivative* $D^\alpha_x h^{(j)}$ *exists and is in* $C([0, T); L^{s_\alpha}_{weak}(\Omega))$ *for some* $s_\alpha \in \mathcal{I}_{2j+|\alpha|}$ *which may depend* j, α, η, ω *and the particular component.*

The same statement holds as well for $(b^{(j)})^\eta$, $(c^{(j)})^\omega$, *and* $d^{(j)}$ *with the upper bound for* $|\alpha|$ *replaced by* $2m-2j-1$, $2m-2j$, *and* $2m-2j-1$ *respectively and* $\mathcal{I}_{2j+|\alpha|}$ *replaced by* $\mathcal{I}_{2j+|\alpha|+1}$, $\mathcal{I}_{2j+|\alpha|+1}$ *and* $\mathcal{I}_{2j+|\alpha|+2}$ *respectively.*

(c) *The bilinear forms* $A^{(0)}, \dots, A^{(m)}$ *given by*

$$\begin{aligned} A^{(j)}(z, w) = \int_\Omega \big[((a^{(j)})^{\eta,\omega} z_{x_\eta}) \cdot w_{x_\omega} &+ ((b^{(j)})^\eta z_{x_\eta}) \cdot w \\ &+ ((c^{(j)})^\omega z) \cdot w_{x_\omega} + (d^{(j)} z) \cdot w\big]\, dx \\ + \int_{\partial\Omega} (f^{(j)} z) \cdot w \, d\sigma & \end{aligned}$$

satisfy the conditions in (a)–(c) *of* Theorem 2.7 *with constant* $C_A(\bar t)$ *determined by* n, m, N, C_{BC} *(see* Definition 5.1), $\bar t$, $\theta(\bar t)$ *and the norms of the coefficients on* $[0, \bar t]$ *occurring in the hypotheses* (5.8), (5.10) *or* (5.11)–(5.14).

Proof. We first show that the boundary integral in the definition of A, if it occurs, can be absorbed into terms in A involving integrals over Ω. To do this we first let $z, w \in C_c^\infty(\Omega)$ and apply (A.26) in Theorem A.26 to obtain

$$(5.16)\quad \int_{\partial\Omega} f^{kl} z^k w^l d\sigma = \int_\Omega \left[S^1(f^{kl}) z^k w^l + S^2(f^{kl})\cdot(w^l \nabla_x z^k + z^k \nabla_x w^l)\right] dx.$$

The middle term on the right, for example, has the same form as the term involving b^η in the definition of A. The reader can easily check via Theorem A.26 that the three hypotheses on the components of f in (5.14) have been correctly tailored to insure that the components of $S^2(f)$ satisfy the three hypotheses in (5.11)–(5.13) imposed on b^η. The same holds for the other two terms in (5.16), which then extends to elements z, $w \in H^1(\Omega)$. The term on the left in (5.16) may therefore be absorbed into the terms in A involving b^η, c^ω and d, thereby affecting only the constant C_A. We may therefore proceed assuming that the boundary integral in the definition of A is omitted.

To prove part (a) we let $h \in L^\infty_{loc}([0,T); L^\infty(\Omega))$ denote a particular component of a particular $a^{\eta,\omega}$. The hypotheses in (5.11) that $D_t^j h_{\mathrm{m}} \in L^{\infty,s}_{(t)loc}$ and $D_t^{j+1} h_{\mathrm{m}} \in L^{2,s}_{(t)loc}$ for some $s \in \mathcal{I}_{2j+1}$ together with Theorem A.24 then show that there is an $h^{(j)} \in wAC^2([0,T); L^s(\Omega))$ whose measurable counterpart is $D_t^j h_{\mathrm{m}}$ and that

$$(5.17)\quad [h^{(j)}]_{\mathrm{m}} = D_t^j h_{\mathrm{m}},$$

which proves (5.15), and whose derivative satisfies $[(h^{(j)})']_{\mathrm{m}} = D_t[h^{(j)}]_{\mathrm{m}}$. Applying (5.17) twice, we conclude that $[(h^{(j)})']_{\mathrm{m}} = D_t^{j+1} h_{\mathrm{m}} = [h^{(j+1)}]_{\mathrm{m}}$. Thus the derivative of $h^{(j)}$ is the $[0,T)$ equivalence class of $h^{(j+1)}$ (or of $h^{(m)}$ if $j = m-1$). This proves the statements in (a) for $a^{\eta,\omega}$, and the corresponding statements for the other coefficients are proved in a similar way.

To prove (b) we let $j \le m-1$ and $|\alpha| \le 2m-2j$ and again denote by h a component of a particular $a^{\eta,\omega}$. Then by the hypotheses (5.12) and (5.13), $D_t^j D_x^\alpha h_{\mathrm{m}} \in L^{\infty,s_\alpha}_{(t)loc}$ for some $s_\alpha \in \mathcal{I}_{2j+|\alpha|}$. Therefore if $\varphi \in C_c^\infty(\Omega)$ and $t_0 \in [0,\bar{t})$,

$$\begin{aligned}\left|\int_\Omega h^{(j)}(t_0) D_x^\alpha \varphi(x) dx\right| &= \lim_{\Delta t\to 0} |\Delta t^{-1} \int_{t_0}^{t_0+\Delta t}\int_\Omega h^{(j)}(t) D_x^\alpha \varphi\, dx dt\Big| \\ &= \Big|\lim_{\Delta t\to 0} \Delta t^{-1} \int_{t_0}^{t_0+\Delta t}\int_\Omega D_x^\alpha D_t^j h_{\mathrm{m}}\, \varphi\, dx dt\Big| \le C(\bar{t})|\varphi|_{L^{s'_\alpha}(\Omega)}\end{aligned}$$

where s'_α is the Hölder conjugate of s_α. Since $s'_\alpha \in [1,\infty)$, $C_c^\infty(\Omega)$ is dense in $L^{s'_\alpha}$ and the above therefore shows that the weak derivative $D_x^\alpha h^{(j)}(t_0)$ exists and is bounded in L^{s_α} by $C(\bar{t})$. The weak continuity of the map $t \to D_x^\alpha h^{(j)}(t)$ then follows from the same density together with the fact that $h^{(j)}$ is a continuous mapping into $L^{s_\alpha}(\Omega)$. This proves the statements in (b) for $a^{\eta,\omega}$ and the corresponding statements for the other coefficients are proved in a similar way.

To prove (c) we need to check that the hypotheses of Theorem 2.7 are satisfied. The first of these, Theorem 2.7(a), follows from (5.11) and (5.12): these are stronger than the conditions in Theorem 3.6, part (d) of which guarantees the requirement in Theorem 2.7(a).

To prove the hypothesis in Theorem 2.7(b) we can (and leave to the reader to) check that, in the notation of Definition 2.6, the bilinear form A_ε is the same as in the definition (5.2) of A but with mollified coefficients: $a^{\eta,\omega}$ is replaced by $a_\varepsilon^{\eta,\omega}$ where

$$a_\varepsilon^{\eta,\omega}(\cdot,t) = \int_{|s-t|<\varepsilon} h_\varepsilon(t-s)a^{\eta,\omega}(\cdot,s+\varepsilon)ds,$$

which defines $a_\varepsilon^{\eta,\omega}$ as an element of $wAC([0,T);L^\infty(\Omega))$, and similarly for the other coefficients. It is then immediate from this and (5.12) that these regularized coefficients satisfy the hypotheses of the elliptic regularity result Theorem 4.2 independently of ε and consequently that the requirement in Theorem 2.7(b) is met.

Next we prove the following statement, which is somewhat stronger than the conditions in Theorem 2.7(ci)–(ciii): for $j=0,\dots,m-1$, $A^{(j)} \in wAC^2(\mathcal{B}(\mathcal{H}_{2j+2}\times\mathcal{H}_1))$ and its derivative $(A^{(j)})'$ is the equivalence class over $[0,T)$ of $A^{(j+1)}$ (or simply of $A^{(m)}$ if $j=m-1$). To prove this we let $z\in\mathcal{H}_{2j+2}$ and $w\in\mathcal{H}_1$. Then by Theorem A.3(b) each summand in $z_{x_\eta}\cdot w_{x_\omega}$ is in $H^{2j+1}\cdot L^2 \subset L^{s'}$ for every s' whose Hölder conjugate s is in $\mathcal{I}_{2j+1}$. In the proof of part (a) of the present lemma we observed that every component of every $(a^{(j)})^{\eta,\omega}$ is in $wAC^2([0,T);L^s(\Omega))$ for some such s with derivative the corresponding component of $(a^{(j+1)})^{\eta,\omega}$. This shows that the mapping $t\to\int_\Omega (a^{(j)}z_{x_\eta})\cdot w_{x_\omega}dx$ is an absolutely continuous function on $[0,T)$ with derivative $\int_\Omega (a^{(j+1)}z_{x_\eta})\cdot w_{x_\omega}dx$. Similar considerations apply to the other terms in $A^{(j)}$, and the mapping $t\to A^{(j)}(z,w)$ is therefore absolutely continuous with derivative $A^{(j+1)}(z,w)$, modulo equivalence classes. This proves that the conditions in Theorem 2.7(ci)–(ciii) are satisfied.

There remains to check that the hypothesis in Theorem 2.7(civ) holds. Thus let $m\geq 2$, $j\in\{1,\dots,m-1\}$, $l\in\{0,\dots,2m-2j-1\}$ and $z\in\mathcal{H}_{2i+l+2}$, as in the statement. Then for each t,

$$A^{(j)}(z,v) = \int_\Omega (\tilde{g}\cdot v + \tilde{G} : \nabla_x v)dx$$

where $\tilde{g} = (b^{(j)})^\eta z_{x_\eta} + d^{(j)}z$ and $(\tilde{G})^{\mu,\omega}$ is the μ-component of the vector $(a^{(j)})^{\eta,\omega}z_{x_\eta} + (c^{(j)})^\omega z$ (recall that the boundary integral involving f has been absorbed). The $\mathcal{V}_l$–norm of the linear functional $v\to A^{(j)}(t)(z,v)$ is therefore bounded by $C(|\tilde{g}|_{H^l} + |\tilde{G}|_{H^{l+1}})$ (see Definition 5.1; we again abbreviate $[H^k(\Omega)]^N$ by H^k, etc.). To bound $|\tilde{G}|_{H^{l+1}}$ we abuse notation somewhat and denote a generic summand in the product $(a^{(j)})^{\eta,\omega}z_{x_\eta}$ by $a^{(j)}z_x$, regarded as the product of scalars. Then for each $t \le \bar{t} < T$ and $|\alpha| \le 2m - 2j$ there is, by the first hypothesis in (5.13), an $s_{\alpha,j} \in \tilde{\mathcal{I}}_{2j+|\alpha|} = \tilde{I}_{(2j+l+1)-(l+1)+|\alpha|}$ such that $|D_x^\alpha a^{(j)}(t)|_{L^{s_{\alpha,j}}} \le C_A(\bar{t})$. Since $l+1 \le 2m-2j$, we therefore have that $a^{(j)} \in \tilde{\mathcal{M}}^{2j+l+1,l+1}$ with the double sum on the right side of the definition in (4.22) (with ζ replaced by $a^{(j)}$) bounded by $C_A(\bar{t})$. Therefore by Lemma 4.5(b) there is a $\delta \in (0,1)$ such that

$$\begin{aligned} |a^{(j)} z_x|_{H^{l+1}} &\le C_A(\bar{t}) |z_x|^{\delta}_{H^{l+1}} |z_x|^{1-\delta}_{H^{2j+l+1}} \\ &\le C_A(\bar{t}) |z|^{\delta}_{H^{l+2}} |z|^{1-\delta}_{H^{2j+l+2}}. \end{aligned}$$

The terms $(c^{(j)})^\omega z$ are bounded in a similar way, and taking the min over the various values of δ occurring in these bounds and noting that $l+2 \le 2j+l$, we obtain that, for $0 \le t \le \bar{t} < T$,

$$|\tilde{G}(t)|_{H^{l+1}} \le C_A(\bar{t}) |z|^{\delta}_{H^{2j+l}} |z_x|^{1-\delta}_{H^{2j+l+2}}$$

for a new $\delta \in (0,1)$ (we apply here the fact that if $0 < a \le b$ then $a^\delta b^{1-\delta}$ is a nonincreasing function of δ). Similar considerations apply to show that $|\tilde{g}(t)|_{H^l}$ satisfies the same bound and therefore so too does the functional $|A^{(j)}(t)(z,\cdot)|_{\mathcal{V}_l}$. This shows that the condition in Theorem 2.7(*civ*) is satisfied and completes the proof of part (c) of the present lemma.

□

Hypotheses on Inhomogeneities. We now formulate hypotheses on the inhomogeneous terms g, G and Γ required for the application of Theorem 2.7 to the problem (5.1) –(5.3):

Lemma 5.4. *Assume that the hypotheses and notations in* Definition 5.1 *are in force and let $T \in (0,\infty]$. Let g, G and Γ be equivalence classes on $[0,T)$ of matrix-valued arrays as described below (5.3) whose components take values in $L^2(\Omega)$ or $L^2(\partial\Omega)$ (references to Γ and $\partial\Omega$ are omitted if $\Omega = \mathbf{R}^n$). Assume the following (in terms of weak derivatives on $\Omega \times (0,T)$):*

$$(5.18)\qquad D_t^j D_x^\alpha g_{\mathfrak{m}} \in \begin{cases} L^{\infty,2}_{(t)loc}, & 0 \le j \le m-1,\ 0 \le 2j+|\alpha| \le 2m-2 \\ L^{2,2}_{(t)loc}, & 0 \le j \le m-1,\ 0 \le 2j+|\alpha| \le 2m-1 \end{cases}$$

and there is an element $g^{(m)} \in L^2_{loc}([0,T);\mathcal{V}_{-1})$ such that for $v \in \mathcal{H}_1$ and $\psi \in C^1_c([0,T))$ (having a one-sided derivative at $t=0$),

$$(5.19)\qquad \int_{\Omega\times[0,T)} D_t^{m-1} g_{\mathfrak{m}}(x,t) \cdot v(x)\,\psi'(t)d(x,t) = \int_{[0,T)} (g^{(m)}(t)\cdot v)\psi(t)\,dt$$

(see the remark following the proof concerning this last hypothesis);

$$(5.20)\qquad D_t^j D_x^\alpha G_{\mathfrak{m}} \in \begin{cases} L^{\infty,2}_{(t)loc}, & 0 \le j \le m-1,\ 0 \le 2j+|\alpha| \le 2m-1, \\ L^{2,2}_{(t)loc}, & 0 \le j \le m,\ 0 \le 2j+|\alpha| \le 2m; \end{cases}$$

and if $\Omega \neq \mathbf{R}^n$,

$$\sum_{|\alpha|\le 2m-2j-1} \langle\Gamma\rangle^{(\alpha,j)}_{L^2,\partial\Omega} \in L^\infty_{loc}([0,T)),\ j=0,\dots,m-1, \tag{5.21}$$
$$\sum_{|\alpha|\le 2m-2j} \langle\Gamma\rangle^{(\alpha,j)}_{L^2,\partial\Omega} \in L^2_{loc}([0,T)),\ j=0,\dots,m$$

(see (5.7)*). Then:*

(a) *For* $j=0,\dots,m-1$ *there are mappings* $g^{(j)}$, $G^{(j)}$ *and* $\Gamma^{(j)}$ *with* $g^{(0)}$, $G^{(0)}$ *and* $\Gamma^{(0)}$ *in the* $[0,T)$*-equivalence classes* g, G *and* Γ, *and there are equivalence classes* $g^{(m)}$, $G^{(m)}$ *and* $\Gamma^{(m)}$ *satisfying the following (in which we abbreviate* $[H^k(\Omega)]^N$ *by* H^k, *etc.):*

- *For* $j \le m-1$, $g^{(j)} \in C([0,T); H^{2m-2j-2}_{weak})$, *and if* $m \ge 2$ *then for* $j \le m-2$, $g^{(j)} \in wAC^2([0,T); H^{2m-2j-3})$ *with derivative the equivalence class of* $g^{(j+1)}$ *in* $L^2_{loc}([0,T); H^{2m-2j-3})$.
- $(g^{(m-1)})^* \equiv \langle g^{(m-1)}, \cdot\rangle_{L^2} \in wAC^2([0,T); \mathcal{V}_{-1})$ *with derivative* $g^{(m)} \in L^2_{loc}([0,T); \mathcal{V}_{-1})$.
- *For* $j \le m-1$, $\Gamma^{(j)} \in wAC^2([0,T); L^2(\partial\Omega))$ *with derivative the equivalence class of* $\Gamma^{(j+1)} \in L^2_{loc}([0,T); L^2(\partial\Omega))$.
- *For* $j \le m-1$,
  $$G^{(j)} \in wAC^2([0,T); H^{2m-2j-2}) \cap C([0,T); H^{2m-2j-1}_{weak})$$
  with derivative $[G^{(j+1)}] \in L^2_{loc}([0,T); H^{2m-2j-2})$. *This statement also holds with either* $S^1(\Gamma^{(j)})$ *or* $S^2(\Gamma^{(j)})$ *(defined in* Theorem A.26*) in place of* $G^{(j)}$.

(b) *The linear functionals* $F^{(0)},\dots,F^{(m)}$ *defined by*

$$F^{(j)}\cdot v = \int_\Omega \left(g^{(j)}\cdot v + G^{(j)} : \nabla_x v\right) + \int_{\partial\Omega} \Gamma^{(j)}\cdot v\, d\sigma \tag{5.22}$$

for $v \in \mathcal{H}_1$ *satisfy the conditions* (2.26)–(2.27) *of* Theorem 2.7 *with constant* $C_F(\bar t)$ *which for given* $\bar t \in (0,T)$ *is determined by bounds for the norms and seminorms on* $[0,\bar t]$ *which are assumed in* (5.18)–(5.21) *to be finite.*

Proof. The first item in (a) and the statement in the fourth item in (a) regarding the $G^{(j)}$ follow from iterations of Theorems A.23 and A.24 as in Corollary A.25. The second item in (a) follows from Theorem A.20(b) and the third from Theorem A.21. Concerning the $\Gamma^{(j)}$ we apply Theorem A.26 to write

$$\int_{\partial\Omega} \Gamma\cdot v\, d\sigma = \int_\Omega \left[S^1(\Gamma)\cdot v + S^2(\Gamma) : \nabla_x v\right] dx$$

and check that the hypotheses on Γ apply in Theorem A.26(e) to show that $S^1(\Gamma)_{\mathrm{m}}$ and $S^2(\Gamma)_{\mathrm{m}}$ satisfy exactly the same hypotheses made for G_{m}, and therefore the same conclusions in the fourth item in (a). This proves (a) and also shows that we can subsume the boundary integral in the definition of F into the terms involving g and G and proceed with this boundary integral omitted.

Next, to prove (b) we recall the definition of $\mathcal{V}_k$ in Definition 5.1 and conclude from the regularity statements in (a) that the equivalence class of $F^{(j)}$ is in $L^\infty_{loc}([0,T);\mathcal{V}_{2m-2j-2})$ for $j \le m-1$ and in $L^2_{loc}([0,T);\mathcal{V}_{2m-2j-1})$ for $j \le m$, as required in (2.26) and (2.27). Finally, the absolute continuity statements in (a) show that, for $v \in \mathcal{H}_1$ and $j \le m-1$,

$$\int_\Omega \big(g^{(j)} \cdot v + G^{(j)} : v\big)dx\Big|_{t_1}^{t_2} = \int_{t_1}^{t_2}\int_\Omega \big(g^{(j+1)}(t) \cdot v + G^{(j+1)}(t) : v\big)dxdt,$$

so that $F^{(j)} \in wAC^2([0,T);\mathcal{V}_{2m-2j-3})$ with derivative $F^{(j+1)}$, as required in Theorem 2.7. This completes the proof of (b).

□

We note that the rather technical hypothesis in (5.19) can be replaced by the simpler but stronger assumption that $D_t^m g_{\mathrm{m}} \in L^{2,q'}_{(t)loc}(\Omega \times [0,T))$ for some q' whose Hölder conjugate q is in $[2,\infty)$ if $n=2$ and in $[2, 2n/(n-2)]$ if $n \ge 3$. For the proof we apply Theorem A.20(a) in place of A.20(b) to show that $g^{(m-1)} \in wAC^2([0,T), L^{q'}(\Omega))$ together with the fact that $H^1(\Omega)$ is continuously included in $L^q(\Omega)$. This stronger assumption is inadequate for certain applications in Chapter 6, however, in which the weaker condition is satisfied but the simpler, stronger condition is not.

§5.2 Higher Order Regularity for Parabolic Systems

In this section we apply the abstract regularity result Theorem 2.7 to the problem (5.1)–(5.3). Three different descriptions of higher–order regularity are given: in the first, the solution and its time derivatives are described as mappings from the time interval into particular Sobolev spaces, mappings which are variously weakly absolutely continuous, continuous or in L^2_{loc}; in the second, the measurable counterpart of the solution and its weak x and t derivatives on the space–time product space are described; and in the third, the continuously differentiable representative is described when the order of regularity is sufficiently high. In the latter case the underlying differential equation and boundary conditions hold in the sense of calculus.

Theorem 5.5. *Assume that the hypotheses and notations in* Definition 5.1 *are in force, let $T \in (0,\infty]$ and let $N \times N$ matrix-valued $[0,T)$–equivalence classes $a^{\eta,\omega}$, b^η, c^ω, d and (if $\Omega \ne \mathbf{R}^n$) f be given satisfying the positive definiteness condition* (5.5) *and the regularity conditions in one of* (5.8), (5.10) *or* (5.11)–(5.14).

Then given g, G, *and (if* $\Omega \neq \mathbf{R}^n$*)* Γ *satisfying* (5.18)–(5.21) *and given* $(\varsigma_0, \dots, \varsigma_m) \in \mathcal{H}_{2m} \times \cdots \mathcal{H}_0$ *satisfying the compatibility conditions*

$$\int_\Omega \varsigma_{j+1} \cdot v\, dx + \sum_{i=0}^{j} \binom{j}{i} A^{(i)}(0)(\varsigma_{j-i}, v) = F^{(j)}(0) \cdot v \tag{5.23}$$

for $j = 0, \dots, m-1$ *and all* $v \in \mathcal{H}_1$*, where* $A^{(i)}$ *and* $F^{(j)}$ *are as in* Lemmas 5.3 *and* 5.4*, the corresponding solution* u *of* (5.1)–(5.3) *in* Theorem 3.6 *satisfies the following:*

(a) **Bochner-Sobolev space description)** *There are mappings* $u, \dots, u^{(m)}$*, where* $u^{(0)} = u$*, satisfying the following:*

(*i*) *For* $j = 0, \dots, m-1$,

$$\begin{aligned} u^{(j)} \in & Lip_{loc}([0,T); [H^{2m-2j-2}(\Omega)]^N) \\ & \cap wAC^2([0,T); [H^{2m-2j-1}(\Omega)]^N) \cap C([0,T); [H^{2m-2j}_{weak}(\Omega)]^N), \end{aligned}$$

$u^{(m)} \in C([0,T); [L^2(\Omega)]^N)$ *and* $(u^{(m)})^* \equiv \langle u^{(m)}, \cdot \rangle_{\mathcal{H}_0} \in wAC^2([0,T); \mathcal{V}_{-1})$*. The derivative of* $u^{(j)}$ *for* $j \le m-1$ *is the equivalence class over* $[0,T)$ *of* $u^{(j+1)}$ *and the derivative of* $(u^{(m)})^*$ *is the functional given by the right side of* (5.26) *below with* $j = m$.

(*ii*) *Given* $\bar{t} \in (0,T)$,

$$\begin{aligned} & \sum_{j=0}^{m} \Big(\sup_{0 \le t \le \bar{t}} |u^{(j)}(t)|^2_{[H^{2m-2j}(\Omega)]^N} + \int_0^{\bar{t}} |u^{(j)}(t)|^2_{[H^{2m-2j+1}(\Omega)]^N} dt \Big) \\ & \qquad \le C C_A(\bar{t}) \Big[\sum_{j=0}^{m} |\varsigma_j|^2_{[H^{2m-2j}(\Omega)]^N} + C_F(\bar{t}) \Big] \end{aligned} \tag{5.24}$$

where $C_A(\bar{t})$ *and* $C_F(\bar{t})$ *are as described in* Lemmas 5.3(c) *and* 5.4(b) *respectively and* C *is a generic positive constant depending on* n, N, m, C_{BC} *and* M_Ω^{2m+1} *(see* Definitions 5.1 *and* 1.7*).*

(*iii*) *The weak forms*

$$\begin{aligned} & \int_\Omega u^{(j)}(t) \cdot v\, dx \\ & \qquad = \int_\Omega \varsigma_j \cdot v\, dx - \sum_{i=0}^{j} \binom{j}{i} \int_0^t A^{(i)}(s)(u^{(j-i)}(s), v) ds \\ & \qquad\qquad + \int_0^t F^{(j)}(s) \cdot v\, ds \end{aligned} \tag{5.25}$$

hold for $v \in \mathcal{H}_1$, $t \in [0,T)$ *and* $j = 0, \dots, m$*; the strong forms*

$$\int_\Omega u^{(j+1)}(t) \cdot v\, dx = -\sum_{i=0}^{j} \binom{j}{i} A^{(i)}(t)(u^{(j-i)}(t), v) + F^{(j)}(t) \cdot v \tag{5.26}$$

hold for $v \in \mathcal{H}_1$, $t \in [0,T)$ *and* $j = 0, \ldots, m-1$; *and the energy equalities*

$$\begin{aligned}&\tfrac{1}{2}|u^{(j)}(t)|^2_{[L^2(\Omega)]^N} + \int_0^t A(s)(u^{(j)}, u^{(j)})ds \\ &-\tfrac{1}{2}|\varsigma_j|^2_{[L^2(\Omega)]^N} + \int_0^t \Big(-\sum_{i=1}^{j} \binom{j}{i} A^{(i)}(u^{(j-i)}, u^{(j)}) + F^{(j)} \cdot u^{(j)} \Big) ds\end{aligned} \tag{5.27}$$

hold for $j = 0, \ldots, m$.

(*iv*) *For* $j = 0, \ldots, m-1$ *and at each fixed* $t \in [0,T)$,

$$u^{(j+1)} = L^{(j)}[u^{(0)}, \ldots, u^{(j)}] + g^{(j)} - \mathrm{Div}_x G^{(j)} \tag{5.28}$$

where

$$\begin{aligned}L^{(j)}[u^{(0)}, \ldots, u^{(j)}] = \sum_{i=0}^{j} \binom{j}{i} \Big[&\big((a^{(i)})^{\eta,\omega} u^{(j-i)}_{x_\eta}\big)_{x_\omega} - (b^{(i)})^{\eta} u^{(j-i)}_{x_\eta} \\ &+ \big((c^{(i)})^{\omega} u^{(j-i)}\big)_{x_\omega} - d^{(i)} u^{(j-i)} \Big].\end{aligned} \tag{5.29}$$

Here the subscripts x_η *and* x_ω *and the notation* Div_x *indicate weak derivatives in* Ω. *Each of the four terms in* (5.28) *is in* $C([0,T); [H^{2m-2j-2}_{weak}(\Omega)]^N) \subseteq C([0,T); [L^2_{weak}(\Omega)]^N)$ *and each of the four terms in the brackets on the right side of* (5.29) *is in* $C([0,T); [L^2_{weak}(\Omega)]^N)$. *In particular, at each* $t \in [0,T)$,

$$u' = L[u] + g - \mathrm{Div}_x G$$

where $L = L^{(0)}$, $u' = [u^{(1)}]_{[0,T)}$ *is the derivative of* $u^{(0)}$ *as a mapping into* $\mathcal{H}_{2m-1}$ *and each term is in* $C([0,T); [H^{2m-2}_{weak}(\Omega)]^N)$.

(*v*)(*For the case that* $\Omega \neq \mathbf{R}^n$) *if* $\mathcal{T} : H^1(\Omega) \to L^2(\partial\Omega)$ *is the trace operator of* Theorem 3.2 *and if* D_x *denotes weak differentiation in* Ω, *then:*

- *For* $j = 0, \ldots, m-1$ *and* $|\alpha| \le 1$, $\mathcal{T} D^\alpha_x u^{(j)} \in C([0,T); [L^2_{weak}(\partial\Omega)]^N)$; *and for* $t \in [0,T)$ *and for every connected component* $\mathcal{C}_l$ *of* $\partial\Omega$, $Q_l \mathcal{T} u^{(j)}(t) = 0$ *as an element of* $[L^2(\mathcal{C}_l)]^N$ (*see* Definition 5.1; *this statement is vacuous if* $N_l = 0$).
- *At each* $t \in [0,T)$ *and for* $j = 0, \ldots, m-1$,

$$L^{(j)}_{\partial\Omega}[u^{(0)}, \ldots, u^{(j)}] - \mathcal{T} G^{(j)} \nu - \Gamma^{(j)} \perp \mathcal{T} v \tag{5.30}$$

for all $v \in \mathcal{H}_1$, *where*

$$\begin{aligned}&L^{(j)}_{\partial\Omega}[u^{(0)}, \ldots, u^{(j)}] \\ &= \sum_{i=0}^{j} \binom{j}{i} \Big[\nu^\omega \mathcal{T}\big((a^{(i)})^{\eta,\omega} u^{(j-i)}_{x_\eta} + (c^{(i)})^\omega u^{(j-i)}\big) + \big(f^{(i)} \mathcal{T} u^{(j-i)}\big) \Big],\end{aligned} \tag{5.31}$$

ν *is the normal vector field on* $\partial\Omega$ (Theorem 3.1(d)) *and* $\perp$ *denotes orthogonality in* $L^2(\partial\Omega)$. *Each of the three terms on the left in* (5.30) *and each of the three terms in the brackets on the right in* (5.31) *is in* $C([0,T);[L^2_{weak}(\partial\Omega)]^N)$.

- *For* $j = 0, \ldots, m-1$, *all* l *and at each* $t \in [0,T)$,

$$L^{(j)}_{\partial\Omega}[u^{(0)}, \ldots, u^{(j)}] - \mathcal{T}G^{(j)}\nu - \Gamma^{(j)} \in (\text{nullspace } Q_l)^{\perp} \tag{5.32}$$

for σ*–a.a.* $x \in \mathcal{C}_l$ *where the null set depends on* t *and on the particular representatives of the functions on the left. In particular, if* $N_l = 0$ *then* $Q_l = 0$ *on* $\mathcal{C}_l$ *and the quantity on the left in* (5.32) *is zero* σ*=a.e. on* $\mathcal{C}_l$*; and if* $N_l = N$ *then* Q_l *is the identity on* $\mathcal{C}_l$ *and* (5.32) *is vacuous.*

(b) **(Weak derivative description)** *The measurable counterpart* u_{m} *of* u *satisfies the following:*

(*i*) *The relation*

$$D_t^j D_x^\alpha u_{\mathrm{m}} = [D_x^\alpha u^{(j)}]_{\mathrm{m}}$$

holds for $j = 0, \ldots, m$ *and* $0 \le |\alpha| \le 2m - 2j + 1$, *where the derivatives on the left are weak derivatives in* $\Omega \times (0,T)$ *and the derivative on the right is the Bochner space derivative* (Definition A.18).

(*ii*) *For* $\bar{t} \in (0,T)$ *the sum*

$$\sum_{j=0}^{m} \Big(\sum_{|\alpha| \le 2m-2j} |D_t^j D_x^\alpha u_{\mathrm{m}}|^2_{L^{\infty,2}_{\Omega\times[0,\bar{t}]}} + \sum_{|\alpha| \le 2m-2j+1} |D_t^j D_x^\alpha u_{\mathrm{m}}|^2_{L^{2,2}_{\Omega\times[0,\bar{t}]}} \Big)$$

is bounded by the expression on the right side of (5.24).

(*iii*) *For* $j = 0, \ldots, m-1$,

$$D_t^{j+1} u_{\mathrm{m}} = L^{(j)}_{\mathrm{m}}\big[u_{\mathrm{m}}, \ldots, D_t^j u_{\mathrm{m}}\big] + D_t^j g_{\mathrm{m}} - D_t^j \mathrm{Div}_x G_{\mathrm{m}} \tag{5.33}$$

where

$$\begin{aligned} &L^{(j)}_{\mathrm{m}}\big[u_{\mathrm{m}}, \ldots, D_t^j u_{\mathrm{m}}\big] \\ &\quad = \sum_{i=0}^{j} \binom{j}{i} \Big(D_{x_\omega}\big[(D_t^i a^{\eta,\omega}_{\mathrm{m}})(D_{x_\eta} D_t^{j-i} u_{\mathrm{m}})\big] - (D_t^i b^{\eta}_{\mathrm{m}})(D_{x_\eta} D_t^{j-i} u_{\mathrm{m}}) \\ &\qquad + D_{x_\omega}\big[(D_t^i c^{\omega}_{\mathrm{m}})(D_t^{j-i} u_{\mathrm{m}})\big] - (D_t^i d_{\mathrm{m}})(D^{j-i} u_{\mathrm{m}}) \Big) \end{aligned} \tag{5.34}$$

where D_t, D_{x_η} *and* D_{x_ω} *denote weak derivatives in* $\Omega \times (0,T)$. *Each of the four terms in* (5.33) *and each of the four terms in the summand on the right in* (5.34) *has weak* $\Omega \times (0,T)$ *derivatives with respect to* x *up to order* $2m - 2j - 2$. *In particular, if* $L_{\mathrm{m}} = L^{(0)}_{\mathrm{m}}$ *then*

$$D_t u_{\mathrm{m}} = L_{\mathrm{m}}[u_{\mathrm{m}}] + g_{\mathrm{m}} - \mathrm{Div}_x G_{\mathrm{m}}$$

in the weak sense on $\Omega \times (0,T)$.

(c) **(Continuous derivative description)** *Let $k_0 \geq 2$ be the integer satisfying*

$$k_0 - 1 \leq n/2 < k_0$$

and assume that $m \geq k_0/2$. Then there is a function $u_{\mathbb{C}} \in C(\overline{\Omega} \times [0,T))$ such that for each t, $u_{\mathbb{C}}(\cdot,t)$ is a member of the Ω–equivalence class $u(t)$ and which satisfies the following:

(*i*) *If j_0 is the integer satisfying*

$$0 \leq j_0 \leq m - k_0/2 < j_0 + 1$$

and if

$$\kappa = 2m - 2j_0 - k_0,$$

which is either 0 or 1, then for $2j + |\alpha| \leq 2j_0 + \kappa$ the calculus derivative $D_x^\alpha D_t^j u_{\mathbb{C}}$ exists on $\Omega \times (0,T)$, its restriction $D_t^j D_x^\alpha u_{\mathbb{C}}(\cdot,t)$ is in the Ω–equivalence class of the weak Ω–derivative $D_x^\alpha u^{(j)}(t)$, and for each $\bar{t} \in (0,T)$ its norm satisfies

$$|D_x^\alpha D_t^j u_{\mathbb{C}}|_{C_{bdd}^{(\lambda,\lambda'_\alpha)}(\Omega\times(0,\bar{t}))} \leq C(\Omega,\, m,\, A,\, F,\, \{\varsigma_j\}, \bar{t},\, \lambda),$$

so that $D_x^\alpha D_t^j u_{\mathbb{C}}$ has a continuous extension to $\overline{\Omega}\times[0,\bar{t}]$. Here $\lambda = k_0 - n/2$ if $k_0 - n/2 < 1$ and can be chosen arbitrarily in $(0,1)$ otherwise, and in either case

$$\lambda'_\alpha = \frac{\lambda}{\lambda + n/2 + |\alpha|} \in (0,1);$$

the constant C on the right above depends on M_Ω^{2m+1} (Definition 1.7), *on $C_A(\bar{t})$ and $C_F(\bar{t})$* (Lemmas 5.3 *and* 5.4) *and on the expression in brackets on the right side of* (5.24) *as well as on λ if $k_0 - n/2 = 1$.*

(*ii*) *Under the stronger hypothesis that $m \geq k_0/2 + 1$ (so that $j_0 \geq 1$), the differential equation and boundary conditions associated to the problem* (5.1)–(5.3) *are satisfied in the sense of continuous differentiability as follows:*

- *There are matrix–valued functions $a_{\mathbb{C}}^{\eta,\omega}$, $b_{\mathbb{C}}^{\eta}$, $c_{\mathbb{C}}^{\omega}$, $d_{\mathbb{C}}$, $g_{\mathbb{C}}$ and $G_{\mathbb{C}}$ whose components are continuous on $\Omega\times(0,T)$, whose Ω–equivalence classes for each fixed t are $(a^{(0)})^{\eta,\omega}(t), \ldots, G^{(0)}(t)$ and which satisfy the following: for $j = 0,\ldots,j_0 - 1$ and for $\bar{t} < T$, $D_t^j g_{\mathbb{C}}$, $D_t^j G_{\mathbb{C}} \in C_{bdd}^{(\lambda,\lambda'_0)}(\Omega \times (0,\bar{t}))$ and $D_t^j \nabla_x G_{\mathbb{C}} \in C_{bdd}^{(\lambda,\lambda'_1)}(\Omega \times (0,\bar{t}))$; and for every ball $B \subset \mathbf{R}^n$, $D_t^j a_{\mathbb{C}}^{\eta,\omega}$, $D_t^j b_{\mathbb{C}}^{\eta}$, $D_t^j c_{\mathbb{C}}^{\omega}$, $D_t^j d_{\mathbb{C}} \in C_{bdd}^{(\lambda,\lambda'_0)}((\Omega\cap B) \times (0,\bar{t}))$ and $\nabla_x D_t^j a_{\mathbb{C}}^{\eta,\omega}$, $\nabla_x D_t^j c_{\mathbb{C}}^{\omega} \in C_{bdd}^{(\lambda,\lambda'_1)}((\Omega\cap B) \times (0,\bar{t}))$.*
- *For $j = 0,\ldots,j_0 - 1$, the differential equation*

$$D_t^{j+1} u_{\mathbb{C}} = L_{\mathbb{C}}^{(j)}[u_{\mathbb{C}},\ldots,D_t^j u_{\mathbb{C}}] + D_t^j g_{\mathbb{C}} - \mathrm{Div}_x D_t^j G_{\mathbb{C}} \tag{5.35}$$

holds on $\Omega \times (0,T)$*, where*

$$
\begin{aligned}
&L_{\mathbb{C}}^{(j)}[u_{\mathbb{C}},\ldots,D_t^j u_{\mathbb{C}}] \\
&= \sum_{i=0}^{j}\binom{j}{i}\Big[\big((D_t^i a_{\mathbb{C}}^{\eta,\omega})(D_t^{j-i}u_{\mathbb{C}})_{x_\eta}\big)_{x_\omega} - (D_t^i b_{\mathbb{C}}^{\eta})(D_t^{j-i}u_{\mathbb{C}})_{x_\eta} \\
&\qquad + \big((D_t^i c_{\mathbb{C}}^{\omega})(D_t^{j-i}u_{\mathbb{C}})\big)_{x_\omega} - (D_t^i d_{\mathbb{C}})(D_t^{j-i}u_{\mathbb{C}})\Big].
\end{aligned}
\tag{5.36}
$$

The x *and* t *derivatives in* (5.35) *and* (5.36) *are in the sense of calculus and each term in these equations has a continuous extension to* $\overline{\Omega} \times [0,\bar{t}]$*. In particular*

$$
\begin{aligned}
(u_{\mathbb{C}})_t = (a_{\mathbb{C}}^{\eta,\omega}(u_{\mathbb{C}})_{x_\eta})_{x_\omega} - b_{\mathbb{C}}^{\eta}(u_{\mathbb{C}})_{x_\eta} + (c_{\mathbb{C}}^{\omega}u_{\mathbb{C}})_{x_\omega} - d_{\mathbb{C}}u_{\mathbb{C}} \\
+ g_{\mathbb{C}} - \mathrm{Div}_x G_{\mathbb{C}}
\end{aligned}
$$

where the subscripts and Div_x *denote calculus derivatives.*

(*iii*) *If* $m \geq k_0/2+1$*, so that* $j_0 \geq 1$*, and if* $\Omega \neq \mathbf{R}^n$*, then the boundary conditions described in* Definition 5.1 *hold pointwise on* $\partial\Omega$ *as follows:*

- *For each connected component* C_l *of* $\partial\Omega$ *and at each* $t \in [0,T)$*,*

$$
Q_l u_{\mathbb{C}}, \ldots, Q_l D_t^{j_0-1} u_{\mathbb{C}} = 0
$$

  pointwise on C_l *(this statement is vacuous if* $N_l = 0$*).*

- *There are continuous functions* $\Gamma_{\mathbb{C}}$ *and* $f_{\mathbb{C}}$ *on* $\partial\Omega \times (0,T)$ *such that for* $j \leq j_0 - 1$ *the calculus derivatives* $D_t^j\Gamma_{\mathbb{C}}$ *and* $D_t^j f_{\mathbb{C}}$ *exist on* $\partial\Omega \times (0,T)$ *and satisfy* $D_t^j\Gamma \in C_{bdd}^{(\lambda,\lambda_0')}(\partial\Omega \times (0,\bar{t}))$ *and* $D_t^j f \in C_{bdd}^{(\lambda,\lambda_0')}((\partial\Omega \cap B)\times(0,\bar{t}))$ *for every* $\bar{t} < T$ *and every ball* $B \subset \mathbf{R}^n$*, and whose* $\partial\Omega$*–equivalence classes with respect to the boundary measure* σ *are* $\Gamma^{(j)}(t)$ *and* $f^{(j)}(t)$ *respectively.*

- *For each connected component* C_l *of* $\partial\Omega$*, for* $j = 0,\ldots,j_0-1$ *and at each* $t \in [0,T)$*,*

$$
L_{\mathbb{C},\partial}^{(j)}[u_{\mathbb{C}},\ldots,D_t^j u_{\mathbb{C}}] - \mathcal{T}(D_t^j G_{\mathbb{C}})\nu - D_t^j\Gamma_{\mathbb{C}} \in (\text{nullspace}\, Q_l)^{\perp}
\tag{5.37}
$$

  pointwise on C_l*,* where

$$
\begin{aligned}
&L_{\mathbb{C},\partial}^{(j)}[u_{\mathbb{C}},\ldots,D_t^j u_{\mathbb{C}}] \\
&\quad = \sum_{i=0}^{j}\binom{j}{i}\Big[\nu^{\omega}\mathcal{T}\big(D_t^i a_{\mathbb{C}}^{\eta,\omega} D_t^{j-i}(u_{\mathbb{C}})_{x_\eta} + D_t^i c_{\mathbb{C}}^{\omega} D_t^{j-i}u_{\mathbb{C}}\big) \\
&\qquad\qquad + D_t^i f_{\mathbb{C}}\mathcal{T}\big(D_t^{j-i}u_{\mathbb{C}}\big)\Big]
\end{aligned}
$$

  and ν *is the outer normal vector* (Theorem 3.1(d))*. In particular, if* $N_l = 0$ *then* $Q_l = 0$ *on* C_l *and the quantity on the left in* (5.37) *is zero pointwise on* C_l*; and if* $N_l = N$ *then* Q_l *is the identity on* C_l *and* (5.37) *is vacuous.*

Note that the statement in (a) that $u^{(j)} \in C([0,T); \mathcal{H}^{2m-2j}_{weak})$ implies in particular that $|u^{(j)}|_{H^{2m-2j}(\Omega)}$ is bounded on compact time intervals, by the uniform boundedness principle. Continuity into H^{m-2j} with the norm topology follows from Theorem 2.7(g) under slightly stronger conditions on the coefficients. The precise formulation is left as an exercise.

Proof of Theorem 5.5(a). The hypotheses of Theorem 2.7 have been checked in Lemmas 5.2–5.4 and the conclusions of that theorem therefore apply to show that (a*i*)–(a*iii*) hold.

To prove the regularity statements in (a*iv*) and (a*v*) we fix $i \le j < m$ and examine in detail the term $(a^{(i)})^{\eta,\omega} u^{(j-i)}_{x_\eta}$ in (5.29). First, $u^{(j-i)}$ maps $[0,T)$ continuously into $\mathcal{H}^{weak}_{2m-2j+2i}$ by (a*i*), hence into $H^{2m-2j+2i}_{weak}$ (by application of orthogonal projection), so that $u^{(j-i)}_{x_\eta} \in C([0,T); H^{2m-2j+2i-1}_{weak})$ (where $H^k = [H^k(\Omega)]^N$, etc.). Next we have from Lemma 5.3(b) that for $|\alpha| \le 2m-2i$, each component of $D^\alpha_x (a^{(i)})^{\eta,\omega}$ is in $C([0,T); L^{s_\alpha}_{weak})$ for some $s_\alpha \in \mathcal{I}_{2i+|\alpha|} = \mathcal{I}_{(2m-2j+2i-1)-(2m-2j-1)+|\alpha|}$, so that $(a^{(i)})^{\eta,\omega}(t)$ is in the multiplier space $\mathcal{M}^{2m-2j+2i-1,2m-2j-1}$ (Definition 4.4). These two observations show that the product $(a^{(i)})^{\eta,\omega} u^{(j-i)}_{x_\eta}$ is a locally bounded map from $[0,T)$ into $H^{2m-2j-1} \subseteq H^1$, and we claim that this map is in $C([0,T); H^1_{weak})$. To see this we let h denote one of its components and apply the strong continuity of $(a^{(i)})^{\eta,\omega}$ into L^s (Lemma 5.3(a)) and the weak continuity of $u^{(j-i)}_{x_\eta}$ into $H^{2m-2j+2i-1}$ to find by triangulation that for any $\varphi \in C^\infty_c(\Omega)$, $\int_\Omega h(t)\,\varphi\,dx$ is continuous in t. Thus if $t_k \to t$ in $[0,T)$ then all H^1_{weak}-subsequential limits of $h(t_k)$, which do exist by the uniform H^1 bound proved above, agree with $h(t)$. This proves the claim and shows as consequences first that $((a^{(i)})^{\eta,\omega} u^{(j-i)}_{x_\eta})_{x_\omega} \in C([0,T); [L^2_{weak}(\Omega)]^N)$ as stated in (a*iv*) and second that the trace of $(a^{(i)})^{\eta,\omega} u^{(j-i)}_{x_\eta}$ is a locally bounded map into $[L^2(\partial\Omega)]^N$. To prove its weak continuity we again let h be one of its components and $\psi \in L^2(\partial\Omega)$. Then by Theorem A.26,

$$\int_{\partial\Omega} \mathcal{T}(h(t))\,\psi\,d\sigma = \int_\Omega \left[h(t) S^1(\psi) + \nabla_x h(t) \cdot S^2(\psi)\right] dx$$

where $S^1(\psi)$ and the components of $S^2(\psi)$ are in $L^2(\Omega)$. The right side here is continuous in t by the previous claim, so that $\mathcal{T}\big((a^{(i)})^{\eta,\omega} u^{(j-i)}_{x_\eta}\big) \in C([0,T); [L^2_{weak}(\partial\Omega)]^N)$ as claimed in the second item in (a*v*).

Similar arguments apply to the other three terms in the brackets in (5.29) and together with Lemma 5.4(a) show that $(a^{(i)})^{\eta,\omega} u^{(j-i)}_{x_\eta}$, $(c^{(i)})^\omega u^{(j-i)}$ and $G^{(j)}$ included in $A^{(i)}(u^{(j-i)}, \cdot)$ and $F^{(j)}$ in (5.26) are in $C([0,T); H^1(\Omega))$ with boundary traces in $C([0,T); L^2(\partial\Omega))$ and that $u^{(j+1)}$, $b^{(i)} u^{(j-i)}_{x_\eta}$ and $d^{(i)} u^{(j-i)}$ are in $C([0,T); L^2(\Omega))$. We can therefore apply the divergence theorem in (5.26) to the first three of these to obtain (5.28)–(5.31). Then since three of the terms in (5.28) are in $C([0,T);H^{2m-2j-2}(\Omega))$ by Lemma 5.4 and (ai), so is the fourth. These considerations complete the proof of (aiv) and the second item in (av), and the first item in (av) follows from the fact that $u^{(j)}$ is weakly continuous into $\mathcal{H}_2 = \mathcal{H}_1 \cap H^2(\Omega)$ for $j \leq m-1$.

To prove the final item in (av) we assume that $N_l < N$ and fix a component $\mathcal{C}_l$ of $\partial\Omega$ intersecting a particular $\mathcal{U}_i$ (in the notation of Definition 1.7). Let ψ denote a representative of the quantity on the left in (5.32) in which specific representatives of all the constituent functions have been chosen, and let $x_0 \in \mathcal{C}_l \cap \mathcal{U}_i$ be a Lebesgue point of ψ so that $\tilde{y}_0 \equiv T_i^{-1}(x_0)$ is a Lebesgue point of $\psi \circ T_i$. Let ξ_0 be an element in the null space of $Q_l(x_0)$ and construct a test function $v \in \mathcal{H}_1$ as follows. First recall from Definition 5.1 that $S_l^{\pm 1} \circ T_i$ are in $C^{2m+1}(\mathcal{C}_l \cap \mathcal{B}_i)$ and that $Q_l S_l^{-1} = [I_{N' \times N'} \;\; 0]$. Thus if $\xi(x) \equiv S_l^{-1}(x) S_l(x_0)\xi_0$ then $Q_l(x)\xi(x) = 0$ in a $\partial\Omega$–neighborhood of x_0 and $\xi \circ T_i$ is C^{2m+1} in a neighborhood of $\tilde{y}_0$. Next let $\delta_h \in C_c^\infty(\mathcal{B}_i)$ be nonnegative with support in a ball of radius h centered at $\tilde{y}_0$ and integrating to one and let $\varphi \in C_c^\infty(\mathbf{R}^n)$ be identically one in a small neighborhood of x_0 with support intersecting Ω in $\mathcal{U}_i$. Then define v mapping an $\mathbf{R}^n$–neighborhood of x_0 into $\mathbf{R}^N$ by $v(x) = \varphi(x)\delta_h(\tilde{y})(\xi \circ T_i)(\tilde{y})$ where $R_i^{-1}(x) = (\tilde{y}, y_n)$. Then v extends to all of Ω and is an element of $\mathcal{H}_1$ with support intersecting Ω in $\mathcal{U}_i$. Therefore by (5.30),

$$\int_{\tilde{B}_i} (\delta_h(\tilde{y})(\psi \cdot \xi) \circ T_i(\tilde{y})\sqrt{1 + |\nabla\psi_i(\tilde{y})|^2}\, d\tilde{y} = 0$$

for h sufficiently small. Letting $h \to 0$ we conclude that $(\psi \cdot \xi)(T_i(\tilde{y}_0)) = 0$, that is, that $\psi(x_0) \cdot \xi_0 = 0$. This proves that $\psi(x_0)$ is orthogonal to the null space of Q_l at x_0 and completes the proof of (av).

Proof of Theorem 5.5(b). Part (i) follows from Theorems A.19 and A.20 and part (ii) follows immediately from these same two theorems and (5.24) via the isometry described in Corollary A.16.

For the proof of (biii) we will apply the fact that the measurable counterpart of a product is the product of the counterparts of the factors:

Lemma 5.6 *Let r, s and p be in $(1,\infty]$ with $r^{-1} + s^{-1} = p^{-1}$ and let $I \subseteq \mathbf{R}$ be a nontrivial interval and $\mathcal{W}$ an open set in $\mathbf{R}^n$. If $f \in L^\infty(I; L^r(\mathcal{W}))$ and $g \in L^\infty(I; L^s(\mathcal{W}))$ then $fg \in L^\infty(I; L^p(\mathcal{W}))$ and $(fg)_{\mathrm{m}} = f_{\mathrm{m}} g_{\mathrm{m}}$.*

PROOF: Since $f_{\mathrm{m}} \in L^{\infty,r}$ and $g_{\mathrm{m}} \in L^{\infty,s}$, their product is an equivalence class of measurable functions on $\Omega \times I$ and is easily seen to be in $L^{\infty,p}$. Corollary A.16 then shows that $f_{\mathrm{m}} g_{\mathrm{m}}$ is the measurable counterpart F_{m} of an element $F \in L^\infty(I; L^p)$ whose representatives are therefore weakly measurable maps into L^p. Choosing specific representatives of f, g, F and their measurable counterparts, we therefore have that for almost all t,

$$f(t)g(t) = f_{\mathrm{m}}(\cdot, t) g_{\mathrm{m}}(\cdot, t) = F_{\mathrm{m}}(\cdot, t) = F(t),$$

so that the $[0,T)$–equivalence class of fg is the same as that of F and therefore is an element of $L^\infty(I;L^p(\mathcal{W}))$. Rearranging these statements we find that $(fg)_{\mathfrak{m}} = F_{\mathfrak{m}} = f_{\mathfrak{m}}g_{\mathfrak{m}}$.

□

Now to prove (b*iii*) we let $\varphi \in [C_c^\infty(\Omega \times (0,T))]^N$ and apply (5.26) with $v = \varphi(\cdot,t)$, which is an element of $\mathcal{H}_1$. Since every term occurring is in $L^\infty_{loc}([0,T);L^2(\Omega))$ by (a*i*), we can integrate over $(0,T)$ and then apply Corollary A.16, Theorems A.19 and A.20 and Lemma 5.6 to compute the term involving $a^{(i)}$ as follows:

$$\begin{aligned}\int_0^T\int_\Omega \left[(a^{(i)})^{\eta,\omega}u_{x_\eta}^{(j-i)}\right]_{x_\omega}\cdot\varphi\,dxdt &= \int_{\Omega\times(0,T)}\left(\left[(a^{(i)})^{\eta,\omega}u_{x_\eta}^{(j-i)}\right]_{x_\omega}\right)_{\mathfrak{m}}\cdot\varphi\\ &= \int_{\Omega\times(0,T)} D_{x_\omega}\left(\left[(a^{(i)})_{\mathfrak{m}}^{\eta,\omega}(u_{x_\eta}^{(j-i)})_{\mathfrak{m}}\right]\cdot\varphi\\ &= \int_{\Omega\times(0,T)} D_{x_\omega}\left[(D_t^i a_{\mathfrak{m}}^{\eta,\omega})(D_t^{j-i}D_{x_\eta}u_{\mathfrak{m}})\right]\cdot\varphi.\end{aligned}$$

The other terms are transformed in a similar way, so that (5.33) and (5.34) follow from (5.28) and (5.29). Finally, Theorems A.19 and A.20 together with the regularity statements below (5.29) in (a*iv*) show that the weak differentiability statements in (b*iii*) hold. This completes the proof of (b).

Proof of Theorem 5.5(c). We begin with the following lemma:

Lemma 5.7. *Let $\mathcal{W}$ be a $C^{0,1}_{bdd}$ domain in $\mathbf{R}^n$* (Definition 1.7), *not necessarily bounded, I an open interval in $\mathbf{R}$ and k_0, λ and λ'_α as in* Theorem 5.5(c*i*). *Assume that $h \in wAC^\infty(I;L^2(\mathcal{W}))$ and that its equivalence class $[h]_I$ is in $L^\infty(I;H^{k_0+i}(\mathcal{W}))$ for some integer $i \geq 0$. Then $h(t) \in H^{k_0+i}(\mathcal{W})$ for every t and there exists $h_{\mathfrak{c}} \in C^{(\lambda,\lambda'_0)}_{bdd}(\mathcal{W}\times I)$ satisfying the following:*

(a) *For $0 \leq |\alpha| \leq i$ the calculus derivative $D_x^\alpha h_{\mathfrak{c}}$ exists on $\mathcal{W}\times I$ and is an element of $C^{(\lambda,\lambda'_\alpha)}_{bdd}(\mathcal{W}\times I)$ with norm depending on $M^{0,1}_{\mathcal{W}}$* (Definition1.7), *λ and the norms of h occurring in the hypotheses. For each $t \in I$ the $\mathcal{W}$–equivalence class of the calculus derivative $D_x^\alpha h_{\mathfrak{c}}(\cdot,t)$ is the weak $\mathcal{W}$–derivative $D_x^\alpha h(t)$.*

(b) *If there is a function $\psi \in C(\mathcal{W}\times I)$ such that $\left[[\psi(\cdot,t)]_{\mathcal{W}}\right]_I$ is the derivative h' of h as a mapping into $L^2(\mathcal{W})$ (equivalently, if $\psi_{\mathfrak{m}} = h'_{\mathfrak{m}}$), then ψ is the calculus derivative $D_t h_{\mathfrak{c}}$ on $\mathcal{W}\times I$.*

PROOF. First, for $\varphi \in C_c^\infty(\mathcal{W})$ and $|\alpha| \leq k_0 + i$,

$$\begin{aligned}\int_{\mathcal{W}} h(t)D_x^\alpha\varphi(x)dx &= \lim_{\Delta t\to 0}\Delta t^{-1}\int_t^{t+\Delta t}\int_{\mathcal{W}} h(s)D_x^\alpha\varphi(x)\,dxds\\ &= (-1)^{|\alpha|}\lim_{\Delta t\to 0}\Delta t^{-1}\int_t^{t+\Delta t}\int_{\mathcal{W}} D_x^\alpha h(s)\varphi(x)\,dxds,\end{aligned}$$

which is bounded in absolute value by $|h|_{L^\infty(I;H^{k_0+i}(\mathcal{W}))}|\varphi|_{L^2(\mathcal{W})}$. The mapping which takes φ to the integral on the left is therefore given by the L^2 inner product of φ with an element of $L^2(\mathcal{W})$, which is therefore the weak derivative $D_x^\alpha h(t)$. This proves that $h(t) \in H^{k_0+i}(\mathcal{W})$ for every t with norm bounded independently of t. It follows from the standard imbedding result Theorem A.6 that for each t there is a representative $h_{\mathbb{c}}(\cdot,t) \in C^{i,\lambda}_{bdd}(\mathcal{W})$ of $h(\cdot,t)$ which is easily seen to satisfy the last statement in (a).

To prove that $h_{\mathbb{c}}$ and its x-derivatives are Hölder continuous in t we fix $x_0 \in \mathcal{W}$ and let $t_1 < t_2$ be points in $\mathcal{I}$. Then if $\varphi_\varepsilon(x) = \varepsilon^{-n}\varphi((x-x_0)/\varepsilon)$ is a smooth function integrating to one and supported in a cone contained in $\mathcal{W}$ with vertex at x_0 and having dimensions independent of x_0, then for $|\alpha| \le i$,

$$\begin{aligned}
\left|D^\alpha h_{\mathbb{c}}(x_0,\cdot)|_{t_1}^{t_2}\right| &= \left|\int_{\mathcal{W}} D_x^\alpha h_{\mathbb{c}}(x,\cdot)|_{t_1}^{t_2}\varphi_\varepsilon(x)dx\right| + O(\varepsilon^\lambda)\\
&= \left|\int_{t_1}^{t_2}\int_{\mathcal{W}} h'\, D_x^\alpha\varphi_\varepsilon\, dxdt\right| + O(\varepsilon^\lambda)\\
&\le C(\varepsilon^{-|\alpha|-n/2}|t_2-t_1| + \varepsilon^\lambda),
\end{aligned}$$

which is bounded by $C|t_2-t_1|^{\lambda'_\alpha}$ for appropriate choice of ε. This proves (a).

To prove (b) we assume that ψ is given as in the statement and let x_0, t_1, t_2 and φ_ε be as above. Then

$$\begin{aligned}
h_{\mathbb{c}}(x_0,\cdot)|_t^{t+\Delta t} &= \lim_{\varepsilon\to 0}\int_{\mathcal{W}} h_{\mathbb{c}}(x,\cdot)|_t^{t+\Delta t}\varphi_\varepsilon(x)dx\\
&= \lim_{\varepsilon\to 0}\int_t^{t+\Delta t}\int_{\mathcal{W}} h'\varphi_\varepsilon dx\, dt\\
&= \lim_{\varepsilon\to 0}\int_{t_1}^{t_2}\int_{\mathcal{W}} \psi\,\varphi_\varepsilon dx\, dt = \int_t^{t+\Delta t}\psi(x_0,t)\, dt,
\end{aligned}$$

so that $\psi(x_0,t)$ is the calculus derivative $D_t h(x_0,t)$.

□

Now to prove (ci) of Theorem 5.5 we let k_0, j_0 and κ be as in the statement and $j \le j_0$. Then from Theorem 5.5(a), $u^{(j)} \in wAC^\infty([0,T);L^2(\Omega))$ as in the first hypothesis of Lemma 5.7 and its equivalence class over $[0,T)$ is in $L^\infty_{loc}([0,T);H^{2m-2j}(\Omega))$. Writing $2m-2j = k_0+i$ where $i = 2(j_0-j)+\kappa \ge 0$, we therefore have that $[u^{(j)}]_{[0,T)} \in L^\infty_{loc}([0,T);H^{k_0+i}(\Omega))$, as in the second hypothesis of Lemma 5.7. There is therefore a continuous function $u_{\mathbb{c}}^{(j)}$ such that for $|\alpha| \le i$, the calculus derivative $D_x^\alpha u_{\mathbb{c}}^{(j)}$ exists and is in $C^{\lambda,\lambda'_\alpha}_{bdd}(\Omega\times(0,\bar{t}))$ for every $\bar{t} < T$ and such that the equivalence class over Ω of $D_x^\alpha u_{\mathbb{c}}(\cdot,t)$ is the weak Ω–derivative $D_x^\alpha u^{(j)}(t)$. Also, if $j \ge 1$ then by Theorem 5.5(a) the derivative of $D_x^\alpha u^{(j-1)}$ (as a wAC mapping into $L^2(\Omega)$) is $[D_x^\alpha u^{(j)}(t)]_{[0,T)}$.

Thus

$$\big(D_x^\alpha u^{(j-1)}\big)' = [D_x^\alpha u^{(j)}]_{[0,T)} = \big[[D_x^\alpha u_{\mathbb{C}}^{(j)}(\cdot,t)]_\Omega\big]_{(0,T))}.$$

Part (b) of Lemma 5.7 therefore applies to show that the calculus derivative $D_t D_x^\alpha u_{\mathbb{C}}^{(j-1)}$ exists on $\Omega \times (0,T)$, is in $C^{\lambda,\lambda'_\alpha}(\Omega \times (0,\bar{t}))$ for every $\bar{t} < T$ and equals the calculus derivative $D_x^\alpha u_{\mathbb{C}}^{(j)}$. Setting $u_{\mathbb{C}} = u_{\mathbb{C}}^{(0)}$, we then find by iteration that for $j \leq j_0$ and $|\alpha| \leq i = 2(j_0 - j) + \kappa$, $D_t^j D_x^\alpha u_{\mathbb{C}} = D_x^\alpha u^{(j)} \in C_{bdd}^{(\lambda,\lambda'_\alpha)}(\Omega \times (0,\bar{t}))$ for every $\bar{t} < T$. This proves the statements in (c*i*).

The existence and regularity statements for $a_{\mathbb{C}}^{\eta,\omega}$, $b_{\mathbb{C}}^{\eta}$, $c_{\mathbb{C}}^{\omega}$, $d_{\mathbb{C}}$, $g_{\mathbb{C}}$ and $G_{\mathbb{C}}$ in the first item in (c*ii*) are proved in exactly the same way as above for $u_{\mathbb{C}}$, *mutatis mutandi.* Note that in every instance in which one of $a^{\eta,\omega}$, b^η, c^ω or d or one of its derivatives is required in (5.11)–(5.13) to be in a Lebesgue space $L^s(\Omega)$, the exponent s is greater than or equal to two, so that $L^s(\Omega) \subset L^2(\Omega \cap B)$ for every ball $B \subset \mathbf{R}^n$. The equality (5.35) then follows directly from (5.28): for example, for $j \leq j_0 - 1$, $i \leq j$ and for fixed t, $D_t^i a_{\mathbb{C}}^{\eta,\omega}(\cdot,t)$ and $D_t^{j-i} u_{\mathbb{C}}(\cdot,t)$ are locally Hölder continuous in Ω and are elements of the Ω–equivalence classes $(a^{(i)})^{\eta,\omega}(t)$ and $u^{(j-i)}(t)$; the same is true for their spatial gradients. Similar considerations apply to the other terms in these equations, and (5.28) therefore holds almost everywhere in Ω for fixed t. Since all the functions involved are Hölder continuous in x and have continuous extensions to $\partial\Omega$, (5.35) holds for all $x \in \overline{\Omega}$. This proves (c*ii*).

Next we prove the statements in (c*iii*) for the case that $\Omega \neq \mathbf{R}^n$. The first item follows directly from the second item in (a*v*): since $D_t^j u_{\mathbb{C}}(\cdot,t)$ is in the Ω–equivalence class $u^{(j)}(t)$, its restriction $\mathcal{T} D_t^j u_{\mathbb{C}}(\cdot,t)$ is in the $\partial\Omega$–equivalence class $\mathcal{T} u^{(j)}(t)$ and therefore satisfies $Q_l(x)\mathcal{T}(D_t^j u_{\mathbb{C}})(x,t) = 0$ for σ–a.a $x \in \mathcal{C}_l$. This then holds for all $x \in \mathcal{C}_l$ because $\mathcal{T} D_t^j u_{\mathbb{C}}(\cdot,t)$ is Hölder continuous.

To prove the second item in (c*iii*) we define $\tilde{\Gamma}_i^{(j)}(\tilde{y},t) \equiv \Gamma^{(j)}(t)(T_i(\tilde{y}))$ for $\tilde{y} \in \mathcal{B}_i$ (in the notation of Definition 1.7) where the $\Gamma^{(j)}$ are as in Lemma 5.4. Then $\tilde{\Gamma}_i^{(j)} \in wAC^2([0,T); L^2(\mathcal{B}_i))$ and its $[0,T)$ equivalence class is in $L_{loc}^\infty([0,T); H^{2m-2j-1}(\mathcal{B}_i))$ by Lemma 5.4(a) and (5.21). The two hypotheses of Lemma 5.7 are therefore satisfied with $k_0 + i = 2m - 2j - 1$, and one easily checks from the definition of κ in (c*i*) that $i \geq 0$ provided $j \leq j_0 - 1$. There are therefore functions $\tilde{\Gamma}_{i,\mathbb{C}}^{(j)}$ in $C_{bdd}^{(\lambda,\lambda'_0)}(\mathcal{B}_i \times (0,\bar{t}))$ with norms bounded independently of i such that $\tilde{\Gamma}_{i,\mathbb{C}}^{(j)}(\cdot,t)$ is in the equivalence class $\Gamma^{(j)}(t)$ on $\mathcal{B}_i$ for each t. Also, by iterating part (b) of Lemma 5.7 we find that $\tilde{\Gamma}_{i,\mathbb{C}}^{(j)} = D_t^j \tilde{\Gamma}_{i,\mathbb{C}}^{(0)}$ in the calculus sense. Now let $\{\varphi_i\}$ be the partition of unity in Theorem 1.8 with $X = \partial\Omega$ and $\delta > 0$ determined by $\underline{r}$, $\underline{s}$, and ε_0 (see Definition 1.7(c)) subordinate to the cover $\{\mathcal{U}_i\}$. Define

$$\Gamma^{(j)}(x,t) = \sum \varphi_i(x)\tilde{\Gamma}_{i,\mathbb{C}}^{(0)}(T_i^{-1}(x),t)$$

where the sum is over i for which $\varphi_i(x) \neq 0$. The regularity statements in the second item in $(ciii)$ for $\Gamma_{\mathfrak{c}}$ are now easily checked, although we note that important use is made of the final item in Definition 1.7(c). The corresponding statements for $f_{\mathfrak{c}}$ are proved in a similar way; we note again, however, that for each space $L^s(\mathcal{B}_i)$ occurring in the hypotheses (5.14), the index s is greater than or equal to two, so that $L^s(\mathcal{B}_i \cap T_i^{-1}(B)) \subseteq L^2(\mathcal{B}_i \cap T_i^{-1}(B))$ for every ball $B \subset \mathbf{R}^n$.

The final item in Theorem 5.5($ciii$) follows from the final item (av), just as in the proof of the first item in (ci), because every function appearing in (5.37) is Hölder continuous. This completes the proof of (c) and of Theorem 5.5.

□

§5.3 Incompatible Data and Initial Layer Regularization

In this section we consider higher order regularity of solutions of the problem (5.1)–(5.3) when the regularity conditions on the coefficients of A in Theorem 5.5 are satisfied but the compatibility conditions (5.23) are not. Specifically, given initial data $\varsigma_0 \in [L^2(\Omega)]^N$ and given inhomogeneous terms g, G and Γ satisfying the hypotheses of the basic existence result Theorem 3.6 on $[0, T)$ but the higher order regularity conditions of Theorem 5.5 only away from $t = 0$, the corresponding solution u of Theorem 3.6 attains the higher order regularity described in Theorem 5.5 instantaneously for $t > 0$ with certain specific rates. The precise results are given in the following theorem, which is a special case of Theorem 2.10:

Theorem 5.8. *Assume that the hypotheses and notations in* Definition 5.1 *are in force, let $T \in (0, \infty]$ and let $N \times N$ matrix-valued $[0, T)$–equivalence classes $a^{\eta,\omega}$, b^η, c^ω, d and (if $\Omega \neq \mathbf{R}^n$) f be given satisfying the positive definiteness condition* (5.5) *and the regularity conditions in one of* (5.8), (5.10) *or* (5.11)–(5.14). *Let $\varsigma_0 \in [L^2(\Omega)]^N$ and let g, G and Γ be given satisfying the following (references to Γ and integrals over $\partial\Omega$ are omitted if $\Omega = \mathbf{R}^n$ and we again abbreviate $[H^k(\Omega)]^N$ by H^k, etc.):*

- *The conditions in* Theorem 3.6(e) *are satisfied.*
- *There is a $\delta \in \{0\} \cup (\frac{1}{2}, \infty)$ such that for $j = 0, \ldots, m-1$ and $k = j+1, \ldots, m$ and on every set $\Omega \times [0, \bar{t}] \subset \Omega \times [0, T)$*

$$
\begin{aligned}
&t^{k+\delta} D_t^j D_x^\alpha g_{\mathrm{m}} \in L^{\infty,2}, \ 0 \le 2j + |\alpha| \le 2k-2, \\
&t^{k+\delta} D_t^j D_x^\alpha G_{\mathrm{m}} \in L^{\infty,2}, \ 0 \le 2j + |\alpha| \le 2k-1, \\
&\sum_{|\alpha| \le 2k-2j-1} t^{k+\delta} \langle \Gamma \rangle^{(\alpha,j)}_{L^2,\partial\Omega} \in L^\infty_{loc}([0,T)).
\end{aligned}
\tag{5.38}
$$

- *For* $j = 0, \ldots, m$ *and* $k = j, \ldots, m$ *and on every set* $\Omega \times [0, \bar{t}]$,

(5.39)
$$\begin{gathered} t^{k+\delta} D_t^j D_x^\alpha g_{\mathrm{m}} \in L^{2,2},\ 0 \leq |\alpha| \leq \max\{2k - 2j - 1), 0\} \\ t^{k+\delta} D_t^j D_x^\alpha G_{\mathrm{m}} \in L^{2,2},\ 0 \leq 2j + |\alpha| \leq 2k, \\ \sum_{|\alpha| \leq 2k-2j} t^{k+\delta} \langle \Gamma \rangle^{(\alpha,j)}_{L^2, \partial\Omega} \in L^2_{loc}([0,T)). \end{gathered}$$

- *For* $j = 0, \ldots, m-1$ *there is an element* $\gamma_{j+1} \in L^2(\Omega)$ *such that for all* $v \in \mathcal{H}_1$,

(5.40)
$$\begin{aligned} \lim_{t \to 0+} t^{-1} \int_0^t \int_\Omega s^{j+1+\delta} \Big[D_t^j \big(g_{\mathrm{m}} + S^1(\Gamma)_{\mathrm{m}} \big) \cdot v \\ + D_t^j \big(G_{\mathrm{m}} + S^2(\Gamma)_{\mathrm{m}} \big) : \nabla_x v \Big] \, dx \, ds \\ = \int_\Omega \gamma_{j+1} \cdot v \, dx \end{aligned}$$

where S^1 *and* S^2 *are the operators defined in* Theorem A.26 *and* D_t^j *denotes a weak derivative on* $\Omega \times (0, T)$ (*see* Exercise 5.5 *for an alternative formulation of this hypothesis*).

Then the solution u *of* (5.1)–(5.3) *described in* Theorem 3.6 *satisfies the following:*

(a)**(Bochner-Sobolev space description)** *The mappings* $F^{(j)}$ *defined in* (5.22) *satisfy the hypotheses of* Theorem 2.10(a)–(c); *the statements in* Theorem 5.5(a) *concerning* $u^{(j)}$ *therefore hold but with* $[0, T)$ *replaced by* $(0, T)$ *in every instance and with the following modifications and additions:*

- *The bound in* Theorem 5.5(a*ii*) *is replaced by*

$$\begin{aligned} \sum_{k=0}^m \sum_{j=0}^k \Big(\sup_{0 < t \leq \bar{t}} t^{2(k+\delta)} |u^{(j)}(t)|^2_{[H^{2k-2j}(\Omega)]^N} \\ + \int_0^{\bar{t}} t^{2(k+\delta)} |u^{(j)}(t)|^2_{[H^{2k-2j+1}(\Omega)]^N} dt \Big) \\ \leq C C_A(\bar{t}) \big[|\varsigma_0|^2_{[L^2(\Omega)]^N} + \sum_1^m |\gamma_j|^2_{[L^2(\Omega)]^N} + \tilde{C}_F \big] \end{aligned}$$

where $C_A(\bar{t})$ *is defined in* Lemma 5.3, $\tilde{C}_F(\bar{t})$ *bounds the sum of the squares of the norms on* $(0, \bar{t})$ *appearing in* (5.38) *and* (5.39) *and* C *is a generic positive constant depending on* n, N, m, C_{BC}, δ *and* M_Ω^{2m+1} (Definitions 5.1 *and* 1.7).

- *The weak form in* (5.25) *is replaced by*

$$\int_\Omega u^{(j+1)}(\cdot)\Big|_{t_1}^{t_2} \cdot v\,dx = -\sum_{i=0}^{j} \binom{j}{i} \int_{t_1}^{t_2} A^{(i)}(s)(u^{(j-i)}(s), v)ds$$
$$+ \int_{t_1}^{t_2} F^{(j)}(s) \cdot v\,ds$$

for $j = 0, \ldots, m-1$, $t_1, t_2 \in (0,T)$ *and* $v \in \mathcal{H}_1$, *where* $A^{(i)}$ *is as in* Lemma 5.2.

- *For* $k = 1, \ldots, m$,

$$t^{k+\delta} u^{(j)}(t) \begin{cases} \rightharpoonup 0 & \text{weakly in } H^{2k-2j}(\Omega),\ 0 \le j < k \\ \to \gamma_j & \text{strongly in } L^2(\Omega),\ j = k \end{cases}$$

as $t \to 0^+$.

(b) **(Weak derivative description)** *The statements in* Theorem 5.5(b) *hold but with the bound in* (*bii*) *replaced by the following: for given* τ, $\bar{t} \in (0,T)$ *with* $\tau < \bar{t}$ *there is a number* C *depending on upper bounds for* τ^{-1}, $\bar{t}$, $C_A(\bar{t})$ (*from* Lemma 5.3), $|\varsigma_0|_{L^2}, |\gamma_1|_{L^2}, \ldots, |\gamma_m|_{L^2}$ *and* $\tilde{C}_F(\bar{t})$ (*from* (a) *above*) *such that for* $j = 0, \ldots, m-1$ *and* $|\alpha| \le 2m - 2j$,

$$|D_t^j D_x^\alpha u_{\mathfrak{m}}|_{L^{\infty,2}(\Omega \times (\tau, \bar{t}))} \le C$$

and for $j = 0, \ldots, m$ *and* $|\alpha| \le 2m - 2j + 1$,

$$|D_t^j D_x^\alpha u_{\mathfrak{m}}|_{L^{2,2}(\Omega \times (\tau, \bar{t}))} \le C.$$

(c) **(Continuous derivative description)** *The statements in* Theorem 5.5(c) *hold but with* $[0,T)$ *replaced by* $(0,T)$ *in each instance, with* $[0,\bar{t}]$ *replaced by* $[\tau, \bar{t}]$ *for arbitrary* $\tau \in (0,\bar{t})$, *and with the bound in* (*cii*) *replaced by the statement that for* $2j + |\alpha| \le 2j_0 + \kappa$ *and for* $0 < \tau < \bar{t} < T$,

$$|D_t^j D_x^\alpha u_{\mathfrak{c}}|_{C_{bdd}^{\lambda,\lambda'_\alpha}(\Omega \times (\tau, \bar{t}))} \le C$$

where C *depends on the same quantities as the constant in* (b) *above as well as on* λ *in the case that* $k_0 - n/2 = 1$. *In addition, for* $j \le j_0$,

$$t^{k+\delta} [D_t^j u_{\mathfrak{c}}(\cdot, t)]_\Omega \begin{cases} \rightharpoonup 0 & \text{weakly in } H^{2k-2j}(\Omega),\ k = j+1, \ldots, m \\ \to \gamma_j & \text{strongly in } L^2(\Omega),\ k = j \end{cases}$$

as $t \to 0^+$.

Proof. We check that the hypotheses of Theorem 2.10 are satisfied. First, (5.38) and (5.39) with $k = m$ show that the hypotheses (5.18)–(5.21) of Lemma 5.4 hold with $(0,T)$ replaced by (τ, T) for arbitrary $\tau \in (0,T)$, and therefore so too do the conclusions of Lemma 5.4 regarding $g^{(j)}$, $G^{(j)}$, $\Gamma^{(j)}$ and $F^{(j)}$. This shows in particular that hypothesis (b) of Theorem 2.10 holds, and (ci) and (cii) are proved exactly as in the proof of Lemma 5.4(b) with the various powers of t inserted (which play no role in these two statements). To check hypothesis $(ciii)$ of Theorem 2.10 we first observe that by (5.39) with $k = j$ the components of $t^{j+\delta}g^{(j)}$ and $t^{j+\delta}G^{(j)}$ are in $L^2_{loc}((0,T); L^2(\Omega))$ and those of $t^{j+\delta}\Gamma^{(j)}$ are in $L^2_{loc}((0,T); L^2(\partial\Omega))$ for $j \leq m$. Consequently $t^{j+1+\delta}F^{(j)} \in wAC^2([0,T); V_{-1})$ for $j \leq m-1$ and therefore for $v \in \mathcal{H}_1$ the difference

$$t^{j+1+\delta}F^{(j)}(t) \cdot v - t^{-1}\int_0^t s^{j+1+\delta}F^{(j)}(s) \cdot v\, ds$$

goes to zero as $t \to 0$. The integral on the right here is

$$\int_0^t s^{j+1+\delta}\Big[\int_\Omega \big(g^{(j)} \cdot v + G^{(j)} : \nabla_x v\big)dx + \int_{\partial\Omega} \Gamma^{(j)} \cdot d\sigma\Big]ds$$

which by Theorems A.20 and A.26 is precisely the double integral on the left in (5.40). It therefore follows that

$$\lim_{t\to 0^+} t^{j+1+\delta}F^{(j)}(t) \cdot v = \int_\Omega \gamma_{j+1} \cdot v\, dx$$

as required in $(ciii)$ of Theorem 2.10. The conclusions of Theorem 2.10 therefore hold and show in particular that for $\tau > 0$, $\{u^{(0)}(\tau), \ldots, u^{(m)}(\tau)\}$ satisfies the compatibility condition (5.23) required for the application of Theorem 5.5 to the initial value problem for (5.1)-(5.3) posed at initial time τ. The conclusions of Theorem 5.5 therefore hold with $[0,T)$ replaced by $[\tau, T)$ in each case.

□

Exercises

(See also exercises 2.11 and 2.12.)

5.1 Rewrite the statements of Lemmas 5.3 and 5.4 and Theorem 5.5 for the case $N = 1$, $n = 2$ or 3 and $m = 1$.

5.2 Assume that $A^{(j)}$ and $F^{(j)}$ are as in Lemmas 5.3 and 5.4 with $\Omega = \mathbf{R}^n$ and let $m \geq 1$. Check that if $\varsigma_0 \in H^{2m}(\mathbf{R}^n)$ then there exist $\varsigma_j \in H^{2m-2j}(\mathbf{R}^n)$, $j = 1, \ldots, m$, satisfying the compatibility conditions (5.23).

5.3 In this problem we formulate the compatibility conditions (5.23) in Theorem 5.5 solely in terms of ς_0 for two additional cases. Assume that the other hypotheses and notations in Theorem 5.5 hold, that the coefficients of A are independent of t and that $\varsigma_0 \in \mathcal{H}^{2m}$ is given. Let L and $L_{\partial\Omega}$ be the operators $L^{(0)}$ and $L^{(0)}_{\partial\Omega}$ defined in (5.29) and (5.31) and let $h^{(j)} = g^{(j)}(0) - \operatorname{Div} G^{(j)}(0)$. Prove that there exist $\varsigma_j \in \mathcal{H}^{2m-2j}$ for which the compatibility conditions (5.23) hold in each of following two cases:

(a) $\mathcal{H}_1 = [H_0^1(\Omega)]^N$, and either $m = 1$ or $m \geq 2$ and

$$L^j[\varsigma_0] + \sum_{i=0}^{j-1} L^{j-1-i}[h^{(i)}] \in [H_0^1(\Omega)]^N, \ j = 1, \dots, m-1.$$

(b) $\mathcal{H}_1 = [H^1(\Omega)]^N$ and

$$L_{\partial\Omega}\big(L^j[\varsigma_0] + \sum_{i=0}^{j-1} L^{j-1-i}[h^{(i)}]\big) = G^{(j)}(0) + \Gamma^{(j)}(0) \text{ on } \partial\Omega$$

for $j = 0, \dots, m-1$ (the sum is absent if $j = 0$).

5.4 Let u be the solution described in Theorem 5.5. Assume that the coefficients of A are constants and show that $u^{(j)} \in C([0,T); \mathcal{H}^{2m-2j})$ for $j = 0, \dots, m$ provided that $g^{(j)} \in C([0,T); [H^{2m-2j-2}]^N)$ and $G^{(j)} \in C([0,T); [H^{2m-2j-2}]^{N\times n})$ for $j = 0, \dots, m-1$.

5.5 Reformulate the hypotheses on Γ in (5.40) of Theorem 5.8 by replacing $S^1(\Gamma)$ and $S^2(\Gamma)$ by appropriate sums of norms on the coordinate domains $\mathcal{B}_i$ of Definition 1.7 in such a way that the argument in the proof of (*ciii*) of that theorem remains valid.

5.6 Let u be the solution described in Theorem 5.8. Show that if $j < k \leq m$ and $2k - 2j - |\alpha| \geq k_0$, then $t^{k+\delta} D_t^j D_x^\alpha u_{\mathbb{C}} \in C_{bdd}^{\lambda,\lambda_0'}(\overline{\Omega} \times [0, \bar{t}])$ for every $\bar{t} < T$ and that its value at $t = 0$ is zero.

CHAPTER 6

Applications to Quasilinear Systems

> Nonlinearity by itself cannot serve as the foundation for a useful mathematical structure.
>
> Ivar Stakgold [**40**]

In this chapter we apply the linear theory of Chapters 3 and 5 to study problems in which coefficients and inhomogeneities depend on the unknown. A basic theory is given in sections 6.1–6.3 for solutions with minimal regularity, including the theorems of Leray and Hopf for the Navier-Stokes equations as well as results for more general problems in which nonlinearities have specific, controlled growth at infinity. More highly regular solutions, derived by iteration from the corresponding regularity theory for linear problems, are considered in sections 6.4–6.7. This discussion includes local-in-time existence with large compatible data, global existence for data near an attracting rest point, invariant regions for parabolic systems, and a large data, global existence theorem in two and three space dimensions for certain systems with symmetry.

§6.1 A Quasilinear System in Hilbert Space

In this section we prove a basic existence theorem for an abstract problem posed in Hilbert spaces, similar to that in (2.1) but with the inhomogeneous term F depending on the unknown function. Thus (2.3) is replaced by

$$\langle u(t), v\rangle_{\mathcal{H}_0} = \langle \varsigma_0, v\rangle_{\mathcal{H}_0} + \int_0^t \big[-A(s)(u(s),v) + F(s,u(s))\cdot v\big]ds \tag{6.1}$$

where A is the bilinear map considered in Theorem 2.1 and F maps $[0,T)\times\mathcal{H}_1$ into $\mathcal{H}_1^*$. The basic result, stated in Theorem 6.2 below, is proved by an approximation and limiting procedure similar to that in the proof of Theorem 2.1, modified to accommodate the nonlinearity. The two key ingredients in the proof are an *estimate*, which leads to compactness of approximate solutions, and a *commutation property*, which allows for the interchange of weak convergence with F. These are stated in Theorem 6.2 as assumptions (hypotheses (d) and (e)) rather than as consequences of assumptions, because their verification typically requires *ad hoc* arguments specific to the particular application, as will be apparent in the examples discussed in sections 6.2 and 6.3.

Before stating the main theorem we address the question of the measurability of the composition $t \to F(t, u(t))$ in (6.1):

Lemma 6.1. *Let $\mathcal{H}$ be a separable Hilbert space, I a nontrivial interval in $\mathbf{R}$ and F a mapping from $I \times \mathcal{H}$ to $\mathcal{H}^*$. Assume that $F(\cdot, w) \cdot v$ is measurable on I for every fixed w and v in $\mathcal{H}$ and that for almost all $t \in I$, $F(t, w) \cdot v$ is a continuous function of $w \in \mathcal{H}$ for every fixed $v \in \mathcal{H}$. If $w : I \to \mathcal{H}$ is weakly measurable on I and $v \in \mathcal{H}$, then the map $\to F(t, w(t)) \cdot v$ is measurable on I.*

Proof. Let $w : I \to \mathcal{H}$ be weakly measurable and let $w^K(t)$ be its projection onto the first K elements of an orthonormal basis $\{\psi_k\}$. Then for almost all t and for all $v \in \mathcal{H}$, $F(t, w^K(t)) \cdot v \to F(t, w(t)) \cdot v$ as $K \to \infty$ by the continuity assumption. It therefore suffices to assume that w maps to the span of $\{\psi_k\}_1^K$, say $w(t) = a_k(t)\psi_k$ (sum over k) where $a = (a_1, \ldots, a_K)$ is measurable in t. Let $\{R_i^h\}_i$ be cubes of side $h > 0$ covering $\mathbf{R}^K$ and fix $a^{i,h} \in R_i^h$ for each i. Then for almost all t,

$$F(t, w(t)) \cdot v = \lim_{h \to 0} \sum_i \chi_{R_i^h}(a(t)) F(t, a_k^{i,h} \psi_k) \cdot v$$

where χ_{R_i} is the characteristic function of R_i^h. Each summand on the right here is measurable in t; hence so is the limit.

□

The following is the main result of this section:

Theorem 6.2. *Let T, $\mathcal{H}_0$, $\mathcal{H}_1$ and A be as in (2.1) and assume also that $A \in L^\infty_{loc}([0, T); \mathcal{B}(\mathcal{H}_1 \times \mathcal{H}_1))$, and let F be a mapping from $[0, T) \times \mathcal{H}_1$ into $\mathcal{H}_1^*$. Assume the following:*

(a) *For fixed w and v in $\mathcal{H}_1$ the mapping $t \mapsto F(t, w) \cdot v$ is Lebesgue measurable on $[0, T)$.*

(b) *There is a set $E \subset [0, T)$ of measure zero such that if $t \in [0, T) - E$ then for each fixed $v \in \mathcal{H}_1$ the mapping $w \mapsto F(t, w) \cdot v$ is continuous from $\mathcal{H}_1$ into $\mathbf{R}$.*

(c) *Given $\bar{t} \in [0, T)$ and $M > 0$ there is a real-valued, locally integrable function ψ on $(0, \bar{t})$ and a nonnegative number $C(M, \bar{t})$ such that for $w \in \mathcal{H}_1$ satisfying $|w|_{\mathcal{H}_0} \leq M$ and for $t \in (0, \bar{t}) - E$,*

$$|F(t, w)|_{\mathcal{H}_1^*} \leq C(M, \bar{t}) \sum_i |w|_{\mathcal{H}_1}^{r_i} + \psi(t)$$

where the sum is finite and each r_i is in $[0, 2)$.

(d) *There is a positive nonincreasing function θ' and locally integrable, nonnegative functions λ', δ_i and γ on $(0, T)$ such that for $\bar{t} \in [0, T)$, $t \in (0, \bar{t}) - E$*

and $w \in \mathcal{H}_1$,

$$-A(t)(w,w) + F(t,w)\cdot w \le -\theta'(\bar t)|w|^2_{\mathcal{H}_1} + \lambda'(t)|w|^2_{\mathcal{H}_0} + \sum_i \delta_i(t)|w|^{p_i}_{\mathcal{H}_0} + \gamma(t) \tag{6.2}$$

where the sum is finite and $p_i > 2$ *for each* i *(compare the final item in* (2.1)*).*

(e) *For all* $\bar t \in (0,T)$ *and* $v \in \mathcal{H}_1$,

$$\int_0^{\bar t} F(t, w^k(t))\cdot v\, dt \to \int_0^{\bar t} F(t, w(t))\cdot v\, dt$$

for any sequence $\{w^k\}_k$ *of weakly measurable functions from* $[0,\bar t]$ *into* $\mathcal{H}_1$ *whose Lebesgue equivalence classes are bounded independently of* k *in* $L^\infty((0,\bar t);\mathcal{H}_0)$ *and in* $L^2((0,\bar t);\mathcal{H}_1)$ *and which converge to a weakly measurable function* w *in the following sense:* $\langle w^k(t), z\rangle_{\mathcal{H}_0} \to \langle w(t), z\rangle_{\mathcal{H}_0}$ *uniformly on* $[0,\bar t]$ *for every* $z \in \mathcal{H}_0$ *and the* $(0,\bar t)$*-equivalence class of* w^k *converges to the equivalence class of* w *weakly in* $L^2((0,\bar t);\mathcal{H}_1)$.

Then there is a local–in–time solution of (6.1) *for given initial data as follows:*

(f) *Given* $\varsigma_0 \in \mathcal{H}_0$, $T_0 \in (0,T)$ *and* $\eta > 0$ *there is a time* $\tau \in (0,T_0]$ *and an element* $u \in C([0,\tau],\mathcal{H}_0^{weak})$ *whose equivalence class is in* $L^2((0,\tau);\mathcal{H}_1)$, *for which the equivalence class on* $[0,\tau]$ *of the mapping* $t \to F(t,u(t))$ *defines an element of* $L^1_{loc}([0,\tau];\mathcal{H}_1^*)$ *and which satisfies* (6.1) *for every* $v \in \mathcal{H}_1$ *and* $t \in [0,\tau]$ *as well as the bound*

$$\sup_{0\le t\le\tau} \tfrac12|u(t)|^2_{\mathcal{H}_0} + \theta'(T_0)\int_0^\tau |u(t)|^2_{\mathcal{H}_1}dt \le (1+\eta)e^{\int_0^{T_0} 2\lambda'(t)\,dt}\Big(\tfrac12|\varsigma_0|^2_{\mathcal{H}_0} + \int_0^{T_0}\gamma(t)\,dt\Big). \tag{6.3}$$

The time τ *is chosen so that* $\sum_i \int_0^\tau \delta_i$ *is sufficiently small depending on* η *and on the exponential and the* ς_0 *and* γ *terms on the right side of* (6.3).

If $\delta_i \equiv 0$ *for every* i *then there is an element* $u \in C([0,T),\mathcal{H}_0^{weak})$ *whose equivalence class is in* $L^2_{loc}([0,T);\mathcal{H}_1)$ *and which satisfies* (6.1) *for every* $v \in \mathcal{H}_1$ *and* $t \in [0,T)$ *as well as the bound in* (6.3) *with* $\eta = 0$, $\tau = T_0$ *and* T_0 *arbitrary in* $[0,T)$; *as a consequence,* $|u(t) - \varsigma_0|_{\mathcal{H}_0} \to 0$ *as* $t \to 0^+$.

(g) *If the hypothesis in* (c) *holds with all* $r_i \le 1$ *and with* $\psi \in L^2_{loc}([0,T))$ *then the solution* u *in* (f) *is in* $C([0,\tau];\mathcal{H}_0)$ *(that is,* $\mathcal{H}_0$ *with the norm topology) and the energy equality*

$$\tfrac12|u(t)|^2_{\mathcal{H}_0} + \int_0^t A(s)(u(s),u(s))\,ds = \tfrac12|\varsigma_0|^2_{\mathcal{H}_0} + \int_0^t F(s,u(s))\cdot u(s)\,ds \tag{6.4}$$

holds for all $t \in [0,\tau]$. *If additionally each* $\delta_i \equiv 0$, *then the solution in* (f) *is in* $C([0,T);\mathcal{H}_0)$ *and the above energy equality holds for all* $t \in [0,T)$.

(h) *If the hypothesis in* (c) *holds with all* $r_i \leq 1$ *and with* $\psi \subset L^2_{loc}([0,T))$, *as in* (g), *and if in addition for every* $M > 0$ *there exist* $C_M > 0$, $s \in [0,2)$ *and* ζ *square integrable on* $[0,\tau]$ *such that for almost all* t,

$$\begin{aligned} &|\big(F(t,w') - F(t,w)\big)\cdot(w'-w)| \\ &\qquad \leq C_M\big[(\zeta(t) + |w'|_{\mathcal{H}_1} + |w|_{\mathcal{H}_1})|w'-w|_{\mathcal{H}_0}\big]^{2-s}|w'-w|^s_{\mathcal{H}_1} \end{aligned} \tag{6.5}$$

for $w, w' \in \mathcal{H}_1$ *with* $\mathcal{H}_0$*–norms bounded by* M, *then solutions of* (6.1) *are unique in the class of elements of* $C([0,\tau];\mathcal{H}_0^w)$ *with equivalence classes in* $L^2((0,\tau);\mathcal{H}_1)$. *And if in addition* $\delta_i \equiv 0$ *for all* i, *then there is a unique solution of* (6.1) *on all of* $[0,T)$.

PROOF. We first treat the case that the δ_i are not all zero. Choose $T_0 \in (0,T)$ and define

$$C_{\lambda'}(T_0) = e^{2\int_0^{T_0}\lambda'(t)\,dt}, \quad C_{data}(T_0) = \tfrac{1}{2}|\varsigma_0|^2 + \int_0^{T_0}\gamma(t)\,dt$$

and

$$\Delta_i(t) = \int_0^t \delta_i(s)\,ds.$$

We then leave to the reader to check that, given $p_i > 2$ and $\eta > 0$ as in the hypotheses, there is a $\tau \in (0,T_0]$ as described in (a) and which will now be fixed, such that the following holds:

If H *is a continuous nonnegative function on* $[0,\tau]$ *satisfying*

$$H(t) \leq C_{\lambda'}(T_0)\Big(C_{data}(T_0) + \sum_i \Delta_i(t)(2H(t))^{p_i/2}\Big)$$

and $H(0) \leq \frac{1}{2}|\varsigma_0|^2_{\mathcal{H}_0}$, *then*

$$H(t) \leq (1+\eta)C_{\lambda'}(T_0)C_{data}(T_0)$$

for all $t \in [0,\tau]$.

The proof is an elementary calculus argument comparing the relative positions of the curves $y = x$ and $y = 1 + \sum_i \Delta_i(t)(2x)^{p_i/2}$ in the (x,y) plane for small, fixed t, together with a scaling.

To construct a solution we let $\{\varphi_k\}_k$ be the orthonormal basis for $\mathcal{H}_0$ described in Theorem A.10(b). Thus if $v \in \mathcal{H}_1$ then the $\mathcal{H}_0$–projections $v^K \equiv \sum_1^K \langle v,\varphi_k\rangle_{\mathcal{H}_0}\varphi_k$ converge to v in $\mathcal{H}_1$ as $K \to \infty$ as well as in $\mathcal{H}_0$, and $|v^K|_{\mathcal{H}_1} \leq 2|v|_{\mathcal{H}_1}$ for all K. Approximate solutions u^K mapping a subinterval of $[0,T)$ into $\mathcal{H}^K \equiv \text{span}\{\varphi_1,\dots,\varphi_K\}$ are then to be found satisfying

$$\begin{aligned}\langle u^K(t),v^K\rangle_{\mathcal{H}_0} = \langle \varsigma_0, v^K\rangle_{\mathcal{H}_0} - \int_0^t A(s)(u^K(s),v^K)\,ds \\ + \int_0^t F(s,u^K(s))\cdot v^K\,ds\end{aligned} \tag{6.6}$$

for all $v^K \in \mathcal{H}^K$. If $u^K(t) = \sum_1^K a_k(t)\varphi_k$ then the $\mathbf{R}^K$-valued function $a = (a_1,\dots,a_K)$ is to satisfy

$$a(t) = a_0 + \int_0^t f(s,a(s))\,ds \tag{6.7}$$

where $(a_0)_k = \langle \varsigma_0,\varphi_k\rangle_{\mathcal{H}_0}$, $f \equiv (f_1,\dots,f_K)$ is defined by

$$f_k(t,a) = -A(t)(w_a,\varphi_k) + F(t,w_a)\cdot\varphi_k$$

and $w_a \equiv \sum_1^K a_k\varphi_k$. We will check that Theorem A.7 applies to guarantee a local-in-time solution $a(t)$. The continuity hypothesis of Theorem A.7 is obvious from assumption (b), and the measurability condition is checked exactly as in the proof of Lemma 6.1. The last requirement is that $|f(t,\cdot)|_{L^\infty(V_j)}$ is integrable on $[0,T_0]$ for each V_j in some increasing sequence $V_j \subset \mathbf{R}^K$ of open sets covering $\mathbf{R}^K$. For this we can take V_j to be the ball of radius j centered at the origin and then apply hypothesis (c) and the fact that $|w_a|$ is bounded by $C|a|$ for a constant C depending on K. The conclusions of Theorem A.7 and Corollary A.8 therefore hold and there is a solution a of (6.7), hence a solution u^K of (6.6), on an interval $[0,t_{max}) \in (0,T_0]$ which is maximal in the sense of Corollary A.8 and which at this point may depend on K.

We will show that t_{max} is at least as large as the time τ defined above, which is independent of K. To do this we note that since the solution a of (6.7) is absolutely continuous, so is $|u^K|^2_{\mathcal{H}_0}$. Proceeding as in the proof of (2.9) in Theorem 2.1 and applying hypothesis (d), we find that, for $t < t_{max}$ and $t \le \tau$,

$$\begin{aligned}\tfrac{1}{2}|u^K(t)|^2_{\mathcal{H}_0} + \theta'(T_0)\int_0^t |u^K(s)|^2_{\mathcal{H}_1}\,ds \\ \le C_{\lambda'}(T_0)\Big(C_{data}(T_0) + \sum_i \int_0^t \delta_i(s)|u^K(s)|^{p_i}_{\mathcal{H}_0}\,ds\Big).\end{aligned} \tag{6.8}$$

Let $H(t')$ be the sup over $t \in [0,t']$ of the left side here and fix $\bar{t} < t_{max}$ and $\bar{t} \le \tau$. Then H is continuous on $[0,\bar{t}]$ and satisfies the hypothesis in the statement in the first paragraph of this proof, hence its conclusion. Thus

$$\tfrac{1}{2}|u^K(t)|^2_{\mathcal{H}_0} + \theta'(T_0)\int_0^t |u^K(s)|^2_{\mathcal{H}_1}\,ds \le (1+\eta)C_{\lambda'}(T_0)C_{data}(T_0) \tag{6.9}$$

for $t \leq \tau$ and $t < t_{max}$. In particular if J is chosen so that $\frac{1}{2}|w_a|^2_{\mathcal{H}_0} > (1+\eta)C_{\lambda'}(T_0)C_{data}(T_0)$ for $a \in \partial V_J$, then $|a(t)|$ cannot exceed J at any such time $t \leq \tau$. Corollary A.8 therefore applies to show that $t_{max} \geq \tau$ as claimed.

The argument is considerably simpler for the case that the δ_i are all zero: the calculus bound in the first paragraph of the proof is avoided and τ is not defined. Instead, we choose $T_0 \in (0, T)$ arbitrarily, then define V as above in terms of T_0. Repeating the above estimates, we find that (6.8) holds with the δ_i terms omitted and therefore that (6.9) holds (with $\eta = 0$) for $t \in [0, \min\{t_{max}, T_0\})$. Corollary A.8 then applies in the same way to show that $t_{max} \geq T_0$ and therefore that $t_{max} = T$ since $T_0 < T$ was arbitrary.

We will give the remainder of the proof only for the case that the δ_i are *not* all zero. The changes required for the improvements stated in the theorem when all δ_i are zero will be obvious.

Since the bound in (6.9) is independent of K, as is τ, we can choose a sequence $K_i \to \infty$ such that u^{K_i} converges weakly in $L^2([0,\tau]; \mathcal{H}_1)$, say to w. We claim that there is a subsequence, still denoted by K_i, and an element $u \in C([0,\tau]; \mathcal{H}_0^{weak})$ such that $\langle u^{K_i}(t), v\rangle_{\mathcal{H}_0} \to \langle u(t), v\rangle_{\mathcal{H}_0}$ uniformly on $[0,\tau]$ for each fixed $v \in \mathcal{H}_1$. To see this we let $t_1 \leq t_2 \in [0,\tau]$, $v \in \mathcal{H}_1$ and v^K its $\mathcal{H}_0$–projection onto $\mathcal{H}^K$. Then by(6.6) and hypothesis (c),

$$
\begin{aligned}
\left|\langle u^K(t_2) - u^K(t_1), v\rangle_{\mathcal{H}_0}\right| &= \left|\langle u^K(t_2) - u^K(t_1), v^K\rangle_{\mathcal{H}_0}\right| \\
&\leq \int_{t_1}^{t_2} \left(|A(u^K, v^K)| + |F(\cdot, u^K)\cdot v^K|\right) \\
&\leq \text{const.} \int_{t_1}^{t_2} \left(|u^K|_{\mathcal{H}_1} + \psi + \sum_i |u^K|^{r_i}_{\mathcal{H}_1}\right)|v|_{\mathcal{H}_1} \\
&\leq \omega_v(|t_2 - t_1|)
\end{aligned}
$$

where ω_v is a continuous function on $[0,\tau]$ which is independent of K and which satisfies $\omega_v(0) = 0$. (We have applied here the bound (6.9), the assumption that $r_i < 2$ for every i and the fact that $|v^K|_{\mathcal{H}_1} \leq 2|v|_{\mathcal{H}_1}$.) Thus for each fixed $v \in \mathcal{H}_1$ the sequence of maps $t \mapsto \langle u^K(t), v\rangle_{\mathcal{H}_0}$ is equicontinuous. We can therefore apply the Ascoli–Arzela theorem together with a diagonal process to arrive at a further subsequence K_i such that, for every v in a countable dense set in $\mathcal{H}_1$, $\langle u^{K_i}(t), v\rangle_{\mathcal{H}_0}$ converges uniformly on $[0,\tau]$. This same result then holds for all $v \in \mathcal{H}_0$ because $\mathcal{H}_1$ is dense in $\mathcal{H}_0$ and the left side above is bounded by const.$|v|_{\mathcal{H}_0}$. The limit is linear in v at each t and is therefore given by $\langle u(t), v\rangle_{\mathcal{H}_0}$ for some $u(t) \in \mathcal{H}_0$. The above bound then shows that $u \in C([0,\tau]; \mathcal{H}_0^{weak})$, and we can apply the argument following (2.11) in the proof of Theorem 2.1 to show that its Lebesgue equivalence class on $[0,\tau]$ is w, which is an element of $L^2([0,\tau,]; \mathcal{H}_1)$. The proof that u satisfies (6.1) is exactly as in the proof of Theorem 2.1 together with a direct application of hypothesis (e) for the term involving F. Observe also that in the case that the δ_i are all zero, (6.3) (with $\eta = 0$) shows that

$\limsup_{t\to 0^+} |u(t)|_{\mathcal{H}_0} \le |\varsigma_0|_{\mathcal{H}_0}$, so that $u(t) \to \varsigma_0$ in the norm of $\mathcal{H}_0$. This completes the proof of (f).

To prove (g) we change notation and now denote by $u^K(t)$ the $\mathcal{H}_0$–orthogonal projection of $u(t)$ onto $\mathcal{H}^K$. Then since each $\langle u(t), \varphi_k\rangle_{\mathcal{H}_0}$ is absolutely continuous in t, we can put $v = \varphi_k$ in (6.1), differentiate, sum over k, and then integrate. The result is that

$$\tfrac{1}{2}|u^K(t)|^2_{\mathcal{H}_0} + \int_0^t A(u, u^K) = \tfrac{1}{2}|u^K(0)|^2_{\mathcal{H}_0} + \int_0^t F(\cdot, u)\cdot u^K.$$

The equality in (g) then follows in the limit as $K \to \infty$. For example, the integrand in $\int_0^t F(s, u(s))\cdot(u^K(s) - u(s))\,ds$ goes to zero almost everywhere because $u \in \mathcal{H}_1$ for almost all s so that the projection $u^K(s)$ converges to $u(s)$ strongly in $\mathcal{H}_1$; also, this integrand is bounded by $C|F(s,u(s))|_{\mathcal{H}_1^*}|u(s)|_{\mathcal{H}_1} \in L^2\times L^2 \subset L^1$ by the hypothesis of (g) and the fact that $|u^K(s)|_{\mathcal{H}_1} \le 2|u(s)|_{\mathcal{H}_1}$. This integral therefore goes to zero by the dominated convergence theorem. The resulting equality in (g) shows in particular that $|u(t)|^2_{\mathcal{H}_0}$ is a continuous function of t and therefore that $u \in C([0,\tau]); \mathcal{H}_0)$ in the norm topology. This proves (g).

Finally to prove (h) we let u and u' be solutions of (6.1) in the given regularity class and proceed as above in the proof of (g) to find that

$$\begin{aligned}\tfrac{1}{2}|(u'-u)(t)|^2_{\mathcal{H}_0} + &\int_0^t A(u'-u, u'-u)\\ &= \int_0^t [F(\cdot,u') - F(\cdot,u)]\cdot(u'-u).\end{aligned}$$

Then by the hypothesis of (h),

$$\begin{aligned}\tfrac{1}{2}|(u'-u)(t)|^2_{\mathcal{H}_0} &+ \int_0^t |u'-u|^2_{\mathcal{H}_1}\\ &\le C\int_0^t \Big[(\zeta + |u'|_{\mathcal{H}_1} + |u|_{\mathcal{H}_1})|u'-u|_{\mathcal{H}_0}\Big]^{2-s} |u'-u|^s_{\mathcal{H}_1}\end{aligned}$$

so that

$$|(u'-u)(t)|^2_{\mathcal{H}_0} \le C\int_0^t \Big[(\zeta^2 + |u'|^2_{\mathcal{H}_1} + |u|^2_{\mathcal{H}_1})|u'-u|^2_{\mathcal{H}_0}\Big].$$

Gronwall's inequality now applies to show that $u'(t) = u(t)$ for every t.

□

It is of interest to know whether a solution can be extended past the time τ given in the theorem and if not to understand the nature of the obstruction. We will say that if u is a solution of (6.1) on $[0,\tau]$ in the sense of Theorem 6.2(f), then a solution $\tilde{u}$ of (6.1) with the same initial value ς_0 on an interval $[0,\tilde{\tau}]$ containing $[0,\tau]$ is an *extension* of u if u and $\tilde{u}$ agree on

$[0, \tau]$ (thus u is an extension of itself). The *maximum time of existence* t_{max} of the solution u relative to $[0, T)$ is then the supremum of times $\tilde{\tau} \in [\tau, T)$ for which such an extension exists on $[0, \tilde{\tau}]$. Notice that t_{max} is an attribute of the particular solution u and not just of the data A, F and ς_0. See in particular Exercise 6.2 for an example in which two different solutions with the same data have different maximal times of existence.

Corollary 6.3. *Let* T, $\mathcal{H}_0$, $\mathcal{H}_1$ *and* A *be as in* Theorem 6.2. *Assume that* F *satisfies hypotheses* (a)–(d) *of* Theorem 6.2 *and that* u *is a solution of* (6.1) *in the sense of* Theorem 6.2(f) *with maximal time of existence* t_{max} *relative to* $[0, T)$. *Then either* $t_{max} = T$ *or*

$$\limsup_{t \to t_{max}^-} \left(|u(t)|^2_{\mathcal{H}_0} + \int_0^t |u|^2_{\mathcal{H}_1} \right) = \infty.$$

Proof. If $t_{max} < T$ and the above limit is finite, then $|u(t)|_{\mathcal{H}_0}$ is uniformly bounded on $[0, t_{max})$ and $|u|^2_{\mathcal{H}_1}$ is integrable on $(0, t_{max})$. The first of these shows that if $t_j \to t_{max}^-$ then $u(t_j)$ has weak-$\mathcal{H}_0$ subsequential limits, and the second together with hypothesis (c) of Theorem 6.2 shows that

$$\int_0^{t_{max}} | - A(u(t), v) + F(t, u(t)) \cdot v| \leq C|v|_{\mathcal{H}_1}$$

for a constant C independent of $v \in \mathcal{H}_1$. Therefore if $u(t_{j_k}) \rightharpoonup w$ in $\mathcal{H}_0$ we can take the limit in

$$\langle u(t_{j_k}), v\rangle_{\mathcal{H}_0} = \langle \varsigma_0, v\rangle_{\mathcal{H}_0} + \int_0^{t_{j_k}} (- A(u, v) + F(t, u) \cdot v)$$

to find that $\langle w, v\rangle_{\mathcal{H}_0}$ is independent of the subsequence $\{t_{j_k}\}$. Since $\mathcal{H}_1$ is dense in $\mathcal{H}_0$, all such weak sequential limits agree and therefore $u(t) \rightharpoonup w$ weakly in $\mathcal{H}_0$ as $t \to t_{max}^-$. Thus if we define $u(t_{max}) = w$, then the extended u is in $C([0, t_{max}]; \mathcal{H}_0^{weak})$. Theorem 6.2 can then be applied to show that there is a nontrivial interval $I = [t_{max}, t_{max} + \tau']$ and a solution $z \in C(I; \mathcal{H}_0^{weak})$ with equivalence class in $L^2(I; \mathcal{H}_1)$ to the problem

$$\langle z(t), v\rangle_{\mathcal{H}_0} = \langle w, v\rangle_{\mathcal{H}_0} + \int_{t_{max}}^t (- A(z, v) + F(t, z) \cdot v)$$

for $v \in \mathcal{H}_1$ and $t \in I$. Concatenating u and z we thus obtain an extension $\tilde{u}$ of u to an interval strictly larger than $[0, t_{max}]$, contradicting the definition of t_{max}.

□

§6.2 Navier–Stokes Equations of Incompressible Flow

In this section we apply Theorem 6.2 to the Navier–Stokes system for viscous, incompressible flow, this being one of the most widely studied models in applied analysis. We present here versions of the first and most basic results in the subject, those of Leray [**29**] and Hopf [**23**] and refer the reader to other texts, particularly those of Galdi [**18**] and Temam [**41**] for more extensive presentations and for references to a very large literature. As we will see, verification of hypotheses (a)–(d) of Theorem 6.2 is nearly immediate but the commutation property (e) is more involved. Its proof depends instead on a crucial approximation result given in Theorem 6.4 below, which is an application of an observation of Friedrichs in [**17**] and which finds application well beyond its use here.

The model is formulated as follows (but see Batchelor [**4**], 3.1–3.3, for example, for a more detailed discussion): let Ω be an open set in $\mathbf{R}^n$, $n = 2, 3$, not necessarily bounded, regarded as the set occupied by a fluid, and let $u(x,t) \in \mathbf{R}^n$ be the velocity of the fluid "particle" which at time t is at position x. The so–called no–slip condition $u(x,t) = 0$ for $x \in \partial\Omega$ is imposed, and the stipulation that $\mathrm{div}_x u(x,t) = 0$ for $x \in \Omega$ expresses the law of conservation of mass under the assumption that the fluid density is constant (unity, without loss of generality). Internal forces are modeled by considering a hypothetical surface element with unit normal ν and applying elementary scaling and frame indifference considerations to describe the force exerted by the fluid immediately on one side of the surface element on the fluid immediately on other side by $\big(\mu(\nabla_x u + \nabla_x^{tr} u) - pI\big)\nu$, where p is the scalar pressure and μ is a positive viscosity constant. External body forces g_i are included, and standard rational mechanics arguments based on Newton's law and the divergence theorem then yield the first equation in the following formal system for the unknown functions u and p:

$$\tag{6.10} \begin{cases} u_t^j + u_{x_k}^j u^k + p_{x_j} = \mu\Delta u^j + \sum_i g_i^j, \ (x,t) \in \Omega \times (0,T), \\ \mathrm{div}_x\, u = 0, \ (x,t) \in \Omega \times (0,T), \\ u = 0, \ (x,t) \in \partial\Omega \times [0,T), \\ u(x,0) = \varsigma_0(x), \ x \in \Omega \end{cases}$$

where ς_0 is the initial velocity, summation over k is understood and Δ is the Laplace operator.

A suitable weak form is derived as follows. Let $\mathcal{H}_1$ be the subset of $\left[H_0^1(\Omega)\right]^n$ whose elements v satisfy $\mathrm{div}_x v = 0$ in the sense of weak derivatives in Ω and let $\mathcal{H}_0$ be the completion of $\mathcal{H}_1$ with respect to the L^2 norm, so that $\mathcal{H}_1$ is dense in $\mathcal{H}_0$. Then for $v \in \mathcal{H}_1$ we multiply the first equation above by v^j and integrate formally with respect to x and t and apply the divergence theorem. The result is that the pressure term p drops out and that u is to satisfy

$$\begin{aligned}\int_\Omega u(t)\cdot v &= \int_\Omega \varsigma_0 \cdot v \\ &+ \int_0^t\int_\Omega \big(-\mu\nabla_x u\cdot\nabla_x v + u^j u^k v^j_{x_k} + \sum_i g_i\cdot v\big)\end{aligned} \tag{6.11}$$

for $t \geq 0$ and $v \in \mathcal{H}_1$. This has the required form (6.1) with $A(w,v) = \int_\Omega \mu\nabla_x w\cdot\nabla_x v$ and

$$F(t,w)\cdot v = \int_\Omega \big(w^j w^k v^j_{x_k} + \sum_i g_i(\cdot,t)\cdot v\big). \tag{6.12}$$

Once a solution u of (6.11) has been constructed, the pressure p is recovered, at least formally, as follows. Taking the x_j-derivative in the first equation in (6.10) and summing over j we find that at almost every t,

$$\Delta p = -(u^j_{x_k}u^k)_{x_j} + \sum_i \operatorname{div}_x g_i.$$

Thus if $n = 3$ and if $u(t) \in \mathcal{H}_1$ for a particular t, then at that t and on Ω, $(u^j_{x_k}u^k) \in L^2\times L^6 \subset L^{3/2}$; and if each $g_i(t) \in [L^{3/2}(\Omega)]^3$ we can anticipate solutions $p(\cdot,t) \in W^{1,3/2}(\Omega)$. Issues of the boundary condition for p, boundary regularity, uniqueness and regularity in t arise, however, and these are beyond the scope of the present discussion. See [**18**] Theorem 2.1 or [**41**] Remark 3.8 for more detailed discussions of the pressure.

We begin the analysis with the following result, which will be applied to verify the commutation assumption in Theorem 6.2(e):

Theorem 6.4. *Let Ω be an open set in $\mathbf{R}^n$, $n \geq 2$, and $\bar t \in (0,\infty)$. Let $\{w^k\}_k$ be a sequence of weakly continuous mappings from $[0,\bar t]$ into $L^2(\Omega)$ and assume that for every $v \in L^2(\Omega)$, $\langle w^k(t), v\rangle_{L^2(\Omega)} \to 0$ uniformly on $[0,\bar t]$ as $k\to\infty$ and that the $[0,\bar t]$–equivalence classes of the w^k are uniformly (in k) bounded in $L^2([0,\bar t]; H^1(\Omega))$. Then $w^k \to 0$ strongly in $L^2([0,\bar t]; L^2(\Omega'))$ for every open $\Omega' \subset \Omega$ whose closure is compact in Ω.*

Proof. We first claim that there is a constant C_n depending on n such that if $K \subset \mathbf{R}^n$ is a cube in $\mathbf{R}^n$ of side $h > 0$ and if $w \in H^1(K)$, then

$$\int_K |w|^2 dx \leq 2h^{-n}\Big(\int_K w\,dx\Big)^2 + C_n h^2\int_K |\nabla_x w|^2 dx. \tag{6.13}$$

It suffices to prove this for the restriction w to K of an element of $C_c^\infty(\mathbf{R}^n)$ (Theorem A.2). For such w and for $x \in K$,

$$\begin{aligned}\Big|w(x) - h^{-n}\int_K w\Big| &= h^{-n}\Big|\int_K\int_0^1 \nabla w(y+s(x-y))\cdot(x-y)\,ds\,dy\Big| \\ &\leq C_n h^{-n/2}\Big(\int_K\int_0^1 |\nabla w(y+s(x-y))|^2 ds\,dy\Big)^{1/2}.\end{aligned}$$

Thus

$$\int_K w(x)^2dx \le 2h^{-n}\Big(\int_K w(y)dy\Big)^2 \\ +C_nh^{-n}\int_K\int_K\int_0^1 |\nabla w(y+s(x-y))|^2 ds\,dy\,dx.$$

We write $\int_0^1 = \int_0^{1/2} + \int_{1/2}^1$ and for the first of these reverse the s and y integrals and make the change of variables $z = y+s(x-y)$, so that $|\det\frac{dy}{dz}| = (1-s)^{-n} \le C_n$. For the second we write $ds\,dy\,dx = dy\,ds\,dx$ so that $|\det\frac{dx}{dz}| = s^{-n} \le C_n$. This proves the claim.

Now if Ω' is compactly contained in Ω and if $h>0$ is sufficiently small, then Ω' can be covered by a finite collection of cubes K_i^h of side h contained in Ω. We let χ_i^h be h^{-n} times the characteristic function of K_i^h and apply (6.13) to w^k to obtain that for almost all t depending on the specific representative,

$$\int_{\Omega'}(w^k)^2 \le \sum_i \int_{K_i^h}(w^k)^2 \\ \le C\Big[\sum_i\Big(\int_\Omega w^k\chi_i^h\Big)^2 + h^2\int_\Omega |\nabla w^k|^2\Big]$$

so that

$$\int_0^{\bar t}\int_{\Omega'}(w^k)^2 \le C\Big[\sum_i\Big(\int_0^{\bar t}\int_\Omega w^k\chi_i^h\Big)^2 + h^2\Big].$$

The integrals on the right go to zero as $k\to\infty$ because each χ_i^h is in $L^2(\Omega)$. The right side can therefore be made arbitrarily small by choosing h small then k large depending on h.

□

We can now give the application of Theorem 6.2 to the Navier–Stokes system (6.11):

Theorem 6.5. *Let Ω be an open set in $\mathbf{R}^2$ or $\mathbf{R}^3$, not necessarily bounded, and let $\mathcal{H}_0$ and $\mathcal{H}_1$ be the subspaces of $[L^2(\Omega)]^n$ and $[H_0^1(\Omega)]^n$ described above preceding* (6.11). *Assume that $\mu>0$ and let $\varsigma_0 \in \mathcal{H}_0$ and (finitely many) g_i be given with $g_i \in L^2_{loc}([0,\infty);[L^{q_i}(\Omega)]^n)$ where $q_i \in (1,2]$ if $n=2$ and $q_i \in [6/5,2]$ if $n=3$. Then:*

(a) *There is an element $u \in C([0,\infty);\mathcal{H}_0^{weak})$ with $[0,\infty)$–equivalence class in $L^2_{loc}([0,\infty);\mathcal{H}_1)$ satisfying* (6.11) *for all $t\ge 0$ and all $v\in\mathcal{H}_1$; u is strongly continuous at $t=0$: $u(t)\to\varsigma_0$ in $L^2(\Omega)$ as $t\to 0^+$; and given $\bar t\in(0,\infty)$ there is a constant $C(\bar t)$ depending on $\bar t$, μ and the q_i if $n=2$*

such that

$$\begin{aligned}(6.14)\qquad &\sup_{0\le t\le \bar t} |u(t)|^2_{[L^2(\Omega)]^n} + \int_0^{\bar t} |\nabla_x u|^2_{[L^2(\Omega)]^n}\,dt \\ &\qquad \le C(\bar t)\big[|\varsigma_0|^2_{[L^2(\Omega)]^n} + \int_0^{\bar t}\sum_i |g_i(t)|^2_{[L^{q_i}(\Omega)]^n}\,dt\big].\end{aligned}$$

(b) *In the case that* $n = 2$ *there is an element* $u \in C([0,\infty);\mathcal{H}_0)$ *(that is,* $L^2(\Omega)$ *with the norm topology) with* $[0,\infty)$*–equivalence class in* $L^2_{loc}([0,\infty);\mathcal{H}_1)$ *satisfying* (6.11) *for all* $t \in [0,T)$ *and all* $v \in \mathcal{H}_1$*. The bound in* (6.14) *holds, as does the energy equality*

$$\begin{aligned}(6.15)\qquad &\tfrac{1}{2}|u(t)|^2_{[L^2(\Omega)]^2} + \mu\int_0^{\bar t} |\nabla_x u|^2_{[L^2(\Omega)]^2}\,dt \\ &\qquad =\tfrac{1}{2}|\varsigma_0|_{[L^2(\Omega)]^2} + \int_0^{\bar t}\int_\Omega \sum_i u\cdot g_i\,dx\,dt.\end{aligned}$$

The solution u *is unique in the set of functions in* $C([0,\infty);\mathcal{H}_0^{weak})$ *with equivalence classes in* $L^2_{loc}([0,\infty);\mathcal{H}_1)$ *which satisfy* (6.11).

PROOF. We apply Theorem 6.2. First, it is obvious that the requirements on $\mathcal{H}_0$, $\mathcal{H}_1$ and A (that is, the hypotheses of Theorem 2.1(b)) are satisfied, as is (a) of Theorem 6.2. We will check that F defined in (6.12) satisfies hypotheses (b)–(d) of Theorem 6.2 for $n = 3$, the case $n = 2$ being similar. For $n = 3$ the space $H_0^1(\Omega)$ is continuously included in $L^6(\Omega)$ (Theorem A.3(a)); thus if $w \in H_0^1$ (we abbreviate $[H_0^1(\Omega)]^n$ by H_0^1, etc.) then by interpolation,

$$(6.16)\qquad |w|_{L^4} \le |w|^{1/4}_{L^2}|w|^{3/4}_{L^6} \le C|w|^{1/4}_{L^2}|w|^{3/4}_{H^1}.$$

Therefore if F is as in (6.12) and w, w', $v \in H_0^1$ then

$$|(F(w') - F(w))\cdot v| \le (|w'|_{L^4} + |w|_{L^4})|w' - w|_{L^4}|v|_{H^1},$$

which goes to zero as $w' \to w$ in H^1 as required in (b) of Theorem 6.2. Next, since the Hölder conjugate of each q_i is in $[2,6]$,

$$|F(t,w)|_{\mathcal{H}_1^*} \le |w|^2_{L^4} + C\sum |g_i(t)|_{L^{q_i}} \le C\big(|w|^{1/2}_{L^2}|w|^{3/2}_{H^1} + \sum |g_i(t)|_{L^{q_i}}\big),$$

which shows that (c) of Theorem 6.2 holds with $r = 3/2$ and $\psi = C\sum |g_i|_{L^{q_i}} \in L^2_{loc}([0,\infty))$. For (d) we note first that if $w \in \mathcal{H}_1$ then

$$(6.17)\qquad \int_\Omega w^j w^k w^j_{x_k} = \int_\Omega \tfrac{1}{2}(|w|^2)_{x_k} w^k = 0$$

by the zero boundary condition and the divergence–free condition. Thus

$$\begin{aligned}-A(w,w) + F(\cdot,w)\cdot w &= \int_\Omega \big(-\mu|\nabla_x w|^2 + \sum g_i(\cdot)\cdot w\big) \\ &\le -\tfrac{1}{2}\mu|w|^2_{H^1} + \mu|w|^2_{L^2} + C(\mu)\Big(\sum |g_i(\cdot)|_{L^{q_i}}\Big)^2,\end{aligned}$$

so that (d) holds with all $\delta_i = 0$ and $\gamma(\cdot) = \sum |g_i(\cdot)|^2_{L^{q_i}}$.

Finally to check (e) of Theorem 6.2 we let $\{w^k\}$, w and $\bar{t}$ be as in the hypothesis. Then for $\varphi \in C_c^\infty(\Omega)$ with compact support $\overline{\Omega'} \subset \Omega$ and for fixed i and j,

$$\Big| \int_0^{\bar{t}} \int_\Omega [(w^k)^i (w^k)^j - w^i w^j] \varphi \Big| \leq C(\varphi) \int_0^{\bar{t}} \int_{\Omega'} (|w^k| + |w|) |w^k - w|$$
$$\leq C(\varphi, \bar{t}) \sup_{0 \leq t \leq \bar{t}} \big(|w^k|_{\mathcal{H}_0} + |w|_{\mathcal{H}_0} \big) \Big(\int_0^{\bar{t}} \int_{\Omega'} |w^k - w|^2 \Big)^{1/2}$$

which goes to zero as $k \to \infty$ by Theorem 6.4. It follows by an application of (6.16) that the double integral on the left goes to zero as $k \to \infty$ for every $\varphi \in L^2(\Omega)$ and in particular for $\varphi = v^i_{x_j}$ if $v \in \mathcal{H}_1$, as required. We have now checked that all the hypotheses (a)–(e) of Theorem 6.2 hold. The conclusions (a) of the present theorem therefore follow from Theorem 6.2(f), including the strong convergence $u(t) \to u(0)$ in L^2 as $t \to 0^+$.

To prove (b) we let $n = 2$ and record that, if $w \in H^1_0(\Omega)$ then

$$|w|^2_{L^4(\Omega)} \leq C(\Omega) |w|_{L^2(\Omega)} |w|_{H^1(\Omega)},$$

which is proved in Theorem A.4. It follows that

$$|F(\cdot, w)|_{\mathcal{H}_1^*} \leq |w|^2_{L^4} + C \sum |g_i(\cdot)|_{L^{q_i}} \leq C(|w|_{L^2} |w|_{H^1} + \sum |g_i(\cdot)|_{L^{q_i}}),$$

which shows that hypothesis (g) of Theorem 6.2 holds with $r = 1$ and $\psi = C \sum |g_i|_{L^{q_i}} \in L^2_{loc}([0, \infty))$. To prove the hypothesis in (h) we let $w, w' \in \mathcal{H}_1$ with $\mathcal{H}_0$-norms bounded by M and compute

$$|(F(\cdot, w') - F(\cdot, w)) \cdot (w' - w)| \leq \int_\Omega (|w'| + |w|) |w' - w| |\nabla_x (w' - w)|$$
$$\leq C \big(|w'|_{L^4} + |w|_{L^4} \big) |w' - w|_{L^4} |w' - w|_{H^1}$$
$$\leq C(M) \Big[(|w'|_{H^1} + |w|_{H^1}) |w' - w|_{L^2} \Big]^{1/2} |w' - w|^{3/2}_{H^1}$$

so that the required bound in (h) of Theorem 6.2 holds with $s = 3/2$. The statements in (b) of the present theorem then follow from the conclusions (g) and (h) of Theorem 6.2.

□

§6.3 Nonlinearities with Polynomial Growth

In this section we consider systems of parabolic equations in which the inhomogeneous terms may depend on the unknown function and are defined globally, with growth at infinity controlled in terms of the dimension of the physical space. Specifically we consider the problem (6.1) in which $\mathcal{H}_0$ and $\mathcal{H}_1$ are subspaces of $[L^2(\Omega)]^N$ and $[H^1(\Omega)]^N$ respectively, the bilinear form A is exactly as in (3.2) with coefficients satisfying the hypotheses of Theorem 3.6, and for $w,\, v \in \mathcal{H}_1$,

$$F(t,w)\cdot v = \int_\Omega \big(g(x,t,w)\cdot v + G(x,t,w) : \nabla_x v\big)dx + \int_{\partial\Omega} \Gamma(x,t)\cdot v\, d\sigma. \tag{6.18}$$

Thus g and G are nonlinear functions of the unknown, but Γ is not. Then given $\varsigma_0 \in \mathcal{H}_1$ the problem is to find a mapping u on an interval $[0,\tau]$ satisfying

$$\langle u(t), v\rangle_{\mathcal{H}_0} = \langle \varsigma_0, v\rangle_{\mathcal{H}_0} - \int_0^t A(s)(u(s),v)ds + \int_0^t F(s,u(s))\cdot v\, ds \tag{6.19}$$

for all $v \in \mathcal{H}_1$.

It will be convenient to regard g and G as measurable *functions* mapping $\Omega\times[0,T)\times\mathbf{R}^N$ into $\mathbf{R}^N$ and $\mathbf{R}^{N\times n}$ respectively, but to retain the formulation of Theorem 3.6 in which the solution u is an element of $C([0,\tau];\mathcal{H}_0^{weak})$ for some $\tau > 0$ with equivalence class in $L^2([0,\tau];\mathcal{H}_1)$. We will show in Theorem 6.6 below that, under suitable hypotheses on g and G and for such u, (6.18) defines F as a weakly measurable, locally integrable mapping from $[0,\tau]$ into $\mathcal{H}_1^*$.

The following gives the application of Theorem 6.2 to the above problem. The required "estimate" referred to in the discussion preceding Lemma 6.1 is derived as a consequence of growth conditions on g and G (hypothesis (a) below) and the "commutation property" is derived from a Hölder-type regularity assumption (hypothesis (b) below) via Theorem 6.4.

Theorem 6.6. *Let $n \geq 2$ and $N \geq 1$, let T, Ω, $\mathcal{H}_0$, $\mathcal{H}_1$, A, λ and θ be as in the hypotheses of* Theorem 3.6, *and let $\Gamma \in L^2_{loc}([0,T); L^2(\partial\Omega))$ be given (this and subsequent references to Γ are omitted if $\Omega = \mathbf{R}^n$). Assume the following:*

(a) *g and G are finite sums $\sum_i g_i$ and $\sum_i G_i$ of functions from $\Omega\times[0,T)\times\mathbf{R}^N$ into $\mathbf{R}^N$ and $\mathbf{R}^{N\times n}$ respectively, and there is a set $E_0 \subset \Omega\times[0,T)$ of measure zero (and therefore a set $E \subset [0,T)$ of measure zero such that for $t \notin E$, $(x,t) \notin E_0$ for almost all $x \in \Omega$) such that:*

- *for $(x,t) \notin E_0$ the maps $w \in \mathbf{R}^N \to g_i(x,t,w)$ and $G_i(x,t,w)$ are continuous;*
- *for all $w \in \mathbf{R}^N$, the maps $(x,t) \in \Omega\times[0,T) \to g_i(x,t,w)$ and $G_i(x,t,w)$ are measurable;*
- *for $\bar t \in (0,T)$ there are measurable functions φ_{ij}^g and φ_i^G, exponents α_i and β_i and a constant $C_F(\bar t)$ such that for $(x,t) \in \Omega\times[0,T) - E_0$ and $w \in \mathbf{R}^N$,*

$$\begin{cases} |g_i(x,t,w) - \sum_j \varphi_{ij}^g(x,t)| \leq C_F(\bar t)|w|^{\alpha_i} \\ |G_i(x,t,w) - \varphi_i^G(x,t)| \leq C_F(\bar t)|w|^{\beta_i} \end{cases} \tag{6.20}$$

satisfying either one of the following sets of conditions (A) *or* (B) (*in which the distinction between* φ_{ij}^g *and* φ_i^G *and their equivalence classes on* $\Omega \times [0,T)$ *is suppressed and spaces* $[L^{p,q}_{(t)loc}(\Omega \times [0,T))]^N$ *are abbreviated* $L^{p,q}_{(t)loc}$):

$$(A) \begin{cases} 1 \le \alpha_i < 1 + 4/n, \\ 1 \le \beta_i < 1 + 2/n, \\ \varphi_{ij}^g \in L^{2,q_{ij}}_{(t)loc} \text{ for some } q_{ij} \in (1,2] \text{ if } n = 2 \text{ and in } [\frac{2n}{n+2}, 2] \\ \qquad \text{if } n \ge 3, \text{ and} \\ \varphi_i^G \in L^{2,2}_{(t)loc}. \end{cases}$$

$$(B) \begin{cases} 1 \le \alpha_i < \begin{cases} 1 + 6/n, & n \le 6, \\ (n+2)/(n-2), & n \ge 7, \end{cases} \\ 1 \le \beta_i < \begin{cases} 1 + 4/n, & n \le 4, \\ n/(n-2), & n \ge 5, \end{cases} \\ \varphi_{ij}^g \in L^{1,q_{ij}}_{(t)loc} \text{ for some } q_{ij} \in (1,2] \text{ if } n = 2 \\ \qquad \text{and in } [\frac{2n}{n+2}, 2] \text{ if } n \ge 3, \\ \varphi_i^G \in L^{1,2}_{(t)loc}, \text{ and} \\ \text{the bound in (6.2) holds for all } w \in \mathcal{H}_1 \text{ and for some } \theta', \lambda', \delta_i, \text{ and } \gamma \\ \text{as described in Theorem 6.2(d) with } F \text{ as in (6.18);} \end{cases}$$

(b) *for each component* h *of* g_i *or* G_i *there are* $\varepsilon > 0$, $q \in [2-\varepsilon, (2-\varepsilon)(1+2/n)]$ ($q \in [2-\varepsilon, 2(2-\varepsilon))$ *if* $n = 2$) *and a measurable function* χ *whose equivalence class is in* $L^{s,s}_{(t)loc}$ *where* $s = 2/(2-\varepsilon)$, *such that for* $(x,t) \in \Omega \times [0,T) - E_0$ *and* $w, w' \in \mathbf{R}^n$,

$$|h(x,t,w') - h(x,t,w) \le \left[\chi(x,t) + C_F(\bar{t})(|w'|^q + |w|^q)\right] |w' - w|^{\varepsilon}.$$

Then:

(c) *Given* $\varsigma_0 \in \mathcal{H}_0$, $\Gamma \in L^2_{loc}([0,T); \partial\Omega)$, $T_0 \in (0,T)$ *and* $\eta > 0$, *there is a* $\tau \in (0, T_0]$ *and an element* $u \in C([0,\tau], \mathcal{H}_0^{weak})$ *with equivalence class in* $L^2([0,\tau]; \mathcal{H}_1)$ *satisfying* (6.19) *for all* $v \in \mathcal{H}_1$ *and* $t \in [0,\tau]$, *in which the equivalence class of* $F(\cdot, u(\cdot))$ *is in* $L^1([0,\tau]; \mathcal{H}_1^*)$. *In addition,*

- *the bound in* (6.3) *holds in case* (B) *with* θ', λ', δ_i, *and* γ *as in the final assumption in* (B), *and in case* (A) *with* $\theta' = \frac{1}{2}\theta(\tau)$, $\lambda' = \lambda(\tau) + C_1$, $\delta_i \le C_1$, *where* C_1 *is a constant depending on* $n, \Omega, C_F(\tau)$, *the exponents* α_i *and* β_i *and the number of summands comprising* g *and* G, *and*

$$\gamma(t) = C(\Omega, n)\Big[\sum_{ij} |\varphi^g_{ij}(\cdot, t)|^2_{[L^{q_{ij}}(\Omega)]^N} + \sum_i |\varphi^G_i(\cdot, t)|^2_{[L^2(\Omega)]^{N\times n}} + |\Gamma(\cdot, t)|^2_{[L^2(\partial\Omega)]^N}\Big].$$

- *If the δ_i referred to in the previous item are known to be zero, then there is an element $u \in C([0,T), \mathcal{H}_0^{weak})$ whose equivalence class is in $L^2_{loc}([0,T); \mathcal{H}_1)$ and which satisfies* (6.19) *for every $v \in \mathcal{H}_1$ and $t \in [0,T)$ as well as the bound in* (6.3) *with $\eta = 0$, $\tau = T_0$ and T_0 arbitrary in $[0,T)$.*
- *More generally, if t_{max} is the maximal time of existence of any such solution u relative to $[0,T)$ as defined above preceding the statement of* Corollary 6.3, *then either $t_{max} = T$ or*

$$\limsup_{t\to t_{max}^-} \Big[|u(t)|^2_{[L^2(\Omega)]^N} + \int_0^t |u(s)|^2_{[H^1(\Omega)]^N} ds\Big] = \infty.$$

- *If $u_{\rm m}$ is any specific representative of the measurable counterpart of such a solution u, then $g(x,t,u_{\rm m}(x,t))$ and $G(x,t,u_{\rm m}(x,t))$ are measurable and locally integrable on $\Omega \times (0,\tau)$, and if $C_c^\infty(\Omega) \subset \mathcal{H}_1$, then the equation*

(6.21)
$$\frac{\partial}{\partial t} u_{\rm m} = \frac{\partial}{\partial x_\omega}(a^{\eta,\omega}_{\rm m} D_{x_\eta} u_{\rm m}) - b^\eta_{\rm m} D_{x_\eta} u_{\rm m} + \frac{\partial}{\partial x_\omega}(c^\omega_{\rm m} u_{\rm m}) - d_{\rm m} u_{\rm m} + g(\cdot,\cdot,u_{\rm m}) - \mathrm{Div}\, G(\cdot,\cdot,u_{\rm m})$$

holds in the sense of distributions on $\Omega\times(0,\tau)$, each term here being either locally integrable on $\Omega\times(0,\tau)$ or the distribution derivative $\frac{\partial}{\partial t}, \frac{\partial}{\partial x_\omega}$, or Div *of a locally integrable function.*

(d) *In case* (A) *there is a solution as described above in* (c) *which is in $C([0,\tau]; [L^2(\Omega)]^N)$ (that is, L^2 with the norm topology) and which satisfies the energy equality* (6.4) *for all $t \in [0,\tau]$. If all $\delta_i = 0$ and the hypotheses of case* (A) *hold, then there is a solution $u \in C([0,T); \mathcal{H}_0)$ satisfying the energy equality for all $t \in [0,T)$.*

(e) *Assume in addition to the hypotheses of case* (A) *that for $\bar{t} \in [0,T)$ there is a constant $C_2(\bar{t})$ such that for $(x,t) \notin E_0$ with $t \le \bar{t}$ and for w, $w' \in \mathbf{R}^N$, $|g(x,t,w') - g(x,t,w)|$ and $|G(x,t,w') - G(x,t,w)|$ are bounded by finite sums of terms of the form $C_2(\bar{t})\xi(x,t)|w' - w|$ where* (*the equivalence class of*) *ξ is in $L^{q_1,q_2}_{(t)loc}$ for g and in $L^{2q_1,2q_2}_{(t)loc}$ for G, where q_1 and q_2 satisfy*

$$n/2 < q_2 \le \infty \text{ and } \frac{(p^*-2)q_2'}{p^*-2q_2'} \le q_1 \le \infty$$

for $n \geq 3$ (q' is the Hölder conjugate of q_2) and $p^* = 2n/(n-2)$); and for $n = 2$, $q_2 \in (1,\infty]$ and $q_1 \in (q_2', \infty]$. Then solutions of (6.19) in $C([0,\tau]; \mathcal{H}_0^{weak})$ having equivalence classes in $L^2([0,\tau]; \mathcal{H}_1)$ are uniquely determined by Γ and ς_0.

For the proof we will need the following consequences of the imbedding theorem in Theorem A.3(a), the latter showing in particular that $H^1(\Omega)$ is continuously included in $L^p(\Omega)$ for $p \in [2,\infty)$ if $n = 2$ and $p \in [2, p^*]$ for $n \geq 3$:

Lemma 6.7. *Let Ω be an open set in $\mathbf{R}^n$, $n \geq 2$, satisfying the cone condition.*

(a) *If $n \geq 3$ and $p \in [2, 2+4/n]$ or if $n = 2$ and $p \in [2,4)$ then there is an $r \in [0,2]$ depending on p such that for all $w \in H^1(\Omega)$,*

$$\int_\Omega |w|^p \leq C|w|^{p-r}_{L^2(\Omega)} |w|^r_{H^1(\Omega)}.$$

The constant C depends on Ω and n if $n \geq 3$ and on Ω and p if $n = 2$.

(b) *Assume that p satisfies*

$$1 \leq p < \begin{cases} 1 + 4/n, & n \leq 4 \\ n/(n-2), & n \geq 5. \end{cases}$$

Then there is a constant C depending on Ω, n and p and a number $r \in [0,2)$ depending on p such that $r \leq p$ and

$$\left(\int_\Omega |w|^{2p} \right)^{1/2} \leq C|w|^{p-r}_{L^2(\Omega)} |w|^r_{H^1(\Omega)}$$

for all $w \in H^1(\Omega)$. The same result holds but with $r \in [0,1)$ under the stronger hypothesis that $p \in [1, 1+2/n)$.

(c) *Assume that*

$$1 \leq p < \begin{cases} 1 + 6/n, & n \leq 5 \\ (n+2)/(n-2), & n \geq 6. \end{cases}$$

Then there is a constant C depending on Ω, n and p, a number $s \in (2n/(n+2), 2]$ and a number $r \in [0,2)$, both depending on p, such that $r \leq p$ and

$$\left(\int_\Omega |w|^{sp} \right)^{1/s} \leq C|w|^{p-r}_{L^2(\Omega)} |w|^r_{H^1(\Omega)}$$

for all $w \in H^1(\Omega)$. The same result holds but with $r \in [0,1)$ under the stronger hypothesis that $1 \leq p < 1 + 4/n$.

Proof. We give the proof of (c) for the case that $n \geq 3$. First, if s and p are positive numbers satisfying $2 \leq sp \leq p^*$, then the L^{sp} norm can be interpolated between the L^2 and L^{p^*} norms, the latter being bounded by $C(\Omega, n)$ times the H^1 norm. Thus for $w \in H^1(\Omega)$,

$$\left(\int_\Omega |w|^{sp}\right)^{1/s} \leq C|w|_{L^2}^{p-r}|w|_{H^1}^r \tag{6.22}$$

where $r = (p-2/s)/(1-2/p^*)$, which shows that $r < 2$ provided $p < 2/s+4/n$. Combining the two assumptions on s and p made so far, we thus have that, given $s > 0$, the bound in (6.22) holds for some $r < 2$ provided

$$2/s \leq p < \min\{(2/s + 4/n), p^*/s\},$$

which we abbreviate $a(s) \leq p < b(s)$. Now, a and b are continuous, decreasing functions of s, so that if s_1 and s_2 are the limits in (c) to be imposed on s, then $\cup_{s_1 < s \leq s_2}[a(s), b(s)) = [a(s_2), b(s_1))$. That is, the statement in (c) holds for p in the latter set, which is easily seen to be the range stated. The other assertions of the lemma are proved in a similar way.

□

PROOF of Theorem 6.6: We will check that the hypotheses of Theorem 6.2 are satisfied. First observe that if $w : \Omega \to \mathbf{R}^N$ is measurable, then $g(x, t, w(x))$ and $G(x, t, w(x))$ are measurable on $\Omega \times [0, T)$. To see this we let $\{R_i^h\}_i$ be nonoverlapping cubes of side $h > 0$ covering $\mathbf{R}^N$, $\chi_{R_i^h}$ their characteristic functions and $w_i^h \in R_i^h$. Then for $(x, t) \notin E_0$,

$$g(x, t, w(x)) = \lim_{h \to 0} \sum_i \chi_{R_i^h}(w(x))g(x, t, w_i^h)$$

by the first item in hypothesis (a). Each summand here is measurable by the second item in (a), hence so is the limit. The same argument applies to G, and the above representation also shows that if w is a measurable function on $\Omega \times I$, where I is a subinterval of $[0, T)$, then so are $g(\cdot, \cdot, w(\cdot, \cdot))$ and $G(\cdot, \cdot, w(\cdot, \cdot))$.

Next we check conditions (a) and (c) of Theorem 6.2. Let $w, v \in H^1$ (we abbreviate $[H^k(\Omega)]^N$ by H^k etc.) so that for $t \notin E$, each $G_i(\cdot, t, w) : \nabla_x v$ is measurable on Ω by the above. We can therefore apply (6.20) together with Lemma 6.7(b) to find that for $t \leq \bar{t} < T$

$$\begin{aligned}\int_\Omega |G_i(\cdot, t, w)||\nabla_x v| \leq \Big(&|\varphi_i^G(\cdot, t)|_{L^2} \\ &+ C(\Omega, n)C_F(\bar{t})|w|_{L^2}^{\beta_i - r_i}|w|_{H^1}^{r_i}\Big)|v|_{H^1}\end{aligned} \tag{6.23}$$

where $r_i < 1$ in case (A) and $r_i < 2$ in case (B). The right side here is locally integrable on $[0, T)$, hence so is the left, and the Fubini and Tonelli theorems therefore apply to show first that the mapping $t \to \int_\Omega G(\cdot, t, w) : \nabla_x v$ is measurable on $[0, T)$, as required in Theorem 6.2(a), and second that the contribution of the G_i term to $|F(t, w)|_{\mathcal{H}_1^*}$ is bounded by the quantity in brackets on the right in (6.23), as required in Theorem 6.2(c) (F is as defined in (6.18)).

The argument for g_i is similar but the bounds are different. First, the hypotheses on α_i together with Lemma 6.7(c) show that there is an $s_i \in (2n/(n+2), 2]$, whose Hölder conjugate s_i' is therefore in $[2, p^*)$ if $n \geq 3$ and in $[2, \infty)$ if $n = 2$, such that for $w, v \in H^1$,

$$\int_\Omega |w|^{\alpha_i}|v| \leq \Big(\int_\Omega |w|^{s_i\alpha_i}\Big)^{1/s_i}|v|_{L^{s_i'}} \leq C|w|_{L^2}^{\alpha_i - r_i}|w|_{H^1}^{r_i}|v|_{L^{s_i'}}$$

where $r_i < 1$ in case (A) and $r_i < 2$ in case (B), and $|v|_{L^{s_i'}} \leq C|v|_{H^1}$. Also, the assumed ranges of the q_{ij} insure that their Hölder conjugates q_{ij}' are also in $[2, p^*]$ if $n \geq 3$ and in $[2, \infty)$ if $n = 2$. Applying these bounds in (6.20) we obtain that for $w, v \in H^1$ and $t \in [0, \bar{t}] - E$,

$$\begin{aligned} \int_\Omega |g_i(\cdot,t,w)|\,|v|dx &\leq \sum_j |\varphi_{ij}^g(\cdot,t)|_{L^{q_{ij}}}|v|_{L^{q_{ij}'}} \\ &\quad + C_F(\bar{t})|w|_{L^2}^{\alpha_i - r_i}|w|_{H^1}^{r_i}|v|_{L^{s_i'}} \\ &\leq C\Big[\sum_j |\varphi_{ij}^g(\cdot,t)|_{L^{q_{ij}}} + C_F(\bar{t})|w|_{L^2}^{\alpha_i - r_i}|w|_{H^1}^{r_i}\Big]|v|_{H^1} \end{aligned} \tag{6.24}$$

(we record that all q_{ij}' and s_i' here are in $[2, \infty)$.) Again, the right side above is locally integrable on $[0, T)$ so that the same conclusions hold for the contributions of the g_i to F as for G_i above, and this shows that hypotheses (a) and (c) of Theorem 6.2 are satisfied.

Next we check hypotheses (b) and (e) of Theorem 6.2. First we let h denote a component of a particular g_i or G_i and fix $t \notin E$ with $t \leq \bar{t} < T$, $w, w' \in H^1(\Omega)$ with $|w'|_{L^2}, |w|_{L^2} \leq M$, and $\varphi \in C_c^\infty(\mathbf{R}^n)$. Then by (b) of the present theorem with $s = 2/(2-\varepsilon)$,

$$\begin{aligned} &\int_\Omega |h(x,t,w') - h(x,t,w)||\varphi(x)|dx \\ &\quad \leq |\varphi|_{L^\infty}\int_\Omega (\chi + C_F(\bar{t}))(|w'|^q + |w|^q)|w' - w|^\varepsilon dx \\ &\quad \leq |\varphi|_{L^\infty}\Big(|\chi(\cdot,t)|_{L^s} + C[|w'|_{L^{sq}}^q + |w|_{L^{sq}}^q]\Big)|w' - w|_{L^2}^\varepsilon \\ &\quad \leq |\varphi|_{L^\infty}\Big(|\chi(\cdot,t)|_{L^s} + C(M,\bar{t})[|w'|_{H^1}^{r/s} + |w|_{H^1}^{r/s}]\Big)|w' - w|_{L^2}^\varepsilon \end{aligned} \tag{6.25}$$

where $r \leq 2$ by Lemma 6.7(a), which has been applied in the last step with $p = sq$. Now to prove that (b) of Theorem 6.2 holds we apply (6.25) with $w' = w_k$ where $w_k \to w$ in H^1. The right side of (6.25) then goes to zero for φ as above, and we can approximate an element $v \in H^1(\Omega)$ by an element $\varphi \in C_c^\infty(\mathbf{R}^n)$ and apply (6.23) and (6.24) to conclude that the same result holds with φ replaced by v if h represents a component of g and by a component of $\nabla_x v$ if h represents a component of G. The result is that

$$\int_\Omega \big[\big(g(\cdot,t,w_k)-g(\cdot,t,w)\big)\cdot v + \big(G(\cdot,t,w_k)-G(\cdot,t,w)\big):\nabla_x v\big]\,dx \to 0$$

as $k\to\infty$, as required.

To prove that hypothesis (e) of Theorem 6.2 holds we let $\{w_k\}$ be a sequence converging to w as described in (e), again let h be a component of a particular g_i or G_i, and first take $\varphi\in C_c^\infty(\Omega)$ with compact support K in Ω. We integrate (6.25) over $(0,\bar t)$ and apply the hypotheses on χ in (b) and the fact that each r in (6.25) is in $[0,2]$:

$$\begin{aligned}
&\int_0^{\bar t}\int_\Omega \big|h(x,t,w_k)-h(x,t,w)\big||\varphi(x)|dx\,dt\\
&\le C(\bar t,\varphi)\Big[1+\Big(\int_0^{\bar t}|(w_k|_{H^1}^r+|w|_{H^1}^r\Big)^{1/s}\Big(\int_0^{\bar t}\int_K |w_k-w|^2\Big)^{\varepsilon/2}\Big] \qquad (6.26)\\
&\le C(\bar t,\varphi)\Big(\int_0^{\bar t}\int_K |w_k-w|^2dx\,dt\Big)^{\varepsilon/2}
\end{aligned}$$

which goes to zero as $k\to\infty$ by Theorem 6.4. Now for the case that h represents a component of g we take $\varphi\in C_c^\infty(\mathbf{R}^n)$, $\tilde\chi\in C_c^\infty(\Omega)$ and $v\in H^1(\Omega)$ and write $\Delta g_k = g(\cdot,\cdot,w_k)-g(\cdot,\cdot,w)$ so that

$$\iint \Delta g_k\cdot v = \iint \Delta g_k\cdot(\tilde\chi\varphi)+\iint \Delta g_k\cdot((1-\tilde\chi)\varphi) + \iint \Delta g_k\cdot(v-\varphi).$$

We can choose $\varphi\in C_c^\infty(\mathbf{R}^n)$ whose restriction to Ω is close to v in $H^1(\Omega)$ so that the third term on the right is small by (6.24), then choose $\tilde\chi$ so that the $L^{q'_{ij}}$ and $L^{s'_i}$ norms of $(1-\tilde\chi)\varphi$ are small (recall that these indices are in $[2,\infty)$) and finally choose k large so that the first term on the right is small, by (6.26). The result is that the integral on the left goes to zero as $k\to\infty$. The same analysis applies to G and is somewhat simpler because the components of $\nabla_x v$ are approximated only in $L^2(\Omega)$. This completes the proof that hypotheses (b) and (e) of Theorem 6.2 are satisfied.

Finally we check hypothesis (d) of Theorem 6.2 in case (A) (this being an *assumption* in case (B)). First let $w\in H^1(\Omega)$, $\bar t<T$ and $t\in[0,\bar t]-E$; then by (6.23) and (6.24),

$$|F(t,w)|_{\mathcal{H}_1^*}\le C(\Omega,n)\big[\tilde\gamma(t)+C_F(\bar t)\sum_l |w|_{L^2}^{p_l-r_l}|w|_{H^1}^{r_l}\big]$$

for various $p_l\ge 1$ and $r_l\in[0,1)$ and where

$$\tilde\gamma(t)=|\Gamma(t)|_{L^2(\partial\Omega)}+\sum_{i,j}|\varphi_{ij}^g(t)|_{L^{q_{ij}}(\Omega)}+\sum_i|\varphi_i^G(t)|_{L^2(\Omega)}.$$

Then by (3.9),

$$\begin{aligned}-A(t)(w,w)+F(t,w)\cdot w &\le -\theta(\bar{t})|w|_{H^1}^2+\lambda(\bar{t})|w|_{L^2}^2\\ &\quad + C(\Omega,n)\Big(\tilde{\gamma}(t)|w|_{H^1}+C_F(\bar{t})\sum_l |w|_{L^2}^{p_l-r_l}|w|_{H^1}^{r_l+1}\Big)\\ &\le -\tfrac{1}{2}\theta(\bar{t})|w|_{H^1}^2+\lambda(\bar{t})|w|_{L^2}^2+C\Big(\tilde{\gamma}(t)^2+\sum_l |w|_{L^2}^{2(p_l-r_l)/(1-r_l)}\Big)\end{aligned}$$

since $r_l < 1$. Observe that $2(p_l - r_l)/(1 - r_l) \ge 2$ for each l, so that (6.2) holds with θ', λ', δ_i and γ as described in the first item in (c) of the present theorem, as required. This proves that hypothesis (d) of Theorem 6.2 is satisfied.

We have now checked all the hypotheses (a)–(e) of Theorem 6.2, so that its conclusion (f) holds, and this together with Lemma 6.1 and Corollary 6.3 proves the statements in (c) of the present theorem with the exception of (6.21) in the final item. The proof of (6.21) is identical to that of (3.12) in Theorem 3.6 except for the weak–form interpretations of the contributions of the nonlinear terms g and G to the rightmost integral in (6.19). For these we observe that if $u_{\mathfrak{m}}$ is any specific representative of the measurable counterpart of u, then for almost all t, $u_{\mathfrak{m}}(\cdot,t)$ is in the Ω-equivalence class $u(t)$, and therefore for $\varphi \in C_c^\infty(\Omega\times(0,\tau))$,

$$\int_0^\tau\int_\Omega g(\cdot,t,u(t))\cdot\varphi = \int_0^\tau\int_\Omega g(\cdot,t,u_{\mathfrak{m}}(\cdot,t))\cdot\varphi = \int_{\Omega\times(0,\tau)} g(\cdot,\cdot,u_{\mathfrak{m}}(\cdot,\cdot))\cdot\varphi,$$

as required for the interpretation in (6.21) in terms of distribution derivatives on $\Omega\times(0,\tau)$ (we have applied here the fact that $g(\cdot,\cdot,u_{\mathfrak{m}}(\cdot,\cdot))\cdot v$ is measurable on $\Omega\times(0,\bar{t})$, which was noted at the very beginning of this proof). The same argument applies to the term involving G. This completes the proof of (c), and (d) is the conclusion (g) of Theorem 6.2 which holds because $r_i < 1$ for all i in case (A), as noted in the first paragraph of this proof, and φ_{ij}^g and φ_i^G have been assumed in case (A) to be in $L^{2,q_{ij}}_{(t)loc}$ and $L^{2,2}_{(t)loc}$ respectively.

There remains to prove (e) of the present theorem. For this we check that the hypothesis in (e) guarantees the condition (6.5) of Theorem 6.2. For the case that $n \ge 3$ we let w, $w' \in H^1$ with L^2 norms bounded by M and write $\Delta w = w' - w$ and $\Delta g = g(\cdot,\cdot,w') - g(\cdot,\cdot,w)$. Then for almost all t the corresponding contribution to the left side of (6.5) is bounded by a sum of terms of the form

$$C\int_\Omega \xi|\Delta w|^2 \le C|\xi|_{L^{q_2}}\Big(\int_\Omega |\Delta w|^{2q_2'}\Big)^{1/q_2'}.$$

The assumption that $q_2 \in (n/2,\infty]$ implies that $2q_2' \in [2,p^*)$; we can therefore interpolate the integral on the right between L^2 and L^{p^*} and apply Theorem A.3(a) to bound the L^{p^*} norm by the H^1 norm. The result is that

$$\int_\Omega \xi|\Delta w|^2 \le C\big(|\xi|_{L^{q_2}}^r|\Delta w|_{L^2}\big)^{2-s}|\Delta w|_{H^1}^s$$

where $s = (2 - 2/q_2')/(1 - 2/p^*)$, which is in $[0, 2)$, and $2r \leq q_1$ by the hypothesis on q_1. This shows that (6.5) holds with $\zeta(t) \equiv |\xi(\cdot, t)|^r_{L^{q'}} \in L^2_{loc}([0, T))$. For $n = 2$ we apply the bound given in Theorem A.4(b) in place of the interpolation; the result is the same, and the argument for G is similar.

□

§6.4 Higher Order Regularity for Quasilinear Systems

In this section we consider problems similar to those posed in §6.3 but with nonlinearities defined only locally in the phase space $\mathbf{R}^N$. Pointwise control of solutions is therefore required, and this suggests that solutions be constructed in the regularity class of Theorem 5.5 with m sufficiently large depending on the dimension n of the physical space.

Specifically, we fix an ensemble (5.4) as in Definition 5.1 with $m > n/2$ and let $T \in (0, \infty]$; then given $\varsigma_0 \in \mathcal{H}_{2m}$ we construct $u : [0, \tau] \subseteq [0, T) \to \mathcal{H}_{2m}$ satisfying

$$\begin{aligned} \langle u(t), v \rangle_{[L^2(\Omega)]^N} &= \langle \varsigma_0, v \rangle_{[L^2(\Omega)]^N} \\ &\quad + \int_0^t \int_\Omega \Big[-(a^{\eta,\omega}(x, t, u) u_{x_\eta}) \cdot v_{x_\omega} + g(x, t, u, \nabla_x u) \cdot v \\ &\qquad + G(x, t, u) : \nabla_x v \Big] \, dx \, dt \end{aligned} \tag{6.27}$$

for all $v \in \mathcal{H}_1$ and $t \leq \tau < T$. We will assume that there are neighborhoods W and W' of zero in $\mathbf{R}^N$ and $\mathbf{R}^{N \times n}$ respectively such that $a^{\eta,\omega}$ and G map $\Omega \times [0, T) \times W$ into $\mathbf{R}^{N \times N}$ and $\mathbf{R}^{N \times n}$ respectively and that g maps $\Omega \times [0, T) \times W \times W'$ into $\mathbf{R}^N$. (A quasilinear boundary term may also be included in (6.27); see Remark 1 below following the statement of Theorem 6.10.)

High–order derivatives of $a^{\eta,\omega}$, g and G will occur, and these are polynomials in the derivatives of the unknown u, complicating the verification of the regularity hypotheses of Theorem 5.5 as well as the description of compatibility conditions. We consider these issues first, then state the main result in Theorem 6.10.

Derivatives of Compositions. The following lemma will facilitate the formulation of the main theorem and will also constitute a significant part of its proof:

Lemma 6.8. *Let Ω be a $C^{0,1}_{bdd}$ domain in $\mathbf{R}^n$, $n \geq 2$, not necessarily bounded* (Definition 1.7) *and I a nontrivial compact interval in $\mathbf{R}$, and let $N \geq 1$ and $m > n/2$. Assume that mappings $w^{(0)} = w, w^{(1)}, \ldots, w^{(m)}$ are given taking values in $[L^2(\Omega)]^N$ and satisfying the following, in which $[H^k(\Omega)]^N$ is abbreviated H^k, etc.:*

$$w^{(j)} \in Lip(I; H^{2m-2j-2}) \cap wAC^2(I; H^{2m-2j-1}) \cap C(I; H^{2m-2j}_{weak})$$

for $j \leq m-1$, $w^{(m)} \in L^\infty(I; L^2) \cap L^2(I; H^1)$, *and the derivative of* $w^{(j)}$ *is the equivalence class over* I *of* $w^{(j+1)}$ *for* $j \leq m-1$. *Assume also that*

$$\sum_{j=0}^{m} \sup_{t \in I} |w^{(j)}(t)|^2_{H^{2m-2j}} \leq M_1 \tag{6.28}$$

and

$$\sum_{j=0}^{m} \int_I |w^{(j)}|^2_{H^{2m-2j+1}} \leq M_2. \tag{6.29}$$

(a) *There is then a bounded continuous function* $w_{\mathbb{C}} \in C(\overline{\Omega} \times I)$ *such that* $w_{\mathbb{C}}(\cdot, t)$ *is a member of the* Ω*–equivalence class* $w(t)$ *for each* t *and the calculus derivatives* $D_x^\alpha D_t^j w_{\mathbb{C}}$ *exist on* $\Omega \times I^\circ$ *for* $2j + |\alpha| \leq 2$ *and have bounded, Hölder continuous extensions to* $\overline{\Omega} \times I$. *The images of* $w_{\mathbb{C}}$ *and* $\nabla_x w$ *are therefore contained in compact sets* K *and* K' *respectively.*

Now assume further that K *and* K' *contain the origins in* $\mathbf{R}^N$ *and* $\mathbf{R}^{N \times n}$ *respectively. The following results then hold for compositions involving involving* w *and* $\nabla_x w$:

(b) *Let* W *be a neighborhood of* K *and assume that a real-valued mapping* $f \in C^{2m+1}_{bdd}(\Omega \times I^\circ \times W)$ *is given whose derivatives have extensions to* $\overline{\Omega} \times I \times K$ *which are Lipschitz with respect to the variable in* K. *Let* $\tilde{f}(t)$ *be the* Ω*-equivalence class* $\tilde{f}(\cdot, t, w(t))$ *as* $w(t)$ *ranges over its* Ω*-equivalence class. Then there are maps* $\tilde{f}^{(0)}, \ldots, \tilde{f}^{(m)}$ *such that* $\tilde{f}^{(0)} = \tilde{f}$, *the equivalence class over* I *of* $\tilde{f}^{(j)}$ *is in* $L^\infty(I; H^{2m-2j}) \cap L^2(I; H^{2m-2j+1})$ *for* $j \leq m$, *and for* $j \leq m-1$,

$$\tilde{f}^{(j)} \in Lip(I; H^{2m-2j-2}) \cap wAC^2(I; H^{2m-2j-1}) \cap C(I; H^{2m-2j}_{weak})$$

with derivative the equivalence class over I *of* $\tilde{f}^{(j+1)}$. *In addition*

$$\sum_{j=0}^{m} \sup_{t \in I} |\tilde{f}^{(j)}(t)|^2_{H^{2m-2j}} \leq \sum_{j=0}^{m} \sup_{t \in I} |(D_t^j f)(\cdot, t, 0)|^2_{H^{2m-2j}} + C(M_1 + \cdots + M_1^{2m}) \tag{6.30}$$

and

$$\sum_{j=0}^{m} \int_I |\tilde{f}^{(j)}(t)|^2_{H^{2m-2j+1}} \leq \sum_{j=0}^{m} \int_I |(D_t^j f)(\cdot, t, 0)|^2_{H^{2m-2j+1}} dt + C(1 + M_1 + \cdots + M_1^{2m}) M_2 \tag{6.31}$$

where the sums on the right are assumed to be bounded and the constant C *depends on the assumed bounds and Lipschitz constants for the derivatives of* f *on* $\overline{\Omega} \times I \times K$ *and on the norm of the imbedding from* H^m *into* L^∞ (Theorem A.3(a)). *Finally, for* $j \leq m$, $\tilde{f}^{(j)}$ *is a polynomial in* $w^{(1)}, \ldots, w^{(j)}$ *whose coefficients are multiples of partial derivatives of* f *evaluated at* $(\cdot, \cdot, w(\cdot))$, *these coefficients being continuous and bounded on*

$\overline{\Omega}\times I$, *and in which the corresponding monomials are in* $L^\infty(I;H^{2m-2j})\cap L^2(I;H^{2m-2j+1})$, *modulo equivalence classes over* I.

(c) *Next let* W' *be a neighborhood of* K' *in* $\mathbf{R}^{N\times n}$ *and assume that a real-valued mapping* $h\in C^{2m}_{bdd}(\Omega\times I^\circ\times W\times W')$ *is given whose derivatives have continuous extensions to* $\overline{\Omega}\times I\times K\times K'$ *which are bounded, continuous, and Lipschitz with respect to the variable in* $K\times K'$. *Let* $\tilde h(t)$ *be the* Ω*-equivalence class* $h(\cdot,t,w(t),\nabla_x w(t))$ *as* $(w(t),\nabla_x w(t))$ *ranges over its* Ω*-equivalence class. Then there are maps* $\tilde h^{(0)},\dots,\tilde h^{(m)}$ *such that* $\tilde h^{(0)}=\tilde h$, *the equivalence class over* I *of* $\tilde h^{(j)}$ *is in* $L^2(I;H^{2m-2j})$ *for* $j\le m$, *and for* $j\le m-1$,

$$\tilde h^{(j)}\in wAC^2(I;H^{2m-2j-2})\cap C(I;H^{2m-2j-1}_{weak})$$

with derivative the equivalence class of $h^{(j+1)}$. *In addition,*

$$\sum_{j=0}^{m-1}\sup_I|\tilde h^{(j)}(t)|^2_{H^{2m-2j-1}}\le\sum_{j=0}^{m-1}\sup_{t\in I}|(D_t^j h)(\cdot,t,0,0)|^2_{H^{2m-2j-1}} + C(M_1+\cdots+M_1^{2m-1}) \tag{6.32}$$

and

$$\sum_{j=0}^{m}\int_I|\tilde h^{(j)}(t)|^2_{H^{2m-2j}}\le\sum_{j=0}^{m}\int_I|(D_t^j h)(\cdot,t,0,0)|^2_{H^{2m-2j}}dt + C(1+M_1+\cdots+M_1^{2m-1})M_2 \tag{6.33}$$

where the constant C *depends on the assumed bounds and Lipschitz constants for the derivatives of* h *on* $\overline{\Omega}\times I^\circ\times K\times K'$ *and on the norms of the imbeddings from* H^m *into* L^∞ *and* H^{m-1} *into* L^4 (Theorem A.3(a)). *Finally, for* $j\le m$, $\tilde h^{(j)}$ *is a polynomial in* $(w^{(1)},\nabla_x w^{(1)}),\dots,(w^{(j)},\nabla_x w^{(j)})$ *whose coefficients are multiples of partial derivatives of* h *evaluated at* $(\cdot,\cdot,w,\nabla_x w)$, *these coefficients being continuous and bounded on* $\overline{\Omega}\times I$, *and in which the corresponding monomials are in* $L^\infty(I;H^{2m-2j-1})\cap L^2(I;H^{2m-2j})$.

Proof. The Hölder continuity and boundedness statements in (a) follow exactly as in the proof of Theorem 5.5(*ci*) and the fact that $m>n/2$, which implies that, in the notation of the latter theorem, $2j_0+\kappa\ge 2$.

We will give the proof of (c), the proof of (b) being similar and somewhat easier. It will be convenient to work at the level of measurable counterparts, applying Theorems A.23 and A.24 and Corollary A.25 to transfer information between the Bochner space statements of the theorem and their weak derivative formulations.

The reader can easily check that $\tilde h_{\mathrm{m}}(\cdot,\cdot)$ is the $\Omega\times I$ equivalence class $h(\cdot,\cdot,w_{\mathrm{m}},\nabla_x w_{\mathrm{m}})$ as w_{m} ranges over its equivalence class. We can therefore fix specific representatives and apply the chain rule ([**19**], Lemma 7.5) to check by induction that

$$
\begin{aligned}
D_t^j D_x^\alpha \tilde{h}_{\mathfrak{m}}(\cdot,\cdot) &= \big(D_t^j D_x^\alpha h\big)(\cdot,\cdot,(0,0)) + \big(D_t^j D_x^\alpha h\big)(\cdot,\cdot,(\cdot))\Big|_{(0,0)}^{(w_{\mathfrak{m}},\nabla_x w_{\mathfrak{m}})} \\
&\quad + \sum \big(\,*\,\big)\big(D_t^{j_1} D_x^{\alpha_1+\sigma_1} w_{\mathfrak{m}}\big)\cdots\big(D_t^{j_l} D_x^{\alpha_l+\sigma_l} w_{\mathfrak{m}}\big)
\end{aligned} \tag{6.34}
$$

where $j_1+\cdots+j_l \le j$, $\alpha_1+\cdots+\alpha_l \le \alpha$, $\sigma_i \in \{0,1\}$, and $\big(\,*\,\big)$ denotes a partial derivative of h evaluated at $(\cdot,\cdot,(w_{\mathfrak{m}},\nabla_x w_{\mathfrak{m}}))$ which by hypothesis is bounded a.e. in $\Omega \times I$ provided $2j+|\alpha| \le 2m$ (the sum is absent if $j=|\alpha|=0$). We need to bound the terms on the right in $L^{\infty,2}$ for $j \le m-1$ and $|\alpha| \le 2m-2j-1$, and in $L^{2,2}$ for $j \le m$ and $|\alpha| \le 2m-2j$. Bounds for the first term on the right are recorded in (6.32) and (6.33), and the second term is bounded pointwise by $C(|w_{\mathfrak{m}}|+|\nabla_x w_{\mathfrak{m}}|)$ by the assumed Lipschitz continuity, and is therefore bounded in $L^{\infty,2}$ and $L^{2,2}$ as required.

Concerning the product Π in the sum on the right in (6.34) we first note that if $2j_i+|\alpha_i|+\sigma_i \le m$, then the corresponding factor $D_t^{j_i} D_x^{\alpha_i+\sigma_i} w_{\mathfrak{m}}$ is in $L^{\infty,\infty}$ because in this case $D_t^{j_i} D_x^{\alpha_i+\sigma_i} w_{\mathfrak{m}}(\cdot,t) \in H^m$ with bound $M_1^{1/2}$ for almost all t by (6.28), and therefore is in L^∞ with bound $CM_1^{1/2}$ by Theorem A.3(a). Now to prove (6.32) we suppose that $2j+|\alpha| \le 2m-1$ and observe that there is then at most one i for which $2j_i+|\alpha_i|+\sigma_i \ge m+1$, so that $\Pi \in L^{\infty,\infty}\cdots L^{\infty,2} \subseteq L^{\infty,2}$ with the bound $C(M_1+\ldots+M_1^{2m-1})^{1/2}$, proving (6.32). To prove (6.33) we suppose that $2j+|\alpha| \le 2m$. If the condition $2j_i+|\alpha_i|+\sigma_i \le m$ fails for at most one value of i, then at most one factor in Π fails to be in $L^{\infty,\infty}$, and therefore $\Pi \in L^{\infty,\infty}\cdots L^{2,2} \subseteq L^{2,2}$ with the bound in (6.33). Otherwise $l=2$ and $2j_i+|\alpha_i|+\sigma_i = m+1$ for $i=1,2$. In this case $D_t^{j_i} D_x^{\alpha_i+\sigma_i} w_{\mathfrak{m}}(\cdot,t) \in H^{m-1} \subset L^4$ for almost all t with bound $CM_1^{1/2}$, and $|\Pi|_{L^{2,2}}^2 \le CM_1M_2$. This proves (6.33), and the corresponding bounds in (6.30) and (6.31) in (b) are proved in a similar way.

The final statements in (b) and (c) regarding $f^{(j)}$ and $h^{(j)}$ are easily deduced from Theorems A.23 and A.24 as in the proof of Corollary A.25.

□

Compatibility Conditions. If solutions of (6.27) are to be obtained in the regularity class of Theorem 5.5 then the compatibility conditions(5.23) must be satisfied. It is important that these be formulated solely in terms of the ς_j appearing in (5.23), independent of properties of a putative solution u. To develop a suitable notation we first illustrate the assertion in (a) above concerning polynomial structure, say of $\tilde{f}^{(2)}$:

$$
\tilde{f}^{(2)} = f_{tt} + 2f_{tw}\cdot w^{(1)} + (f_{ww}w^{(1)})\cdot w^{(1)} + f_w \cdot w^{(2)}
$$

where subscripts denote partial derivatives evaluated at $(\cdot,\cdot,w)$, f_w and f_{tw} are $N\times 1$ and f_{ww} is $N\times N$. The expression on the right is indeed a polynomial in $w^{(1)}, w^{(2)}$ and will be denoted by $f_2(\cdot,t,w(t))[w^{(1)},w^{(2)}]$. More generally $\tilde{f}^{(j)}(t)$ will be denoted $f_j(\cdot,t,w(t))[w^{(1)}(t),\ldots,w^{(j)}(t)]$, so that f_j is a mapping from $\Omega\times I\times W$ into the set of polynomials of degree j over $\mathbf{R}^N$

(the brackets are omitted if $j = 0$). Thus whereas the notation $\tilde{f}^{(j)}$ suppresses the dependence on w, the more satisfactory notation $f_j(\cdot,\cdot,w)[w^{1)},\ldots,w^{(j)}]$ makes clear that w is an *argument* of f_j and that the mapping f_j itself is independent of w. Applying this notation to the components of $a^{\eta,\omega}$ and G we obtain matrices denoted $(a^{\eta,\omega})_j$ and G_j whose components are mappings from $[0,T)\times\Omega\times W$ into the set of polynomials of degree j on $\mathbf{R}^N$, and to the components of g to obtain $N\times 1$ arrays g_j of mappings from $[0,T)\times\Omega\times W\times W'$ into the set of polynomials of degree j on $\mathbf{R}^N\times\mathbf{R}^{N\times n}$.

Now suppose that u is a solution of (6.27) in the sense of Theorem 5.5 on an interval $[0,\tau]$. That is, $u=u^{(0)},\ldots,u^{(m)}$ satisfy the regularity conditions stated in Theorem 5.5(a) and for $v\in\mathcal{H}_1$ and $t\in[0,\tau]$,

$$\langle u(t),v\rangle_{L^2(\Omega)} = \langle\varsigma_0,v\rangle_{L^2(\Omega)} + \int_0^t\int_\Omega \Big[-(\tilde{a}^{\eta,\omega}u_{x_\eta})\cdot v_{x_\omega} + \tilde{g}\cdot v + \tilde{G}:\nabla_x v\Big]\,dx\,ds$$

where $\tilde{a}^{\eta,\omega}(x,s)=a(x,s,u(x,s))$ and similarly for $\tilde{g}$ and $\tilde{G}$ (we have replaced the mapping w in Lemma 6.8 by u). The compatibility conditions (5.23) must then be satisfied: if $\varsigma_j\equiv u^{(j)}(0)$ then for $j\le m-1$ and $v\in\mathcal{H}_1$,

$$\langle\varsigma_{j+1},v\rangle_{L^2} = \int_\Omega\Big[-\sum_{i=0}^{j}\binom{j}{i}\big((\tilde{a}^{\eta,\omega})^{(i)}(x,0)(\varsigma_{j-i})_{x_\eta}\big)\cdot v_{x_\omega} + \tilde{g}^{(j)}(x,0)\cdot v + \tilde{G}^{(j)}(x,0):\nabla v\Big]\,dx.$$

In light of the above discussion these conditions can be be written

$$\begin{aligned}\langle\varsigma_{j+1},v\rangle_{L^2} = \int_\Omega\Big[&-\sum_{i=0}^{j}\binom{j}{i}\big(a_i^{\eta,\omega}(\cdot,0,\varsigma_0)[\varsigma_1,\ldots,\varsigma_i](\varsigma_{j-i})_{x_\eta}\big)\cdot v_{x_\omega}\\ &+\Big(g_j(\cdot,0,\varsigma_0,\nabla_x\varsigma_0)[(\varsigma_1,\nabla_x\varsigma_1),\ldots,(\varsigma_j,\nabla\varsigma_j)]\Big)\cdot v\\ &+\Big(G_j(\cdot,0,\varsigma_0)[\varsigma_1,\ldots,\varsigma_j]\Big):\nabla_x v\Big]\,dx\end{aligned}\tag{6.35}$$

(again, the brackets are omitted if i or j is zero) which, while somewhat more complicated, involve only the initial values $\varsigma_0,\ldots,\varsigma_m$ and the partial derivatives of $a^{\eta,\omega}$, g and G evaluated at ς_0 or $(\varsigma_0,\nabla\varsigma_0)$, as intended.

It will be convenient in what follows to adopt the following terminology:

Definition 6.9. *Fix an ensemble* (5.4) *as in* Definition 5.1 *and a nontrivial interval I in* $\mathbf{R}$. *A mapping $w: I\to\mathcal{H}_{2m}$ is m-regular on I with respect to the spaces $\mathcal{H}_0,\ldots,\mathcal{H}_{2m+1}$ if there are mappings $w^{(0)}=w,\ldots,w^{(m)}$ satisfying the following four conditions: for $j=0,\ldots,m-1$,*

$$w^{(j)}\in Lip(I;\mathcal{H}_{2m-2j-2})\cap wAC^2(I;\mathcal{H}_{2m-2j-1})\cap C(I;\mathcal{H}^{weak}_{2m-2j})$$

and $w^{(m)} \in C(I;\mathcal{H}_0)$; *for* $j = 0,\ldots,2m$ *the equivalence class over* I *of* $w^{(j)}$ *is in* $L^2(I;\mathcal{H}_{2m-2j+1})$; *the derivative of* $w^{(j)}$ *for* $j < m$ *is the equivalence class over* I *of* $w^{(j+1)}$; *and* $(w^{(m)})^* \equiv \langle w^{(m)},\cdot\rangle_{\mathcal{H}_0} \in wAC^2(I;\mathcal{H}_1^*)$.

For example, the solution described in Theorem 5.5 is m-regular on $[0,\bar{t}]$ for every $\bar{t} \in (0,T)$ with respect to the particular spaces $\mathcal{H}_j = \mathcal{H}_1 \cap [H^j(\Omega)]^N$, the final condition being a consequence (5.25) with $j = m$. Observe that the functions considered in Lemma 6.8 satisfy all but this final condition and thus narrowly fail to be m-regular.

We can now state the main result of this section:

Theorem 6.10. *Fix an ensemble* (5.4) *and constant* C_{BC} *as in* Definition 5.1 *with* $m > n/2$, *let* W *and* W' *be neighborhoods of the origins in* $\mathbf{R}^N$ *and* $\mathbf{R}^{N\times n}$ *respectively and fix* $\bar{t} > 0$. *Assume the following:*

(a) *Mappings* $a^{\eta,\omega}$ *and* G *in* $C^{2m+1}_{\mathrm{bdd}}(\Omega \times (0,\bar{t}) \times W)$ *are given taking values in* $\mathbf{R}^{N\times N}$ *and* $\mathbf{R}^{N\times n}$ *respectively with derivatives having continuous extensions to* $\overline{\Omega} \times [0,\bar{t}] \times K$ *for every compact* $K \subset W$ *which are bounded, continuous and Lipschitz with respect to the variable in* K. *And a mapping* g *in* $C^{2m}_{\mathrm{bdd}}(\Omega \times (0,\bar{t}) \times W \times W')$ *is given taking values in* $\mathbf{R}^N$ *whose derivatives have continuous extensions* $\overline{\Omega} \times [0,\bar{t}] \times K \times K'$ *for all compact* $K \subset W$ *and compact* $K' \subset W'$ *which are bounded, continuous and Lipschitz with respect to the variable in* $K \times K'$.

    The derivatives and Lipschitz constants included here are assumed to be bounded by a constant $C_{sys}(K,K')$.

(b) *The following regularity conditions are satisfied* (*in which* $j \geq 0$ *in each instance and* $L^{p,2}$ *denotes* $[L^{p,2}(\Omega \times [0,\bar{t}])]^{r\times s}$ *for various* p, r *and* s):

$$D_t^j D_x^\alpha g_{\mathfrak{m}}(\cdot,\cdot,0,0) \in \begin{cases} L^{\infty,2}, & j \leq m-1,\ 2j + |\alpha| \leq 2m-1, \\ L^{2,2}, & j \leq m,\ 2j + |\alpha| \leq 2m; \end{cases} \tag{6.36}$$

$$D_t^j D_x^\alpha G_{\mathfrak{m}}(\cdot,\cdot,0) \in \begin{cases} L^{\infty,2}, & j \leq m,\ 2j + |\alpha| \leq 2m, \\ L^{2,2}, & j \leq m,\ 0 \leq 2j + |\alpha| \leq 2m+1; \end{cases} \tag{6.37}$$

and for $(j,\alpha) \neq (0,0)$,

$$D_t^j D_x^\alpha a_{\mathfrak{m}}^{\eta,\omega}(\cdot,\cdot,0) \in \begin{cases} L^{\infty,2}, & j \leq m,\ 2j + |\alpha| \leq 2m, \\ L^{2,2}, & j \leq m,\ 0 \leq 2j + |\alpha| \leq 2m+1. \end{cases} \tag{6.38}$$

    The norms of the derivatives in (6.36)–(6.38), *although independent of* K *and* K', *are assumed to be bounded by the aforementioned* $C_{sys}(K,K')$ *for any compact subsets* K *and* K' *of* W *and* W'.

(c) *There is a positive number* θ *such that for all* $(x,t,w) \in \Omega \times [0,\bar{t}] \times W$,

$$(M^{tr} a_{\mathfrak{m}}^{\eta,\omega}(x,t,w)M)^{\omega,\eta} \geq 2\theta(\bar{t})|M|^2 \tag{6.39}$$

for all $N \times n$ matrices M.

(d) *Initial data $(\varsigma_0, \dots, \varsigma_m) \in \mathcal{H}_{2m} \times \dots \times \mathcal{H}_0$ is given satisfying the compatibility conditions* (6.35), *and the continuous counterparts of ς_0 and $\nabla_x \varsigma_0$* (*see* Lemma 6.8(a)) *take values in compact sets K and K' a positive distance ε from ∂W and $\partial W'$ respectively* (*see also* Remark 2 *below concerning this hypothesis*).

(e) *There is an m-regular function $u_{data} : [0, \bar{t}] \to \mathcal{H}_{2m}$ such that: $u^{(j)}_{data}(0) = \varsigma_j$ for $j \leq m$, $(u_{data}, \nabla_x u_{data})$ takes values in $W \times W'$ on $\overline{\Omega} \times [0, \bar{t}]$, and $|u_{data}(t) - \varsigma_0|^2_{H^{2m}} \leq \delta$ for $t \in [0, \bar{t}]$ where $\delta > 0$ is chosen so that if $z \in H^{2m}$ with $|z|^2_{[H^{2m}(\Omega)]^N} \leq \delta$ then $|z|_{[L^\infty(\Omega)]^N}, |\nabla_x z|_{[L^\infty(\Omega)]^{N \times n}} \leq \frac{1}{2}\varepsilon$* (*see* Remark 3 *below concerning this hypothesis*).

There is then an m-regular solution u to the problem (6.27) *on an interval $[0, \tau]$, described in terms of constants C_{solve} and C_{time} as follows:*

- *C_{solve} and C_{time} depend on n, N, m and on upper bounds for C_{BC}, $C_{sys}(K_{\varepsilon/2}, K'_{\varepsilon/2})$, where $K_{\varepsilon/2}$ is the closure of $K + B_{\varepsilon/2}(0)$ and similarly for $K'_{\varepsilon/2}$, and on upper bounds for θ^{-1}, ε^{-1} and $\sum_j |\varsigma_j|^2_{\mathcal{H}_{2m-2j}}$* (*$C_{solve}$ and C_{time} are otherwise independent of the particular solution u and its data*)*; and*

$$\tau \geq \min\left\{\bar{t},\, C_{time}^{-1},\, \sup_{\zeta \in \mathcal{H}_1} \frac{\delta - C_{time}|\varsigma_m - \zeta|^2_{L^2}}{|\zeta|^2_{\mathcal{H}_1}}\right\}. \tag{6.40}$$

- *u is m-regular on $[0, \tau]$ with respect to the spaces $\mathcal{H}_0, \dots, \mathcal{H}_{2m+1}$; the compositions $g(\cdot, \cdot, u, \nabla_x u)$, $G(\cdot, \cdot, u)$ and $a^{\eta,\omega}(\cdot, \cdot, u)$ satisfy the hypotheses of* Theorem 5.5*; and u is a solution of* (6.27) *on $[0, \tau]$ and satisfyies the conclusions in* (a)–(c) *of* Theorem 5.5 *with*

$$A(t)(z, v) = \int_\Omega \left[(a^{\eta,\omega}(x, t, u) z_{x_\eta}\right) \cdot v_{x_\omega} dx$$

$$F(t) \cdot v = \int_\Omega \left[g(x, t, u, \nabla_x u) \cdot v + G(x, t, u) : \nabla_x v\right] dx$$

$$L(t)[z] = \left(a^{\eta,\omega}(x, t, u(x, t)) z_{x_\eta}\right)_{x_\omega}$$

and if $\Omega \neq \mathbf{R}^n$,

$$L_\partial(t)[z] = \nu^\omega \mathcal{T}\left[a^{\eta,\omega}(\cdot, t, u(\cdot, t)) z_{x_\eta}\right]$$

where ν is the unit outer normal on $\partial\Omega$ and $\mathcal{T}$ is the trace operator.

- *The following bounds are saisfied:*

$$\sum_{j=0}^{m}\Big(\sup_{0\le t\le\tau}|u^{(j)}(t)|^2_{H^{2m-2j}}+\int_0^\tau|u^{(j)}(t)|^2_{H^{2m-2j+1}}dt\Big)$$
$$\le C_{solve}\Big[\sum_{j=0}^{m}|\varsigma_j|^2_{H^{2m-2j}}+\sum_{0}^{m-1}\big(|\gamma_j^g|^2_{H^{2m-2j-1}}+|\gamma_j^G|^2_{H^{2m-2j}}\big)\Big] \tag{6.41}$$

where $\gamma_j^g=g_j(\cdot,0,\varsigma_0,\nabla_x\varsigma_0)\big[(\varsigma_1,\nabla_x\varsigma_1),\ldots\big]$, $\gamma_j^G=G_j(\cdot,0,\varsigma_0)\big[\varsigma_1,\ldots\big]$ *(see* (6.35)*) and* $[H^k(\Omega)]^N$ *is abbreviated* H^k *etc.*
- *For* $j=0,\ldots,m$, $u^{(j)}\in C([0,\tau];\mathcal{H}_{2m-2j})$, *that is,* $\mathcal{H}_{2m-2j}$ *with the norm topology.*

The solution u *is unique in the following sense: if* $u'\in C([0,\tau'];\mathcal{H}_0)$ *with equivalence class in* $L^2([0,\tau'];\mathcal{H}_1)$ *and with* $(u'_{\mathrm{m}},\nabla_x u'_{\mathrm{m}})$ *taking values in* $W\times W'$ *a.e., and if* (6.27) *holds for all* $v\in\mathcal{H}_1$ *and* $t\in[0,\tau']$ *with* u *replaced by* u', *then* $u'=u$ *on their common domain (see also* Remark 4 *below).*

Remarks:

(1) A quasilinear boundary term $\int_0^t\int_{\partial\Omega}f(x,t,u)\cdot v\,d\sigma dt$ may be included in (6.27) under suitable hypotheses on f: the proof requires that if w is in the regularity class under consideration and if S^1 and S^2 are the functionals defined in Theorem A.26, then $S^1(f(\cdot,\cdot,w))$ and $S^2(f(\cdot,\cdot,w))$ should satisfy the same regularity conditions as those satisfied respectively by $g(\cdot,\cdot,w,\nabla_x w)$ and $G(\cdot,\cdot,w)$. This allows for absorption of the boundary integral into the integral over Ω, exactly as in the proofs of Theorems 4.2 and 5.5. It suffices that f have derivatives up to order $2m+1$ on a neighborhood of $\partial\Omega\times[0,\bar{t}]\times\overline{W}$ which are bounded, continuous and Lipschitz with respect to the variable in W. Details and proofs are left to the interested reader.

(2) If either W or W' is the entire space, then the corresponding condition in (d) on the distance from the image of ς_0 or $\nabla_x\varsigma_0$ to ∂W or $\partial W'$ is omitted. If both, then the dependence of constants on ε^{-1} is omitted as is the sup on the right side of (6.40).

(3) Hypothesis (e) is obviously necessary for the validity of the theorem since the solution itself, if one exists, satisfies (e). We list here four important cases in which a suitable u_{data} is known to exist; a construction for the general case of the theorem remains elusive, however:

- $\Omega=\mathbf{R}^n$: the proof is given in Exercise 6.4;
- $\varsigma_j\in\mathcal{H}_{2m+1}$ for all j: in this case the obvious Taylor polynomial in t satisfies the required conditions;

- there is a time-independent bilinear form A on $\mathcal{H}_1 \times \mathcal{H}_1$ satisfying the hypotheses of Theorem 5.5 with corresponding differential operators L and $L_{\partial\Omega}$ as defined in section 3.1 such that $L^i\varsigma_j$ and $L_{\partial\Omega}L^i\varsigma_j$ satisfy essential and natural boundary conditions for certain i and j; the construction of u_{data} for this case is by induction on m and is outlined in Exercise 6.5;
- $n = 2$ or 3, $m = 2$, $a^{\eta,\omega} = a^{\eta,\omega}(x,t)$, $g = g(x,t,u)$, and G is absent: the proof, which is outlined Exercise 6.6, is somewhat lengthy and depends on the relative strength of the Sobolev imbeddings $H^k \to L^p$ in low dimensions.

(4) Concerning the final statement of the theorem, we leave to the reader to check that the given hypotheses on u' are sufficient to insure that the terms in (6.27) are well-defined with u' in place of u.

PROOF. We give the proof for the case that the boundaries of W and W' are nonempty, the minor modifications required for the other cases being obvious. Throughout the proof we abbreviate $[H^k(\Omega)]^N$ by H^k, etc., and $L^p([0,\tau]; H^k)$ by $L^p(H^k)$ when the value of τ is clear from context.

Let C_0 be the term in brackets on the right side of (6.41):

$$C_0 = \sum_{j=0}^{m} |\varsigma_j|^2_{H^{2m-2j}} + \sum_{0}^{m-1} \left(|\gamma_j^g|^2_{H^{2m-2j-1}} + |\gamma_j^G|^2_{H^{2m-2j}}\right),$$

and define

$$M_1 = 2(C_0 + \delta).$$

Note that C_0 and δ are bounded in terms of the same quantities as C_{solve} and C_{time} by Lemma 6.8 and hypothesis (b); the same is therefore true for M_1.

Step 1. Preliminary bounds. Let $\tau \in (0, \bar{t}]$ and $w = w^{(0)}, \ldots, w^{(m)}$ be as in the hypotheses of Lemma 6.8 on $I = [0,\tau]$ with $w^{(j)}(0) = \varsigma_j$ for $j = 0, \ldots, m$. Assume that (6.28) holds with the above M_1, that (6.29) holds with some finite M_2 and that the continuous representatives of w and $\nabla_x w$ take values in $K_{\varepsilon/2}$ and $K'_{\varepsilon/2}$ respectively. Define $\tilde{a}^{\eta,\omega}(t) = a(\cdot, t, w(t))$, $\tilde{g}(t) = g(\cdot, t, w(t)), \nabla_x w(t))$ and $\tilde{G}(t) = G(\cdot, t, w(t))$ for $t \in I$. Then:

There are maps $(\tilde{a}^{\eta,\omega})^{(0)}, \ldots, (\tilde{a}^{\eta,\omega})^{(m)}$ with $(\tilde{a}^{\eta,\omega})^{(0)}(t) = a^{\eta,\omega}(\cdot, t, w(t))$ and for which the associated bilinear forms $\tilde{A}^{(0)}, \ldots, \tilde{A}^{(m)}$ defined by

$$\tilde{A}^{(j)}(t)(z,v) = \int_\Omega \left((\tilde{a}^{\eta,\omega})^{(j)}(t) z_{x_\eta}\right) \cdot v_{x_\omega}$$

satisfy the hypotheses of Theorem 2.7 *and therefore of* Theorem 5.5 *on I.*

The proof consists in first applying Lemma 6.8(b) to $a(\cdot,\cdot,w) - a(\cdot,\cdot,0)$ to show that for $0 < 2j + |\alpha| \le 2m$, $D_x^\alpha \tilde{a}^{(j)}(\cdot)$ is bounded in $L^\infty([0,\tau]); L^2)$ (modulo equivalence class) by a constant depending on the same quantities as for

$C_{sys}(K_{\varepsilon/2}, K'_{\varepsilon/2})$ and on M_1, and then applying these bounds in Lemma 5.3. The details are straightforward and left to the reader. It follows that solutions of corresponding linear problems (2.3) as described in Theorem 2.7 satisfy the bound in (2.28) and that the multiplier $C(\bar{t})\exp\left(\int_0^{\bar{t}} \beta^2\right)$, which we denote by $\frac{1}{2}C_{solve}$, is as described in the statement of the present theorem and does not depend on M_2.

Now define

$$M_2 = C_{solve}(K_{\varepsilon/2}, K'_{\varepsilon/2})C_0$$

and assume that w also satisfies (6.29) with this value of M_2. We claim that:

There are mappings $\tilde{g}^{(0)}, \ldots, \tilde{g}^{(m)}$ and $\tilde{G}^{(0)}, \ldots, \tilde{G}^{(m)}$ with $\tilde{g}^{(0)}(t) = g(\cdot, t, w(t), \nabla_x w(t))$ and $\tilde{G}^{(0)}(t) = G(\cdot, t, w(t))$ and for which the associated functionals $\tilde{F}^{(0)}, \ldots, \tilde{F}^{(m)}$ defined by

$$\tilde{F}^{(j)}(t) \cdot v = \int_\Omega \left(\tilde{g}^{(j)}(t) \cdot v + \tilde{G}^{(j)}(t) : \nabla_x v\right)$$

satisfy the hypotheses of Theorem 2.7 *and therefore of* Theorem 5.5. *In particular,*

$$\begin{aligned}(6.42)\qquad &\sum_{j=0}^{m-1} |\tilde{g}^{(j)}(t)|^2_{L^\infty(H^{2m-2j-1})} + \sum_{j=0}^{m} |\tilde{G}^{(j)}(t)|^2_{L^\infty(H^{2m-2j})} \\ &\quad + \sum_{j=0}^{m} \left(|\tilde{g}^{(j)}(t)|^2_{L^2(H^{2m-2j})} + |\tilde{G}^{(j)}(t)|^2_{L^2(H^{2m-2j+1})}\right) \\ &\quad \le C,\end{aligned}$$

and as a consequence

$$\begin{aligned}(6.43)\qquad &\sum_{j=0}^{m-1} |\tilde{F}^{(j)}|^2_{L^\infty(\mathcal{V}_{2m-2j-2})} + \sum_{j=0}^{m} |\tilde{F}^{(j)}|^2_{L^2(\mathcal{V}_{2m-2j-1})} \\ &\qquad \le \sum_{0}^{m-1} \left(|\gamma_j^g|^2_{H^{2m-2j-1}} + |\gamma_j^G|^2_{H^{2m-2j}}\right) + C\tau^{1/2},\end{aligned}$$

where γ_j^g and γ_j^G are as defined following (6.41) *and C is a generic positive constant which depends on the same quantities as C_{solve}.*

For the proof we first note that the statements concerning $\tilde{g}^{(j)}$ and $\tilde{G}^{(j)}$ and the bounds in (6.42) are immediate from Lemma 6.8 and that for (6.43) it suffices to prove that

$$\sum_0^{m-1}\Big(|\tilde g^{(j)}|^2_{L^\infty(H^{2m-2j-2})}+|\tilde G^{(j)}|^2_{L^\infty(H^{2m-2j-1})}+|\tilde g^{(j)}|^2_{L^2(H^{2m-2j-1})}\Big)$$

$$+|(g^{(m)})^*|^2_{L^2(\mathcal{H}_1^*)}+\sum_0^m|\tilde G^{(j)}|^2_{L^2(H^{2m-2j})}$$

$$\le\sum_0^{m-1}\big(|\gamma_j^g|^2_{H^{2m-2j-1}}+|\gamma_j^G|^2_{H^{2m-2j}}\big)+C\tau^{1/2}.$$

We label the terms on the left (i)-(v). First, (6.42) shows that (iii) and (v) are bounded by $C\tau$. To bound (iv) we fix $v\in\mathcal{H}_1$ and $t\in[0,\tau]$ and observe that the highest order term in the expansion (corresponding to (6.34)) of $\langle\tilde g^{(m)}(t),v\rangle_{L^2}$ has the form $\int_\Omega k(x,t,w(t),\nabla_x w(t))\cdot\big(\nabla_x w^{(m)}(t)\,v\big)$ where k is a partial derivative of g of order one. We apply the divergence theorem and the fact that k and $D_x^2w(t)$ are bounded in $L^\infty(\Omega)$ (since $D_x^2w(t)\in H^{2m-2}\subseteq H^m\subset L^\infty$) to bound this integral by

$$C\Big(|w^{(m)}(t)|_{L^2(\partial\Omega)}+|w^{(m)}(t)|_{L^2(\Omega)}\Big)|v|_{\mathcal{H}_1}$$

$$\le C\Big(|w^{(m)}(t)|^{1/2}_{L^2(\Omega)}|w^{(m)}(t)|^{1/2}_{H^1(\Omega)}+|w^{(m)}(t)|_{L^2(\Omega)}\Big)|v|_{\mathcal{H}_1}$$

(we have applied the trace theorem for $W^{1,1}$ integrands to $|w^{(m)}|^2$). Thus $|(\tilde g^{(m)})^*(t)|_{\mathcal{H}_1^*}\le C(1+|w^{(m)}(t)|^{1/2}_{\mathcal{H}_1})$ and (iv) is bounded by $C\tau^{1/2}$. We record this result for later reference, including a bound for $\tilde G$ from (6.42):

$$|(\tilde g^{(m)})^*|^2_{L^2(\mathcal{H}_1^*)},\ |\tilde G^{(m)}|^2_{L^2(L^2)}\le C\tau^{1/2}. \tag{6.44}$$

Next we examine term (i). By considering actions on test functions we find that for $j\le m-1$ and $|\alpha|\le 2m-2j-2$ and for all $t\in[0,\tau]$,

$$|D_x^\alpha\tilde g^{(j)}(t)|_{L^2}\le|g_j(\cdot,0,\varsigma_0,\nabla_x\varsigma_0)[\varsigma_1,\dots,\varsigma_j]|_{H^{2m-2j-2}}$$

$$+\int_0^t|D_x^\alpha\tilde g^{(j+1)}|_{L^2}.$$

The integral here is bounded by $C\tau^{1/2}$ by (6.42). Squaring and summing over α we then obtain

$$|\tilde g^{(j)}(t)|^2_{H^{2m-2j-2}}\le|\gamma_j^g|^2_{H^{2m-2j-2}}+C\tau$$

where γ_j^g is as in (6.41). A similar argument applies to $\tilde G^{(j)}$ in term (ii). Summing over j we then find that terms (i)–(v) together are bounded by the right side of (6.43) and therefore that (6.42) holds as claimed.

Step 2: Construction of an invariant set. Now fix $\tau\in(0,\bar t]$ and let $\mathcal{S}_\tau$ denote the set of mappings $w:[0,\tau]\to\mathcal{H}_{2m}$ which are m-regular on $[0,\tau]$ with respect to the spaces $\mathcal{H}_0,\dots,\mathcal{H}_{2m+1}$ and which satisfy the following:

$$w^{(j)}(0) = \varsigma_j,\ j = 0, \ldots, m, \tag{6.45}$$

$$(w(t), \nabla w_x(t)) \in K_{\varepsilon/2} \times K'_{\varepsilon/2} \text{ on } \Omega \text{ for all } t \in [0, \tau], \tag{6.46}$$

$$\sum_{j=0}^{m} \sup_{t \in [0,\tau]} |w^{(j)}(t) - \varsigma_j|^2_{H^{2m-2j}(\Omega)} \leq \delta \tag{6.47}$$

and

$$\sum_{j=0}^{m} \int_0^\tau |w^{(j)}(t)|^2_{H^{2m-2j+1}(\Omega)} dt \leq M_2. \tag{6.48}$$

We will show that for τ sufficiently small, $\mathcal{S}_\tau$ is invariant for simple iteration applied to (6.27) as described below in Step 3. Observe first that, by the choice of δ, (6.47) implies (6.46), which is therefore redundant. Also, the triangle inequality, (6.47) and the definitions of C_0 and M_1 show that if $w \in \mathcal{S}_\tau$ then

$$\sum_{j=0}^{m} \sup_{t \in [0,\tau]} |w^{(j)}(t)|^2_{H^{2m-2j}} \leq M_1. \tag{6.49}$$

Finally, by reducing $\bar{t}$ if necessary, we may assume that $u_{data} \in \mathcal{S}_\tau$ for every $\tau \leq \bar{t}$, so that all such $\mathcal{S}_\tau$ are nonempty.

Now let $w \in \mathcal{S}_\tau$, define $\tilde{a}^{\eta,\omega}(t)$, $\tilde{g}(t))$ and $\tilde{G}(t)$ as in Step 1 and consider the following linear problem for an unknown $u : [0, \tau] \to \mathcal{H}_{2m}$:

$$\begin{aligned} \langle u(t), v \rangle_{L^2} = \langle \varsigma_0, v \rangle_{L^2} + \int_0^t \int_\Omega & \big[\big(\tilde{a}^{\eta,\omega}(s) u_{x_\eta}(s) \big) \cdot v_{x_\omega} \\ & + \tilde{g}(s) \cdot v + \tilde{G}(s) : \nabla_x v \big] dx\, ds \end{aligned} \tag{6.50}$$

for all $v \in \mathcal{H}_1$ and $t \in [0, \tau]$. We have already checked in Step 1 that $\tilde{a}$, $\tilde{g}$ and $\tilde{G}$ satisfy the regularity conditions required for the application of Theorems 2.7 and 5.5, and the compatibility conditions (5.23) hold because w satisfies (6.45). Theorem 5.5 therefore applies to show that there is a unique m-regular solution u defined on $[0, \tau]$. We will prove that if τ is sufficiently small consistent with (6.40) then $u \in \mathcal{S}_\tau$. First, Theorem 5.5 shows that u satisfies (6.45), and (6.46) follows from (6.47); we therefore need only check (6.47) and (6.48). For the latter we recall the definitions of C_{solve} and $\tilde{F}^{(j)}$ from Step 1 and apply (2.28) and (6.43) to obtain that

$$\begin{aligned}
&\sum_{j=0}^{m}\Big(\sup_{[0,\tau]}|u^{(j)}(t)|^2_{H^{2m-2j}}+\int_0^\tau|u^{(j)}|^2_{H^{2m-2j+1}}\Big)\\
&\qquad\le \tfrac{1}{2}C_{solve}\Big[\sum_{j=0}^{m}|\varsigma_j|^2_{H^{2m-2j}}+\sum_{j=0}^{m-1}|F^{(j)}|^2_{L^\infty(\mathcal{V}_{2m-2j-2})}\\
&\qquad\qquad+\sum_{j=0}^{m}|F^{(j)}|^2_{L^2(\mathcal{V}_{2m-2j-1})}\Big]\\
&\qquad\le \tfrac{1}{2}C_{solve}(C_0+C\tau^{1/2})\le C_{solve}C_0=M_2
\end{aligned}\tag{6.51}$$

if τ is sufficiently small, consistent with (6.40). In particular, u satisfies (6.48).

There remains to prove that u satisfies (6.47). To do this we let $\beta_j(t)=u^{(j)}(t)-\varsigma_j$ and apply the strong form (5.26) of (6.50) to obtain that for $t\le\tau$, $j\le m-1$ and all $v\in\mathcal{H}_1$,

$$\langle\beta_{j+1},v\rangle_{L^2}+\sum_{i=0}^{j}\binom{j}{i}\tilde{A}^{(j-i)}(t)(\beta_i,v)=H_j\cdot v\tag{6.52}$$

where the $\tilde{A}^{(i)}$ are as in Step 1 and

$$\begin{aligned}
H_j(t)\cdot v=\Big[-&\sum_{i=0}^{j}\binom{j}{i}\tilde{A}^{(j-i)}(\cdot)(\varsigma_i,v)\\
&+\int_\Omega\big(\tilde{g}^{(j)}(\cdot)\cdot v+\tilde{G}^{(j)}(\cdot):\nabla_x v\big)\Big]\Big|_0^t.
\end{aligned}\tag{6.53}$$

We checked in Step 1 that the $\tilde{A}^{(i)}$ satisfy the hypotheses of Theorems 2.7 and 5.5 and therefore also of Lemma 2.9. Thus from (2.32), for each t,

$$\sum_{j=0}^{m-1}|\beta_j|^2_{H^{2m-2j}}\le C\Big(|\beta_m|^2_{L^2}+\sum_{j=0}^{m-1}|H_j|^2_{\mathcal{V}_{2m-2j-2}}\Big).\tag{6.54}$$

The contribution to $|H_j|_{\mathcal{V}_{2m-2j-2}}$ in (6.53) from $\tilde{A}^{(j-i)}$ is bounded as follows. If $t\le\tau$ and $v\in\mathcal{H}_1$ then

$$\tilde{A}^{(j-i)}(t)(\varsigma_i,v)\big|_0^t=\int_0^t\int_\Omega\big((\tilde{a}^{(j-i+1)})^{\eta,\omega}(s))(\varsigma_i)_{x_\eta})\big)\cdot v_{x_\omega}\,ds,\tag{6.55}$$

so that

$$\Big|A^{(j-i)}(t)(\varsigma_i,\cdot)\big|_0^t\Big|_{\mathcal{V}_{2m-2j-2}}\le C\int_0^\tau|\tilde{a}^{(j-i+1)})^{\eta,\omega}(s))(\varsigma_i)_{x_\eta})|_{H^{2m-2j-1}}$$

(recall Definition 5.1). We checked in Step 1 that the bounds (6.30) and (6.31) apply to the components of $\tilde{a}-a(\cdot,\cdot,0)$, so that in particular $\tilde{a}^{j-i+1}\in L^2(H^{2m-2j+2i-1})$ with norm bounded by C; and $\varsigma_{j-i}\in H^{2m-2j}$ by hypothesis (and is constant in time). Their product is therefore in $L^2([0,\tau];H^{2m-2j-1})$

by Theorem A.3(d), and the contribution to $|H_j|_{\mathcal{V}_{2m-2j-2}}$ of the term in question is therefore bounded by $C\tau^{1/2}$. Applying (6.42) in a similar way to bound the $\tilde{g}^{(j)}$ and $\tilde{G}^{(j)}$ terms in (6.53) and substituting into (6.54), we then obtain that at each t,

$$\sum_{j=0}^{m-1} |\beta_j|^2_{H^{2m-2j}} \leq C\big(|\beta_m|^2_{L^2} + \tau\big). \tag{6.56}$$

To bound β_m we apply (5.25) of Theorem 5.5 to write

$$\langle u^{(m)}(t), v\rangle_{L^2} = \langle \varsigma_m, v\rangle_{L^2} + \int_0^t \big(- \tilde{A}^{(0)}(u^{(m)}, v) + \Phi \cdot v\big) ds$$

where

$$\Phi(s) \cdot v = - \sum_{i=0}^{m-1} \binom{m}{i} \tilde{A}^{(m-i)}(s)(u^{(i)}, v) + \int_\Omega \big(\tilde{g}^{(m)}(s) \cdot v + \tilde{G}^{(m)}(s) : \nabla_x v\big).$$

Now if $\zeta \in \mathcal{H}_1$ is an L^2-approximation of $u^{(m)}$ then

$$\langle u^{(m)}(t) - \zeta, v\rangle_{L^2} = \langle \varsigma_m - \zeta, v\rangle_{L^2} + \int_0^t \big[- \tilde{A}^{(0)}(u^{(m)} - \zeta, v) + \Phi(s) \cdot v - \tilde{A}^{(0)}(\zeta, v)\big] ds$$

so that by Theorem 2.1,

$$\sup_{t\in[0,\tau]} |u^{(m)}(t) - \zeta|^2_{L^2} \leq C\Big(|\varsigma_m - \zeta|^2_{L^2} + \int_0^\tau |\Phi|^2_{\mathcal{H}_1^*} + \tau|\zeta|^2_{\mathcal{H}_1}\Big).$$

The contributions to the integral on the right from $\tilde{g}^{(m)}$ and $\tilde{G}^{(m)}$ are bounded by $C\tau^{1/2}$ by (6.44), and an argument similar to that given above for $\tilde{A}^{(j-i)}$ following (6.55) shows that the contribution from $\tilde{A}^{(m-i)}$ is bounded by $C\tau$. Triangulating $\beta_m = u^{(m)} - \varsigma_m$ via ζ, we then obtain from (6.54) that

$$\sum_{j=0}^{m} \sup_{t\leq\tau} |\beta_j(t)|^2_{H^{2m-2j}} \leq C\Big(|\varsigma_m - \zeta|^2_{L^2} + \tau|\zeta|^2_{\mathcal{H}_1} + \tau^{1/2}\Big),$$

which is less than δ for appropriate choice of ζ and for τ small consistent with (6.40) (observe that the $j = m$ term is included on the left). This proves that u satisfies the condition in (6.47) and completes the proof that $u \in \mathcal{S}_\tau$.

We have now established that for all positive τ sufficiently small consistent with (6.40), the function $w \mapsto u$ defined by (6.50) maps $\mathcal{S}_\tau$ to $\mathcal{S}_\tau$.

Step 3: Iteration. Now let τ be as above and define a sequence $\{w_k\}$ by taking $w_0 = u_{data} \in \mathcal{S}_\tau$ and $w^{k+1} \in \mathcal{S}_\tau$ the solution of

$$\langle w^{k+1}(t), v\rangle_{L^2} = \langle \varsigma_0, v\rangle_{L^2} + \int_0^t \int_\Omega \Big[\big(\tilde{a}_k^{\eta,\omega}(s) w^{k+1}_{x_\eta}(s)\big)\cdot v_{x_\omega} + \tilde{g}_k(s)\cdot v + \tilde{G}_k(s) : \nabla_x v\Big]\,dx\,ds \tag{6.57}$$

where $\tilde{a}_k^{\eta,\omega}(\cdot,\cdot) = a^{\eta,\omega}(\cdot,\cdot,w^k)$ as in Step 2, and similarly for $\tilde{g}_k$ and $\tilde{G}_k$. Subtracting the equations satisfied by w^{k+1} and w^k we find that their difference Δw^{k+1} satisfies

$$\langle \Delta w^{k+1}(t), v\rangle_{L^2} = \int_0^t\int_\Omega \Big(- (a^{\eta,\omega}(x,t,w^k)\Delta w^{k+1}_{x_\eta})\cdot v_{x_\omega} + \Delta g_k \cdot v + \Delta G_k : \nabla_x v - (\Delta(\tilde{a}_k)^{\eta,\omega} w^k_{x_\eta})\cdot v_{x_\omega}\Big)\,dx\,dt$$

for $v \in \mathcal{H}_1$. The bounds in Theorem 2.1(b) (or Theorem 3.6) together with the assumed Lipschitz continuity of a, g, and G and the fact that $\nabla_x w^k \in L^\infty([0,\tau];L^\infty)$ then show that

$$|\Delta w^{k+1}(t)|^2_{L^2} + \int_0^t\int_\Omega |\nabla_x \Delta w^{k+1}|^2 \le C\int_0^t\int_\Omega \Big[\big(|\Delta w^k| + |\nabla_x \Delta w^k|\big)|\Delta w^{k+1}| + |\Delta w^k||\nabla_x \Delta w^{k+1}|\Big].$$

Applying Gronwall's inequality we find that if

$$\mathcal{E}^{k+1} = \sup_{t\in[0,\tau]} |w^{k+1}(t) - w^k(t)|^2_{L^2} + \int_0^T |w^{k+1}(t) - w^k(t)|^2_{H^1}$$

and if τ is sufficiently small consistent with (6.40) then $\mathcal{E}^{k+1} \le \lambda \mathcal{E}^k$ for some $\lambda \in (0,1)$. The sequence w^k therefore converges in the norm implied by $\mathcal{E}$. That is, $w^k(t)$ converges in $L^2(\Omega)$, say to $w(t)$, uniformly for $t \in [0,\tau]$, so that $w \in C([0,\tau];L^2)$; and the equivalence classes over $[0,\tau]$ of the w^k converge in $L^2([0,\tau];\mathcal{H}_1)$ to that of w.

It follows from the uniform bound in (6.48), translated to the measurable counterparts of the w^k, that for $2j + |\alpha| \le 2m+1$ the weak derivatives $D_t^j D_x^\alpha w_{\mathrm{m}}$ of the measurable counterpart of w exist on $\Omega \times (0,\tau)$ and inherit the bounds

$$\begin{aligned} &\sum_{2j+|\alpha|\le 2m} |D_t^j D_x^\alpha w_{\mathrm{m}}|^2_{L^{\infty,2}} \le M_1,\\ &\sum_{2j+|\alpha|\le 2m} |D_x^\alpha(D_t^j w_{\mathrm{m}} - \varsigma_j)|^2_{L^{\infty,2}} \le \delta,\\ &\sum_{2j+|\alpha|\le 2m+1} |D_t^j D_x^\alpha w_{\mathrm{m}}|^2_{L^{2,2}} \le M_2 \end{aligned} \tag{6.58}$$

where $L^{p,2} = [L^{p,2}(\Omega\times[0,\tau])]^N$. For example, to prove the second of these we fix a particular representative w_{m} and let $\{\varphi_i\}_i \subset C_c^\infty(\Omega)$ be countable and dense in $L^2(\Omega)$. There is then a null set E such if $t \notin E$ then t is a Lebesgue point of $\int_\Omega D_x^\alpha(D_t^j w_{\mathrm{m}}(\cdot,t)-\varsigma_j)\varphi_i$ for all j and α with $2j+|\alpha|\leq 2m$ and all i. Thus if $\psi^h(t)$ is a smooth nonnegative function supported on an interval of length h in $[0,\tau]$ containing t and integrating to one, then the latter integral is equal to

$$\begin{aligned}\pm \lim_{h\to 0}\int_{\Omega\times(0,\tau)} & w_{\mathrm{m}} D_x^\alpha D_t^j(\varphi_i\psi^h) - \int_\Omega D_x^\alpha \varsigma_j\cdot\varphi_i \\ &= \lim_{h\to 0}\lim_{k\to\infty}\int_{\Omega\times(0,\tau)}(D_x^\alpha D_t^j w_{\mathrm{m}}^k)(\varphi_i\psi^h) - \int_\Omega D_x^\alpha\varsigma_j\cdot\varphi_i \\ &= \lim_{h\to 0}\lim_{k\to\infty}\int_0^\tau\int_\Omega D_x^\alpha\big[(w^k)^{(j)}-\varsigma_j\big]\varphi_i\psi^h.\end{aligned}$$

Therefore if elements $\varphi^{j,\alpha}\in\{\varphi_i\}_i$ are chosen for each j and α, then

$$\begin{aligned}\Big|\sum_{2j+|\alpha|\leq 2m} & \int_\Omega D_x^\alpha(D_t^j w_{\mathrm{m}}(\cdot,t)-\varsigma_j)\varphi^{j,\alpha}\Big| \\ &= \lim_{h\to 0}\lim_{k\to\infty}\Big|\sum_{2j+|\alpha|\leq 2m}\int_0^\tau\int_\Omega D_x^\alpha\big[(w^k)^{(j)}-\varsigma_j\big]\varphi^{j,\alpha}\psi^h\Big| \\ &\leq \overline{\lim}_{k\to\infty}\Big(\sum_{j=0}^m |(w^k)^{(j)}-\varsigma_j|^2_{L^\infty(H^{2m-2j})}\Big)^{1/2}\Big(\sum_{j,\alpha}|\varphi^{j,\alpha}|^2_{L^2}\Big)^{1/2} \\ &\leq \delta^{1/2}\Big(\sum_{j,\alpha}|\varphi^{j,\alpha}|^2_{L^2}\Big)^{1/2}\end{aligned}$$

by (6.47). This proves the second bound in (6.58); the others are proved in a similar way.

The bounds in (6.58) are exactly the conditions stated in Corollary A.25(a); there are therefore mappings $w^{(0)}=w,\ldots,w^{(m)}$ satisfying (b) of that corollary. In particular, $w\in C([0,\tau],\mathcal{H}_{2m}^{weak})$ with $|w(\cdot)-\varsigma_0|^2_{\mathcal{H}_{2m}}\leq\delta$ on $[0,\tau]$, so that $(w,\nabla_x w)$ takes values in $K\times K'$. Thus w satisfies the hypotheses in Step 1 and therefore its conclusions: if $\tilde a^{\eta,\omega}(\cdot,\cdot)=a^{\eta,\omega}(\cdot,\cdot,w)$ and similarly for $\tilde g$ and $\tilde G$, then $\tilde a^{\eta,\omega}$, $\tilde g$, and $\tilde G$ satisfy the regularity hypotheses of Theorem 5.5. We can also infer that

$$\begin{aligned}\langle w(t),v\rangle_{L^2} = \langle\varsigma_0,v\rangle_{L^2} + \int_0^t\int_\Omega & \big[(\tilde a^{\eta,\omega}(s)w_{x_\eta}(s))\cdot v_{x_\omega} \\ & + \tilde g(s)\cdot v + \tilde G(s):\nabla_x v\big]\,dx\,ds\end{aligned}\tag{6.59}$$

for all $v\in\mathcal{H}_1$. The proof consists in taking $\lim_{k\to\infty}$ in (6.57), the details being nearly identical to those leading to the bound for $\mathcal{E}^{k+1}$ above. In particular w is a solution of (6.27), but at this point only in the sense of the basic existence result Theorem 2.1 (or Theorem 3.6). To prove that w is m-regular we have to check the compatibility conditions $w^{(j)}(0)=\varsigma_j$ for

$j = 0, \ldots, m-1$. These hold because $(w^k)^{(j)}(0) = \varsigma_j$ for all k and $w^{(j)}$ and $(w^k)^{(j)}$ are in $C^{1/2}_{bdd}([0,\tau];\mathcal{H}_1)$ with modulus independent of k, which follows from (6.58). Thus both the compatibility and the regularity hypotheses of Theorem 5.5 are satisfied and we can therefore conclude that w is m-regular, $w \in \mathcal{S}_\tau$ and that the other descriptions given in Theorem 5.5 and listed in the statement of the present theorem hold (with $u = w$).

The penultimate statement of the theorem, that $u^{(j)} \in C([0,\tau];\mathcal{H}_{2m-2j})$, is proved by checking the hypotheses of Theorem 2.7(g). For example, the requirement that $\tilde{F}^{(j)} \in C([0,\tau];\mathcal{V}_{2m-2j-2})$ for $j \le m-1$ follows by applying (6.33) and (6.31) of Lemma 6.8 to $\tilde{g}$ and $\tilde{G}$ respectively to obtain that $\tilde{g}^{(j+1)} \in L^2([0,\tau];H^{2m-2j-2})$ and $\tilde{G}^{(j+1)} \in L^2([0,\tau];H^{2m-2j-1})$ so that $\tilde{g}^{(j)} \in wAC^2([0,\tau];H^{2m-2j-2})$ and $\tilde{G}^{(j)} \in wAC^2(([0,\tau];H^{2m-2j-1})$, which are somewhat stronger than is required. The hypothesis on $\tilde{a}^{\eta,\omega}$ is proved in a similar way and the ellipticity condition follows because the hypotheses of Theorem 5.5, which we have checked in Step 1, include those of Theorem 4.2.

Finally, to prove uniqueness we let u' be as in the statement and subtract the equations satisfied by u and u' to obtain that, with the obvious meanings,

$$\langle \Delta u(t), v\rangle_{L^2} = \int_0^t\int_\Omega \big(- (a^{\eta,\omega}(x,t,u')\Delta u_{x_\eta}) \cdot v_{x_\omega} + \Delta g \cdot v + \Delta G : \nabla_x v - (\Delta a^{\eta,\omega} u_{x_\eta}) \cdot v_{x_\omega}\big) dx\, dt$$

for $v \in \mathcal{H}_1$ and t in the common domain. The bounds in Theorem 2.1(b) (or Theorem 3.6) together with the assumed Lipschitz continuity of a, g and G and the fact that $\nabla_x u \in L^\infty([0,\tau];L^\infty)$ then apply to show that

$$|\Delta u(t)|^2_{L^2} + \int_0^t\int_\Omega |\nabla_x \Delta u|^2 \le C \int_0^t\int_\Omega \big(|\Delta u|^2 + |\Delta u||\Delta u_x|\big)$$

from which we conclude via Gronwall's inequality that $u' = u$ on the common domain.

□

We conclude this section with a result describing the maximal time of existence of a solution of (6.27). Recall Corollary 6.3, which applies to problems for which uniqueness may fail and for which the maximal existence time is a property of a particular solution, not just of the data. The opposite is true here, although the proof is similar: the solution is shown to extend past a time for which the stated blowup criteria fail. The main point of interest in the present case is the derivation of a time-independent lower bound away from zero for the approximation term involving ζ in (6.40).

Corollary 6.11. *Fix an ensemble* (5.4) *as in* Definition 5.1 *with* $m > n/2$, *let* $T \in (0,\infty]$ *and assume that sets* W *and* W' *and mappings* a, g, *and* G *are given satisfying hypotheses* (a), (b) *and* (c) *of* Theorem 6.10 *on* $[0,\bar{t}]$ *for every* $\bar{t} \in (0,T)$, *and finally that hypotheses* (d) *and* (e) *concerning initial data hold. Let* t_{max} *be the supremum of the set of times* $\tau \in (0,T)$ *for* *which a corresponding m-regular solution u of (6.27) exists on $[0, \tau]$ with $(u, \nabla_x u)$ taking values in $W \times W'$. Then at least one of the following is true:* $t_{max} = T$,

$$\lim_{t \to t_{max}^-} \sum_{j=0}^{m} \left(\sup_{0 \le \sigma \le t} |u^{(j)}(s)|^2_{H^{2m-2j}} + \int_0^t |u^{(j)}|^2_{H^{2m-2j+1}} \right) = \infty,$$

or

$$\lim_{t \to t_{max}^-} \operatorname{dist}(\mathcal{I}(u(t)), \partial W) \wedge \operatorname{dist}(\mathcal{I}(\nabla_x u(t)), \partial W')$$

exists and is zero ($\mathcal{I}$ here denotes the image of Ω under the mapping $u(t)$ or $\nabla_x u(t)$; the term involving ∂W or $\partial W'$ is omitted if W or W' is the whole space, and the third alternative is omitted altogether if both.)

PROOF. If all three alternatives fail then $t_{max} \in (0, T)$, there is a number $M < \infty$ such that the sum in the second alternative is bounded by M for every $t < t_{max}$, and there is a sequence of times $t_i \to t_{max}^-$ such that the minimum in the third alternative is bounded below by a positive number ε for every $t = t_i$ (because the negation of the third alternative is that this term has at least one nonzero limit point).

We will show that Theorem 6.10 guarantees a positive existence time τ which is independent of i for each of the problems (6.27) posed at initial time t_i with initial data $(u^{(0)}(t_i), \ldots, u^{(m)}(t_i))$. These problems will then be solvable on $[t_i, t_i + \tau]$, and this together with the uniqueness result in Theorem 6.10 will then yield a solution on $[0, t_{max} + \tau)$, thus contradicting the maximality of t_{max}.

First note that the above bounds in terms of M and ε show that there are compact subsets K and K' of W and W' such that the continuous representatives of $u^{(0)}(t_i)$ and $\nabla_x u^{(0)}(t_i)$ take values in K and K' for all i. Therefore if we choose $\bar{t} \in (t_{max}, T)$ then $\bar{t}$, M and ε determine the constants C_{sys} and C_{time} (see their descriptions preceding (6.40)) independently of i. We thus conclude from (6.40) that there is a positive existence time τ which applies to each of the problems posed at initial time t_i provided that for every i there is an element $\zeta_i \in \mathcal{H}_1$ with

$$C_{time} |u^{(m)}(t_i) - \zeta_i|^2_{L^2} \le \delta/2 \text{ and } |\zeta_i|^2_{H^1} \le M' \tag{6.60}$$

for some constant M'.

To construct ζ_i we note that $u^{(m)}$ is m-regular on $[0, t]$ for every $t < t_{max}$ and consequently $(u^{(m)})^*$ is in $wAC^2([0, t_{max}); \mathcal{H}_1^*)$ with derivative in $L^2_{loc}([0, t_{max}); \mathcal{H}_1^*)$. The action of this derivative on an element $v \in \mathcal{H}_1$ is given by the right side of (5.26) with $j = m - 1$, applied to the problem (6.27). The bounds associated with the constant M can then be applied to show that the derivative of $(u^{(m)})^*$ is in fact in $L^2([0, t_{max}]; \mathcal{H}_1^*)$. We denote this derivative by $\langle z(\cdot), \cdot \rangle_{\mathcal{H}_1}$ and conclude that $z \in L^2([0, t_{max}]; \mathcal{H}_1)$.

Now let $\{\varphi_k\}$ be an L^2-orthonormal basis as described in Theorem A.10: each φ_k is in $\mathcal{H}_1$ and if $v \in \mathcal{H}_1$ then its L^2-projection onto the first J elements is bounded in the $\mathcal{H}_1$ norm by $2|v|_{\mathcal{H}_1}$ and converges to v in $\mathcal{H}_1$ as $J \to \infty$. The relation

$$\langle u^{(m)}(t_i), \varphi_k \rangle_{L^2} = \langle \varsigma_m, \varphi_k \rangle_{L^2} + \int_0^{t_i} \langle z(t), \varphi_k \rangle_{\mathcal{H}_1} dt$$

then shows that

$$\sum_{J+1}^{\infty} \langle u^{(m)}(t_i), \varphi_k \rangle_{L^2}^2 \leq 2 \sum_{J+1}^{\infty} \langle \varsigma_m, \varphi_k \rangle_{L^2}^2 + 2t_{max} \int_0^{t_{max}} \sum_{J+1}^{\infty} \langle z(t), \varphi_k \rangle_{\mathcal{H}_1}^2 dt.$$

The integrand on the right goes to zero for almost all t as $J \to \infty$ and is dominated by $2|z(t)|_{\mathcal{H}_1}^2$, which is integrable. This integral therefore goes to zero. We can therefore choose J sufficiently large that the first inequality in (6.60) holds with ζ_i the L^2-orthogonal projection of $u^{(m)}(t_i)$ onto $\{\varphi_1, \ldots, \varphi_J\}$; the second will then hold for a constant M' determined by J.

□

§6.5 Global Existence and Stability of Steady-States

In this section we consider m-regular solutions of (6.27) with initial data close to that of a known global solution when the functions g and G enforce a certain dissipativity. A typical example is that in which the known solution is a steady-state, that is, a time-independent solution of the associated differential equations and boundary conditions. The main result is given in Theorem 6.12 in which $u \equiv 0$ is a steady-state and the dissipativity assumptions guarantee that solutions corresponding to nearly initial data converge to zero as $t \to \infty$ (if $T = \infty$). More general stability problems can then by treated by applying Theorem 6.12 to the difference between a known global solution and a solution with nearby initial data with the conclusion that the latter is also global and coalesces with the former as $t \to \infty$. A steady-state for which this holds is therefore an "attracting rest point" or a "stable steady-state " of the underlying differential equations and boundary conditions. We will give only a brief, informal discussion (below following the proof of Theorem 6.12) of the application to these more general settings, the details typically being specific to the specific problem as in the example in Exercise 6.9.

To motivate the analysis let us suppose that g and G are as in (6.27) and that $g(x, t, 0, 0) = 0$ and $G(x, t, 0) = 0$ for all x and t so that $u \equiv 0$ is a solution of (6.27) with $\varsigma_0 = 0$. Then letting $P \in \mathbf{R}^{N \times n}$ denote the variable in W' we can write

(6.61) $$g(x, t, w, P) = g_{lin}(x, t, w, P) + g_{quad}(x, t, w, P)$$

where g_{lin} consists of the linear terms in the Taylor expansion of $g(x, t, \cdot, \cdot)$ about $(w, P) = (0, 0)$ and g_{quad} is the quadratic remainder term. Specifically,

(6.62) $$g_{lin}^k(x, y, w, P) = g_{w^i}^k(x, t, 0, 0)w^i + g_{P^{il}}^k(x, t, 0, 0)P^{il},$$

where subscripts denote partial derivatives and summation over repeated indices is understood, and

$$(6.63)\quad g^k_{quad}(x,t,w,P) = \Big[\int_0^1\int_0^1 g^k_{w^iw^l}(x,t,s_1s_2w,s_1s_2P)ds_1ds_2\Big]w^iw^l + \dots$$

and similarly for G, G_{lin}, and G_{quad}. Observe that if $g \subset C^{2m+2}_{bdd}$ on a given domain, then $g_{lin} \in C^{2m+1}_{bdd}$ and $g_{quad} \in C^{2m}_{bdd}$; similarly, if $G \in C^{2m+3}_{bdd}$ then $G_{lin} \in C^{2m+2}_{bdd}$ and $G_{quad} \in C^{2m+1}_{bdd}$. In particular, g_{quad} and G_{quad} will satisfy the regularity conditions in Lemma 6.8(c) and (b) respectively.

We will impose sign conditions on the derivatives defining g_{lin} and G_{lin} sufficient to insure that the bilinear form

$$(6.64)\qquad B(t)(z,v) \equiv \int_\Omega \Big(g_{lin}(x,t,z,\nabla_x z)\cdot v + G_{lin}(x,t,z) : \nabla_x v\Big)dx$$

(which *is* bilinear in (z,v)) is dissipative in a certain sense, and then to combine B with the bilinear form

$$A(t,u)(z,v) \equiv \int_\Omega (a^{\eta,\omega}(\cdot,t,u)z_{x_\eta})\cdot v_{x_\omega}$$

to write (6.27) in the form

$$(6.65)\quad \begin{aligned}\langle u(t),v\rangle_{L^2} &= \langle \varsigma_0,v\rangle_{L^2} + \int_0^t \big[-A(s,u)(u,v) + B(s)(u,v)\big]ds\\ &+ \int_0^t\int_\Omega \big[g_{quad}(x,s,u,\nabla_x u)\cdot v + G_{quad}(x,t,u) : \nabla_x v\big]dx\,ds\end{aligned}$$

in which the $-A+B$ term is strongly dissipative and the g_{quad} and G_{quad} terms are of secondary importance. An explicit expression for B is given below in (6.68) followed by a brief discussion of the required conditions on the partials of g and G.

These ideas are formalized in the following theorem in which hypotheses (a) and (b) are routine regularity conditions, (c) guarantees that $u \equiv 0$ is a steady-state solution, (d) is the important dissipativity hypothesis and (e) is the assumption that a, g and G vary slowly in time relative to the dissipation. Observe that the assertion of global existence of solutions with small, consistent initial data is made only subject to hypothesis (e) of Theorem 6.10 (recall the discussion in Remark 3 following the statement of that theorem).

Theorem 6.12. *Fix an ensemble* (5.4) *as in* Definition 5.1 *with* $m > n/2$, *let* $T \in (0,\infty]$ *and assume the following:*

(a) *Mappings* $a^{\eta,\omega}$ *are given satisfying the hypotheses in* Theorem 6.10(a) *for all* $\bar{t} < T$ *and with* W *replaced by the ball* B_{r_0}, *and there is a positive* θ *such that* (6.39) *holds for all* $t \in (0,T)$.

(b) *Mappings* g *and* G *are given satisfying the hypotheses in* Theorem 6.10(a) *for all* $\bar{t}$ *with* W *and* W' *replaced by balls* B_{r_0} *and* B_{s_0} *and* m *replaced by* $m+1$.

The derivatives and Lipschitz constants included in these hypotheses are assumed to be bounded by a constant C_{sys} which is independent of $\bar{t}$, and B_{r_0} and B_{s_0} are the balls of positive radii r_0 and s_0 centered at the respective origins.

(c) $g(x,t,0,0) = 0$ *and* $G(x,t,0) = 0$ *for all* $(x,t) \in \Omega \times [0,T)$.

(d) *There are constants* $\theta' \in (0,\theta)$ *and* $\gamma > \theta - \theta'$ *such that if* B *is the bilinear form defined in* (6.64) *then*

$$B(t)(z,z) \leq \theta' |\nabla_x z|^2_{[L^2(\Omega)]^{N\times n}} - \gamma |z|^2_{[L^2(\Omega)]^N}.$$

(e) *The number* ζ *defined by*

$$\zeta^{1/2} \equiv \sup \sum_{\eta,\omega} |D_t^j a^{(\eta,\omega)}(x,t,0)| + \sup |D_t^j g(x,t,0,0)|$$

$$+ \sup |D_t^j G(x,t,0)|,$$

where each sup *is over* $(x,t) \in \Omega \times [0,T)$ *and* $j = 1, \ldots, m$, *satisfies* $2\zeta < 2\lambda \equiv \gamma - (\theta - \theta')$.

Then given $\varepsilon \in (0, r_0 \wedge s_0)$ *there are positive numbers* κ^{-1} *and* C_{solve}, *upper bounds for which depend on* n, m, N *and on upper bounds for* C_{BC}, C_{sys}, λ^{-1}, ε^{-1} *and* $\theta - \theta')^{-1}$ *such that the following holds: Given initial data* $(\varsigma_0, \ldots, \varsigma_m) \in \mathcal{H}_{2m} \times \ldots \mathcal{H}_0$ *for which*

- *the compatibility conditions* (6.35) *are satisfied*
- $(\varsigma_0, \nabla_x \varsigma_0)$ *takes values in* $B_{r_0-\varepsilon} \times B_{s_0-\varepsilon}$
- $\displaystyle\sum_{j=0}^{m} |\varsigma_j|^2_{H^{2m-2j}(\Omega)} \leq \kappa$
- *hypothesis* (e) *of* Theorem 6.10 *holds for some* $\bar{t} > 0$,

the corresponding m*-regular solution of* (6.27) *described in* Theorem 6.10 *has maximal existence time* $t_{max} = T$ *relative to* $B_{r_0} \times B_{s_0}$ *in the sense of* Corollary 6.11. *That is, there is an* m*-regular solution* u *of* (6.27) *with these data satisfying the conclusions of* Theorem 6.10 *on* $[0,\tau]$ *for every* $\tau < T$. *In particular, the bound in* (6.41) *holds with* C_{solve} *replaced by* $C_{solve}e^{-\lambda\tau}$. *Thus if* $T = \infty$, *then* $u^{(j)} \to 0$ *in* $H^{2m-2j}(\Omega)$ *as* $t \to \infty$ *for all* $j = 0 \ldots, m$.

PROOF. It will be convenient to denote by $\|w\|^2_\tau$ the squared norm on the left side of (6.41) with w in place of u. The first step is to reexamine the multiplier $C(\bar{t})e^{\int_0^{\bar{t}} \beta^2}$ on the right side of (2.28) in the application of Theorem 2.7 to the bilinear form $A(\cdot, w) - B(\cdot)$:

Step 1: Dissipation Estimate. We claim the following: *there is a constant* $M_0 \in (0,1]$ *such that if* w *is* m*-regular on an interval* $[0,\bar{t}] \subset (0,T)$ *with* $\|w\|^2_{\bar{t}} \leq M_0$, *then* $(w, \nabla_x w)$ *takes values in* $B_{r_0 - \frac{1}{2}\varepsilon} \times B_{s_0 - \frac{1}{2}\varepsilon}$ *and if* $\tilde{A}(t) \in$

$\mathcal{B}(\mathcal{H}_1 \times \mathcal{H}_1)$ is the bilinear form

$$\tilde{A}(t)(z,v) = \int_\Omega \left((a^{\eta,\omega}(x,t,w)z_{x_\eta}\right) \cdot v_{x_\omega} dx - B(t)(z,v)$$

where B is as in (6.64), *then $\tilde{A}$ satisfies the regularity hypotheses of* Lemma 5.3 *and therefore of* Theorem 5.5. *In addition there is a constant C_{solve} which is independent of $\bar{t}$ such that if z is an m-regular solution of* (2.3) *with A and u replaced by $\tilde{A}$ and z and with inhomogeneous term F and compatible initial data $(\varsigma_0', \ldots, \varsigma_m')$ then*

$$\|z\|_{\bar{t}}^2 \leq C_{solve} e^{-\lambda \bar{t}} \Big[\sum_{j=0}^{m} |\varsigma_j'|^2_{\mathcal{H}_{2m-2j}} + \sum_{j=0}^{m-1} \sup_{0 \leq t \leq \bar{t}} |F^{(j)}(t)|^2_{\mathcal{V}_{2m-2j-2}} + \sum_{j=0}^{m} \int_0^{\bar{t}} |F^{(j)}|^2_{\mathcal{V}_{2m-2j-1}} \Big]. \tag{6.66}$$

The first statement above holds because $\nabla_x w(t) \in H^{2m-1} \subset H^m \subset L^\infty$ with norm bounded by $CM_0^{1/2}$, which can be taken arbitrarily small, and similarly for w. For the second we examine the multiplier $C(\bar{t})e^{\int_0^t \beta^2(s)ds}$ on the right side of (2.28) in Theorem 2.7, first applied to the shifted bilinear form $\tilde{A}_\lambda = \tilde{A} - 2\lambda\langle\cdot,\cdot\rangle_{L^2}$. Note that the hypothesis in (d) of the present theorem and the definition of λ show that $\tilde{A}_\lambda$ satisfies the boundedness and coerciveness conditions in (a) of Theorem 2.7, which is therefore applicable. The description in (f) of that theorem shows that in the present case $C(\bar{t})$ is independent of $\bar{t}$ and therefore may be designated C_{solve}. To estimate β, which is defined in (c*i*) of Theorem 2.7, we fix $i \in \{1, \ldots, m\}$ and $(z,v) \in \mathcal{H}_{2i} \times \mathcal{H}_1$ and bound $|\tilde{A}_\lambda^{(i)}(z,v)|$ under the assumption that $\|w\|_T$ is small. The dominant term in $\tilde{A}_\lambda^{(i)}(z,v)$ is seen to be

$$\begin{aligned}\int_\Omega &\left(D_t^i a^{\eta,\omega}(x,t,w)z_{x_\eta}\right) \cdot v_{x_\omega} \\ &= \int_\Omega \left[(D_t^i a^{\eta,\omega})(x,t,0)z_{x_\eta} + (D_t^i a^{\eta,\omega})(x,t,\cdot)\Big|_0^w z_{x_\eta}\right] \cdot v_{x_\omega} \\ &\qquad + \sum \int_\Omega O\big(|w^{(i_1)}| \cdots |w^{(i_l)}||\nabla_x z||\nabla_x v|\big)\end{aligned}$$

where $i_1 + \cdots + i_l = i$ is as in (6.34). The contribution of the first term on the right is bounded by $\left(\sup_{\Omega\times[0,T)}\right) \sum_{\eta,\omega} |D_t^j a^{(\eta,\omega)}(x,t,0)|\big)|z|_{\mathcal{H}_1}|v|_{\mathcal{H}_1}$ and the second can be absorbed into the third by the assumed Lipschitz continuity of $D_t^i a^{\eta,\omega}$. The summand in the third term is bounded by

$$\left(\int_\Omega |\Pi|^2 |\nabla_x z|^2\right)^{1/2} |v|_{\mathcal{H}_1}$$

where Π is the product of the $|w^{(i_j)}|$. Consider two cases: if $2i \geq m+1$ then $|\nabla_x z|_{L^\infty} \leq |\nabla_x z|_{H^m} \leq C|\nabla_x z|_{H^{2i-1}}$; it follows that $|\Pi|_{L^2} \leq C\|w\|_{\bar{t}}$, as in the derivation of (6.30) from (6.28), because we can take $\|w\|_{\bar{t}} \leq 1$. The above term is therefore bounded in this case by $C\|w\|_{\bar{t}}|z|_{H^{2i}}|v|_{H^1}$. In the second case $2i \leq m$ and therefore each $2i_j \leq m$ and $w^{(i_j)} \in H^{2m-2i_j} \subseteq H^m \subset L^\infty$, and the same bound holds. Treating the $D_t^i g_{lin}$ and $D_t^i G_{lin}$ terms in the integral representing $\tilde{A}_\lambda^{(i)}(z,v)$ in a similar way we conclude that

$$|\tilde{A}_\lambda^{(i)}(z,v)| \leq (\zeta^{1/2} + C\|w\|_{\bar{t}})|z|_{H^{2i}}|v|_{\mathcal{H}_1}$$

and therefore that $\beta \leq (\zeta^{1/2} + C\|w\|_{\bar{t}})$. Consequently if $\|w\|_t^2 \leq M_0$ then the multiplier $C(\bar{t})e^{\int_0^t \beta^2(s)ds}$ on the right side of (2.28), applied to $\tilde{A}_\lambda$, is bounded by $C_{solve}e^{(\zeta+CM_0)t}$ and the corresponding multiplier for $\tilde{A}$ is bounded by $C_{solve}e^{(\zeta+CM_0-2\lambda)t}$ (Exercise 2.1), which is bounded by $C_{solve}e^{-\lambda t}$ if M_0 is chosen sufficiently small. This proves the claim.

Step 2: Estimates for g_{quad} and G_{quad}. We claim the following: *The functions g_{quad} and G_{quad} defined in* (6.63) *satisfy the hypotheses in* Lemma 6.8(c) *and* (b) *respectively with $W \times W'$ replaced by $B_{r_0} \times B_{s_0}$ and with m as in the present theorem. And if w is as in* Step 1 *and $\tilde{F} : [0,T) \to \mathcal{H}_1^*$ is defined by*

$$\tilde{F}(t) \cdot v = \int_\Omega \left(g_{quad}(x,t,w,\nabla_x w) \cdot v + G(x,t,w)_{quad} : \nabla_x v\right)$$

then $\tilde{F}$ satisfies the regularity hypotheses (2.26) *and* (2.27) *of* Theorem 2.7 *and there is a constant C_F independent of $t \in (0,T)$ and w such that*

$$\sum_{j=0}^{m-1} |\tilde{F}^{(j)}|^2_{L^\infty((0,t);\mathcal{V}_{2m-2j-2})} + \sum_{j=0}^{m} |\tilde{F}^{(j)}|^2_{L^2((0,t);\mathcal{V}_{2m-2j-1})} \leq C_F\|w\|_t^4. \tag{6.67}$$

The regularity statements here are obvious from (6.63) for g_{quad} and the corresponding representation for G_{quad}. To prove (6.67) we apply Lemma 6.8(c) with $h = g_{quad}$, with M_1 and M_2 in (6.30) and (6.31) bounded by $\|w\|_t^2 \leq 1$ and with $l \geq 2$ in the sum in (6.34). The proof for G_{quad} is similar. These observations prove the claim.

We now complete the proof of the theorem as follows. Fix the constants $C_{solve} \geq 1$ and C_F from Steps 1 and 2 and choose $\kappa > 0$ so that $(4C_{solve}^2 C_F)\kappa < 1$ and the roots $s_\pm$ of $C_{solve}C_F s^2 - s + C_{solve}\kappa$ are real and positive with $s_- < M_0$. It then follows that $\kappa < s_-$ and that if $s_1 \in (s_-, s_+ \wedge M_0)$ then the graph of the line $y = s$ in the s–y plane lies strictly below that of $y = C_{solve}(C_F s^2 + \kappa)$ for $s \in (0, s_-)$ and strictly above for $s \in (s_-, s_1]$.

Now let u be the solution of (6.27) of Theorem 6.10 with initial data as described in the hypotheses of the present theorem and assume that the maximal time of existence t_{max} of u with respect to $B_{r_0} \times B_{s_0}$ is strictly less than T. Let $s(t) \equiv \|u\|_t^2$, which is increasing and continuous in t by the penultimate conclusion of Theorem 6.10. Then $s(0) \leq \kappa$ by hypothesis and so $s(0) \leq \kappa < s_- < s_1$. There is therefore a first time $t_1 < T$ such that $s(t) < s_1$ on $[0, t_1)$ and $s(t_1) = s_1$ (otherwise $s(t) < s_1 \leq M_0$ for

$t \in [0, t_{max})$ and by Step 1 all three alternatives in Corollary 6.11 would be precluded, contradicting the maximality of t_{max}). The bounds (6.66) and (6.67) therefore apply on $[0, s_1]$ and show that

$$s_1 = \|u\|_{t_1}^2 \le C_{solve}(\kappa + C_F s_1^2),$$

which violates the last conclusion in the preceding paragraph. This proves that $t_{max} = T$.

□

Before considering the application of Theorem 6.12 to more general problems we examine the dissipativity assumption in (d) of the theorem. We have that for $z \in H^1(\Omega)$,

$$B(t)(z,z) = \int_\Omega \Big[(g_w(x,t,0,0)z) \cdot z) + \Big(g^k_{Pil}(x,t,0,0) + G^{il}_{wk}(x,t,0) \Big) z^i_{x_l} z^k \Big] \tag{6.68}$$

where g_w is the $N \times N$ Jacobian matrix of g with respect to w, superscripts denote components and subscripts denote partial derivatives. Thus (d) holds if $g_w(\cdot,\cdot,0,0)$ is uniformly negative definite on $\Omega \times [0,T)$, say $(g_w z) \cdot z \le -C^{-1}|z|^2$ for $z \in \mathbf{R}^N$, and the partial derivatives of g and G in the second and third terms in the integrand on the right above are small relative to C. The latter holds in particular if g is independent of P and G of w.

The application of Theorem 6.12 to more general settings is straightforward in principle if not in detail. Suppose that $\tilde{u}$ is a known, global m-regular solution of (6.27) with initial data $(\tilde{\varsigma}_0, \dots, \tilde{\varsigma}_m)$, as in Theorem 6.10. Suppose also that nearby compatible initial data $(\varsigma_0, \dots, \varsigma_m)$ is given and that u is the corresponding local-in-time solution of Theorem 6.10. We can then consider the application of Theorem 6.12 to the difference $z = u - \tilde{u}$, which satisfies

$$\langle z(t), v \rangle_{L^2} = \langle \varsigma_0 - \tilde{\varsigma}_0 \rangle_{L^2} + \int_0^t \int_\Omega \big[-(\tilde{a}^{\eta,\omega}(x,t,z) z_{x_\eta}) \cdot v_{x_\omega} + h(x,t,z,\nabla_x z) \cdot v + H(x,t,z) : \nabla_x v \big] dx\, dt$$

for $v \in \mathcal{H}_1$. Here $\tilde{a}^{\eta,\omega}(x,t,z) = a^{\eta,\omega}(x,t,\tilde{u}(x,t) + z)$ and explicit expressions for h and H are easily computed. These expressions show first that $h(x,t,0,0) = 0$ and $H(x,t,0) = 0$, so that $z \equiv 0$ is a steady-state solution, and second that the dissipativity hypothesis in (d) holds if the Jacobian matrix $g_w(\cdot,\cdot,\tilde{u},\nabla_x \tilde{u})$ is uniformly negative definite and if the partial derivatives $a^{\eta,\omega}_w(\cdot,\cdot,\tilde{u})$, $g_P(\cdot,\cdot,\tilde{u},\nabla_x\tilde{u})$ and $G_w(\cdot,\cdot,\tilde{u})$ are small, as in the paragraph above. These smallness conditions hold in particular if a and G are independent of w and g is independent of P. If Theorem 6.12 can be shown to apply then the conclusion will be that u is also globally defined and coalesces with $\tilde{u}$ as $t \to \infty$. This application of Theorem 6.12 requires two additional hypotheses, however, which cannot be overlooked: first, the domains of regularity of the nonlinear functions $a^{\eta,\omega}$, g and G must be respected, and second, the continuous differentiability hypotheses in (a) and (b) will require a high degree of regularity of the known solution $\tilde{u}$, as in Theorem 5.5(c) and Theorem 4.2(c). A representative example is outlined in Exercise 6.9.

§6.6 Invariant Regions for Quasilinear Systems

In this section we establish sufficient conditions for a region Σ in u-space to be *invariant* for m-regular solutions of (6.27). This means that if the initial function ς_0 takes values in Σ then so too does the solution $u(\cdot,t)$ for as long as it is defined. The existence of such a set is of particular interest in applications in which the solution loses its physical meaning outside Σ, such as when its components represent nonnegative concentrations or densities. Also, the existence of an invariant set can serve as the basis for precluding the blow-up alternative in Corollary 6.11, as will be seen in Theorem 6.14 in the next section.

The key idea is illustrated by the same question for systems of ordinary differential equations $\dot{u}(t) = g(u(t))$, where $g : W \subseteq \mathbf{R}^N \to \mathbf{R}^N$ and an initial condition $u(0) = \varsigma_0 \in W$ is given. If

$$\Sigma = \{u \in W \text{ such that } \Phi(u) < 0\}$$

where Φ is C^1 and $\nabla_u \Phi(u) \cdot g(u) < 0$ for $u \in \partial\Sigma$, then Σ is invariant. The proof is that if $\varsigma_0 \in \Sigma$ and if there is a minimal time $t_0 > 0$ at which $u \in \partial\Sigma$, then $\frac{d}{dt}\Phi(u(t))\big|_{t_0} < 0$, contradicting the minimality of t_0.

We will extend this idea to m-regular solutions of systems (6.27) with the restriction that g is linear in $\nabla_x u$ and $\mathcal{H}_j = [(H^j \cap H_0^1)(\Omega)]^N$. In this case the divergence theorem can be applied to the term involving G, bringing the underlying differential equation to the form

$$u_t = \left(a^{\eta,\omega}(x,t,u)u_{x_\eta}\right)_{x_\omega} + g_1(x,t,u) + g_2^i(x,t,u)u_{x_i} \tag{6.69}$$

where g_1 is $\mathbf{R}^N$-valued and each g_2^i is $R^{N\times N}$-valued. The corresponding weak form is therefore

$$\begin{aligned}\langle u(t), v\rangle_{L^2} = \langle \varsigma_0, v\rangle_{L^2} + \int_0^t\int_\Omega \Big[&-(a^{\eta,\omega}(x,t,u)u_{x_\eta})\cdot v_{x_\omega}\\ &+\left(g_1(x,t,u) + g_2^i(x,t,u)u_{x_i}\right)\cdot v\Big].\end{aligned} \tag{6.70}$$

The following theorem finds wide application in spite of the complexity of its statement. In many cases invariance is more easily established directly by adapting its proof rather than by checking its hypotheses, as in the first three examples given at the end of this section. On the other hand the theorem can be quite useful for identifying *candidates* for invariant sets when these are not otherwise suggested by the application, as in the fourth example.

Theorem 6.13. *Fix an ensemble* (5.4) *as in* Definition 5.1 *with* $m > n/2$ *and* $\mathcal{H}_1 = [H_0^1(\Omega)]^N$ *and let* $T \in (0,\infty]$ *and* W *be a neighborhood of the* *origin in* $\mathbf{R}^N$. *Let mappings* g_1 *and* g_2^i *with components in* C^{2m}_{bdd} *and* $a^{\eta,\omega}$ *with components in* C^{2m+1}_{bdd} *be given, all on* $(\Omega \times (0,T) \times W)$, *with* g_1 *taking values in* $\mathbf{R}^N$ *and* g_2^i *and* $a^{\eta,\omega}$ *taking values in* $\mathbf{R}^{N\times N}$. *Assume that the derivatives of these mappings up to the given orders have continuous extensions to* $\overline{\Omega} \times [0, \bar{t}] \times K$ *for every compact* $K \subset W$ *and* $\bar{t} \in (0,T)$ *which are bounded, continuous and Lipschitz with respect to the variable in* K, *and finally that the* $a^{\eta,\omega}$ *satisfy the positivity condition* (6.39) *for every* $\bar{t} \in (0,T)$.

Suppose also that a C^2 *mapping* $\Phi : W \to \mathbf{R}$ *is given, taking both positive and negative values in* W *and satisfying* $\Phi(0) < 0$ *and* $\nabla_u \Phi \neq 0$ *on* $\Phi^{-1}(0)$. *Define* $\Sigma = \{w \in W : \Phi(w) < 0\}$, $\partial\Sigma = \{w \in W : \Phi(w) = 0\}$ *and* $\overline{\Sigma} = \Sigma \cup \partial\Sigma$. *Assume that the following conditions hold:*

(a) *There are mappings* $\lambda^{\eta,\omega} : \Omega \times (0,T) \times W \to \mathbf{R}$, $1 \le \eta, \omega \le n$, *such that*

- *for* $t \in (0,T)$, $\lambda^{\eta,\omega}(\cdot, t, \cdot) \in C^1(\Omega \times W)$
- for all η and ω, $(a^{\eta,\omega})^{tr} \nabla_u \Phi = \lambda^{\omega,\eta} \nabla_u \Phi$ *on* $\Omega \times (0,T) \times W$
- *if* S *is a nonnegative symmetric* $n\times n$ *matrix, then the scalar* $\lambda^{\eta,\omega} S^{\eta,\omega}$ *is nonnegative on* $\Omega \times (0,T) \times W$
- *for all (constant)* $N \times n$ *matrices* M,
$$\left(M^{tr} \Phi'' a^{\eta,\omega} M\right)^{\omega,\eta} \ge 0$$
on $\Omega \times (0,T) \times W$.

(b) *Either*

(*i*) $g_1 \cdot \nabla_u \Phi < 0$ *on* $\Omega \times (0,T) \times \partial\Sigma$

or

(*ii*) $g_1 \cdot \nabla_u \Phi \le 0$ *on* $\Omega \times (0,T) \times \partial\Sigma$ *and there is a (constant)* $N \times N$ *matrix* P *such that* $(Pu) \cdot \nabla_u \Phi < 0$ *on* $\partial\Sigma$.

(c) *There are maps* $\mu^i : \Omega \times (0,T) \times \partial\Sigma$ *such that*
$$(g_2^i)^{tr} \nabla_u \Phi = \mu^i \nabla_u \Phi$$
on $\Omega \times (0,T) \times \partial\Sigma$.

Let u *be an* m*-regular solution of* (6.70) *on an interval* $[0, t_{max})$ *taking values in* W *and with compatible initial data* ς_0 *and let* $u_{\mathbb{C}}$ *be its continuous representative (recall the description in* Lemma 6.8(a)). *Under hypothesis* (b*i*): *if* $\varsigma_0(x) \in \Sigma$ *for all* $x \in \Omega$, *then* $u_{\mathbb{C}}(x,t) \in \Sigma$ *for all* $(x,t) \in \overline{\Omega} \times [0, t_{max})$. *Under either hypothesis* (b*i*) *or* (b*ii*): if $\varsigma_0(x) \in \overline{\Sigma}$ for all $x \in \Omega$, then $u_{\mathbb{C}}(x,t) \in \overline{\Sigma}$ for all $(x,t) \in \overline{\Omega} \times [0, t_{max})$.

PROOF. We denote the continuous counterpart $u_{\mathbb{C}}$ by u. Recall from Lemma 6.8(a) that u and its (calculus) derivatives u_t, u_{x_j} and $u_{x_j x_k}$ have Hölder continuous extensions to every $\overline{\Omega} \times [0, \bar{t}]$. Also, Theorem 6.10 invokes (5.35) in Theorem 5.5(c) to show that (6.69) holds pointwise in $\Omega \times (0, t_{max})$ and therefore on $\overline{\Omega} \times [0, t_{max})$. We will prove the second statement in the conclusion of the theorem, that $\overline{\Sigma}$ is invariant; the proof of the first is similar (see Exercise 6.10).

Assume first that (bi) holds and let u be as in the statement with $u(x, 0) \in \overline{\Sigma}$ for all x. Define

$$\bar{t} = \sup\{t < t_{max} : u(x, s) \in \overline{\Sigma} \text{ for all } (x, s) \in \overline{\Omega} \times [0, t]\}$$

(notice that $t = 0$ is a competitor in the above set). We have to show that $\bar{t} = t_{max}$. Suppose to the contrary that $0 \leq \bar{t} < t_{max}$. There is then a sequence $\{(x_k, t_k)\} \in \overline{\Omega} \times (\bar{t}, t_{max})$ such that $t_k \to \bar{t}^+$ and $\varphi(x_k, t_k) \equiv \Phi(u(x_k, t_k)) > 0$ for all k. Note that there is a compact set $K \subset \Omega$ such that $\varphi < 0$ on $(\overline{\Omega} - K) \times [0, t_1]$. This follows because $u(\cdot, t) = 0$ on $\partial\Omega$, $u(x, t) \to 0$ as $|x| \to \infty$ if Ω is unbounded, $\Phi(0) < 0$ and φ is Hölder continuous. Thus $x_k \in K$ for all k, so that, modulo a subsequence, x_k converges to an element $\overline{x} \in K \subset \Omega$. It follows that $\varphi(\overline{x}, \bar{t}) = 0$ and we claim further that $\varphi_t(\overline{x}, \bar{t}) \geq 0$. This is obvious if $\bar{t} > 0$ and in any case follows from the observation that the inequalities $\varphi(x_k, \bar{t}) \leq 0 < \varphi(x_k, t_k)$ show that there is a $\tau_k \in (\bar{t}, t_k)$ with $\varphi_t(x_k, \tau_k) > 0$, and $(x_k, \tau_k) \to (\overline{x}, \bar{t})$. Noting also that the C^2 function $\varphi(\cdot, \bar{t})$ has a relative maximum at $\overline{x}$, we then conclude that at $(\overline{x}, \bar{t})$, $\varphi = 0, \varphi_t \geq 0$, $\varphi_{x_j} = 0$ for all j and the second derivative matrix φ'' is nonpositive.

On the other hand we can take the dot product of both sides of (6.69) with $\nabla_u \Phi(u(\overline{x}, \bar{t}))$ to find that the term on the left, which is φ_t, is nonnegative, the term involving g_1 is strictly negative by hypothesis (bi), and the terms involving g_2^i are all zero by hypothesis (c). The remaining term is

$$\begin{aligned}
\left(\nabla_u \Phi \cdot (a^{\eta,\omega} u_{x_\eta})\right)_{x_\omega} &- \Phi_{u^i u^j} u^j_{x_\omega} (a^{\eta,\omega})^{ik} u^k_{x_\eta} \\
=&(\lambda^{\eta,\omega} \varphi_{x_\eta})_{x_\omega} - \left(\Phi'' \nabla_x u\right)^{i\omega} \left(a^{\eta,\omega} \nabla_x u\right)^{i\eta} \\
=&\lambda^{\eta,\omega} \left(\varphi''\right)^{\eta,\omega} - \left((\nabla_x u)^{tr} \Phi'' a^{\eta,\omega} \nabla_x u\right)^{\omega,\eta},
\end{aligned}$$

which is nonpositive by the last item in hypothesis (a). These results contradict equality in (6.69), however, so that $\bar{t} = t_{max}$, as required.

Now suppose that hypothesis (bii) holds in place of (bi) and let u and $\bar{t}$ be as in the second paragraph of this proof. Suppose again that $\bar{t} < t_{max}$ and let (x_k, t_k) be as above with $t_k \to \bar{t}^+$ and $\varphi(x_k, t_k) > 0$. Consider the problem (6.70) posed at initial time $\bar{t}$ but with g_1 replaced by

$$h_1(x, t, w) = g_1(x, t, w) + \delta(t - \bar{t})^{2(m+1)} P w,$$

with initial data $(u(\cdot, \bar{t}), \ldots, D_t^m u(\cdot, \bar{t}))$ and $\delta \in (0, 1)$. Hypotheses (a)–(e) of Theorem 6.10 are then satisfied (note in particular that u itself provides the required element u_{data} in (e)). There is therefore an m-regular solution u^δ defined on an interval $[\bar{t}, \bar{t} + t^\delta_{max})$, and the description in (6.40) shows that there is a positive lower bound τ for t^δ_{max} independent of δ. The result just proved with hypothesis (bi) in effect applies to this modified system, so that u^δ takes values in $\overline{\Sigma}$ on $\overline{\Omega} \times [\bar{t}, \bar{t} + \tau]$. On the other hand, it is easy to show that $u^\delta(\cdot, t) \to u(\cdot, t)$ as $\delta \to 0$, say in $L^2(\Omega)$, for $t \in [\bar{t}, \bar{t} + \tau]$ (see the last paragraph in the proof of Theorem 6.10, for example). We conclude that $\Phi(u(x, t)) \leq 0$ on $\overline{\Omega} \times [\bar{t}, \bar{t} + \tau]$, which contradicts the supposition that $\Phi(u(x_k, t_k)) > 0$ for k large. It follows that $\bar{t} = t_{max}$, as required. □

We consider several representative examples:

Example 1. $N = 1$. In this case $a^{\eta,\omega}$, g_1, g_2^i and Φ are scalar-valued and, apart from the regularity hypotheses preceding (a), conditions (a) and (c) hold automatically: the third item in (a) follows from the spectral representation of S and Remark 2 following Theorem 3.6, and the fourth is equivalent to (6.39). Thus if zero is an interior point of a compact subset $[\underline{u}, \overline{u}]$ of W and $\Phi(u) = (u - \underline{u})(u - \overline{u})$, then (b$ii$) holds with $P = -I$ provided

$$g_1(\cdot,\cdot,\overline{u}) \leq 0 \leq g_1(\cdot,\cdot,\underline{u}).$$

The theorem then applies to show that $[\underline{u}, \overline{u}]$ is invariant for m-regular solutions. That is, the standard "maximum principle" holds for scalar equations (6.69) satisfying the hypotheses of the theorem.

Example 2. Example 1 generalizes to the reaction-diffusion system

$$u_t^i + w^i(x,t,u) \cdot \nabla_x u^i = (d^i(x,t,u)u^i_{x_j})_{x_j} + F^i(x,t,u) \tag{6.71}$$

posed on $\Omega \times [0,T)$ with initial data ς_0, as in the theorem. Here u, w^i and F take values in $\mathbf{R}^N$ and the d^i are scalar-valued. In a typical application u^i represents the concentration of the i-th species in a mixture which in the absence of spatial dependence undergoes chemical reactions governed by an ordinary differential equation $\dot{u} = F$. On the other hand if the i-th species is assumed to convect in a region Ω along the integral curves of a known vector field w^i and to undergo molecular diffusion at the rate d^i, then $\dot{u}^i$ is replaced by the convective time-derivative on the left in (6.71) and both diffusion and reaction are represented on the right. The more common boundary condition in this setting, that the normal derivative of u^i is zero on $\partial\Omega$, is not accommodated in Theorem 6.13, however. Instead we renormalize the nominal range of u^i to be the interval $[-1,1]$ and take $\mathcal{H}_1 = H_0^1(\Omega)$ (which includes the case that $\Omega = \mathbf{R}^n$). The corresponding weak form is then (6.70) with $a^{\eta,\omega} = \delta_{\eta\omega} d^\eta I$ (Kronecker δ here), $g_1 = F$ and g_2^i the diagonal $N \times N$ matrix whose diagonal entries are the components of $-w^i$.

It is then easy to check that if the regularity hypotheses preceding (a) of Theorem 6.13 are satisfied and $\Phi(u) = (u^i - 1)(u^i + 1)$ for a fixed i, then (a), (bii) and (c) hold with $P = -I$ provided

$$F^i(\cdot,\cdot,e_i) \leq 0 \leq F^i(\cdot,\cdot,-e_i)$$

where e_i is the i-th standard basis vector in $\mathbf{R}^N$. If this condition holds for all i, then the set $[-1,1]^N$ is invariant for m-regular solutions, consistent with the interpretation of its components as normalized concentrations.

Example 3. Here we consider a simplification of the Navier-Stokes system considered in section 6.2 in which the motion of a fluid of constant density occupying a region $\Omega \subseteq \mathbf{R}^n$ is described by its velocity field $u(x,t)$ (so that $n = N$), which is to satisfy

$$u_t^i + \nabla_x u^i \cdot u = \big(\varepsilon(x,t,u)u^i_{x_j}\big)_{x_j} + F^i(x,t,u)$$

together with the condition that $u = 0$ on $\partial\Omega$. The weak form is then exactly as in (6.70) with $\mathcal{H}_1 = [H_0^1(\Omega)]^n$, $a^{\eta,\mu} = \delta_{\eta\mu}\varepsilon I$, $g_1 = F$, and $g_2^i = -u^i I$. We leave to the reader to check that if the regularity hypotheses preceding (a) are satisfied and if $\Phi(u) = \frac{1}{2}(|u|^2 - R^2)$ where $R > 0$ is constant, then (a), (b*ii*) and (c) hold with $P = -I$ provided $F(\cdot,\cdot,u)\cdot u \le 0$ for $|u| = R$. If so, then the ball of radius R centered at the origin of u-space is invariant for m-regular solutions.

Example 4. The p-system is the set of equations

$$\begin{aligned} v_t - u_x &= 0 \\ u_t + p(v)_x &= 0 \end{aligned}$$

for unknown functions v and u of $x \in \mathbf{R}$ and $t \ge 0$, where p is a given function of v (the reader need not be concerned that we have assumed that $n \ge 2$ throughout the text). These equations model the flow of a compressible fluid or gas occupying a thin tube along the real axis for which v, u and p represent specific volume (the reciprocal of density), velocity and pressure, and x is the so-called Lagrangean coordinate, which is constant along fluid-particle trajectories. (Thus the time derivatives on the left above correspond to the convective derivatives on the left sides in Examples 2 and 3.) It is generally assumed that $p \in C^2((0,\infty))$ with $p' < 0 < p''$. See [**8**] Sect. I.18 for the underlying rational mechanics and [**39**] Sect. 17§A for the fairly rich mathematical theory, disjoint from that of this text, of this deceivingly simple system.

The p-system can be "regularized," both for theoretical and computational purposes, by the addition of so-called artificial viscosity terms εv_{xx} and εu_{xx} to the right sides, where ε is a positive constant. Making this change and replacing v by $v+1$ and writing $c(v) = \sqrt{-p'(v+1)}$, we obtain the system

$$\begin{bmatrix} v \\ u \end{bmatrix}_t + \begin{bmatrix} 0 & -1 \\ -c^2(v) & 0 \end{bmatrix}\begin{bmatrix} v \\ u \end{bmatrix}_x = \varepsilon \begin{bmatrix} v \\ u \end{bmatrix}_{xx} \tag{6.72}$$

for which the origin of v–u space is in the domain of regularity. The eigenvectors of the transpose of the matrix on the left are $\begin{bmatrix} \pm c \\ 1 \end{bmatrix}$, so that if

$$\Phi(v,u) = u - \int_{v_0}^{v} c(s)ds$$

where $v_0 \in (-1,0)$ is fixed, then $\nabla_{(v,u)}\Phi$ is an eigenvector, and hypotheses (a) and (c) of Theorem 6.13 hold with $W = \{(v,u) : v > -1\}$. The term g_1 is zero in this example, so we check condition (b*ii*) with $P = -I$: if $\Phi(v,u) = 0$ then at (v,u),

$$\nabla_{(v,u)}\Phi \cdot \left(P\begin{bmatrix} v \\ u \end{bmatrix}\right) = c(v)v - u = c(v)v - \int_{v_0}^{v} c(s)ds.$$

If $v > v_0$ then the integral here is bounded below by $c(v)(v - v_0)$, since c is decreasing, and the right side here is therefore negative. For $v \in (-1, v_0)$ the derivative of the right side with respect to v is $c'(v)v$, which is positive, so that its value is bounded above by its value $c(v_0)v_0$ at $v = v_0$, which again is negative. The ($n = 1$ version of) Theorem 6.13 together with a similar analysis for the second eigenvector therefore shows that the set

$$\{(v,u) \in \mathbf{R}^2 : v \geq v_0 \text{ and } -\int_{v_0}^{v} c(s)ds \leq u \leq \int_{v_0}^{v} c(s)ds\}$$

is invariant for (6.72).

§6.7 A Global Existence Result for Systems in $\mathbf{R}^2$ and $\mathbf{R}^3$

In this section we prove global existence of 2-regular solutions of the quasilinear system (6.70) in two and three space dimensions when the $a^{\eta,\omega}$ depend only on x and satisfy the symmetry conditions in section 3.4 and when the solution in question is known to take values in an invariant set $\mathcal{S}$ in which the derivatives of $a^{\eta,\omega}$, g_1 and g_2^i are pointwise bounded. The proof consists in precluding the two blow-up alternatives of Corollary 6.11 by chaining in a particular order three *a priori* bounds established in previous sections (Theorems 3.7, 4.2 and 5.5) and by exploiting the strength of the Sobolev imbeddings $H^k \to L^p$ in low dimensions. The reader can check that the required invariant region hypothesis of the following theorem is satisfied in all four examples in the previous section. The first three of these, posed with initial conditions taking values in $\mathcal{S}$, therefore have global solutions if $n = 2$ or 3 and if the other relatively routine conditions of the theorem below are met.

The solution u to be found is to satisfy

$$\text{(6.73)} \quad \begin{aligned} \langle u(t), v\rangle_{L^2} = \langle \varsigma_0, v\rangle_{L^2} + \int_0^t\int_\Omega \Big[&-\big(a^{\eta,\omega}(x)u_{x_\eta}\big)\cdot v_{x_\omega} \\ &+\big(g_1(x,t,u) + g_2^i(x,t,u)u_{x_i}\big)\cdot v \end{aligned}$$

for $v \in \mathcal{H}_1$. The global existence result is as follows:

Theorem 6.14. *Fix an ensemble* (5.4) *as in* Definition 5.1 *with* $n = 2$ *or* 3 *and* $m = 2$, *let* $T \in (0,\infty]$ *and let* W *be a neighborhood of the origin in* $\mathbf{R}^N$. *Assume the following:*

(a) *Mappings* $a^{\eta,\omega}$, $0 \leq \eta,\omega \leq n$, *from* Ω *into the set of* $N \times N$ *matrices are given, satisfying the hypotheses of* Theorem 5.5, *that is, the positive definiteness condition* (5.5)*, any one of the sets of regularity conditions* (5.8), (5.10) *or* (5.11)–(5.13) *(in which all t-derivatives are now zero) and the symmetry conditions $(a^{\eta,\omega})^{tr} = a^{\omega,\eta}$ for all η, ω.*

(b) *g_1 and g_2 are C^4 mappings from $\Omega \times (0,T) \times W$ into $\mathbf{R}^N$ and $\mathbf{R}^{N\times n}$ respectively and g_1 satisfies* (6.36) *with $m = 2$.*

(c) *There is a closed set $\mathcal{S} \subset W$ which is star-shaped with respect to $0 \in \mathcal{S}$ such that*

(*i*) *$\mathcal{S}$ is invariant for* (6.73) *in the sense that if u is a 2-regular solution on $[0,\bar{t}]$ with $u(x,0) \in \mathcal{S}$ for all $x \in \Omega$, then $u(x,t) \in \mathcal{S}$ for all $(x,t) \in \Omega \times [0, t_{max})$ where t_{max} is its maximal time of existence with respect to W and T in the sense of* Corollary 6.11.

(*ii*) *For every $\bar{t} \in (0,T)$ there is a constant $C_{sys}(\bar{t})$ which bounds all the derivatives and Lipschitz constants of $a^{\eta,\omega}$, g_1 and g_2 referred to above in* (a) *and* (b) *on $\Omega \times [0,\bar{t}] \times \mathcal{S}$, as well as θ^{-1} (which is now independent of t; see* (5.5)*) and the $L^{\infty,2}(\Omega \times [0,\bar{t}])$ and $L^{2,2}(\Omega \times [0,\bar{t}])$ norms of $g_1(x,t,0)$ in* (6.36) *with $m = 2$.*

Let initial data $(\varsigma_0, \varsigma_1, \varsigma_2)$ for (6.73) *be given satisfying the compatibility conditions* (6.35) *with $m = 2$ and with ς_0 taking values in $\mathcal{S}$. Then if the corresponding solution described in* Theorem 6.10 *exists (which is true if hypothesis* (e) *of that theorem is satisfied) then it is global: its maximal time of existence with respect to W and T in the sense of* Corollary 6.11 *is T.*

PROOF. We give the proof for $n = 3$, the proof for $n = 2$ being similar. Spaces $[L^p(\Omega)]^N$ will be abbreviated L^p, etc., and the bound

$$|w|_{L^4} \leq C(\Omega)|w|_{L^2}^{1/4}|w|_{H^1}^{3/4}, \tag{6.74}$$

which follows from the imbedding $H^1 \to L^6$ for $N = 1$ (Theorem A.3(a)) via interpolation, will be applied throughout, as will the Bochner-Sobolev space description in Theorem 5.5(a).

Let u be as in the statement and t_{max} its maximal time of existence with respect to W and T. We will assume that $t_{max} < T$ and prove that

$$\sum_{j=0}^{2}\Big(\sup_{0\leq t<t_{max}} |u^{(j)}(t)|_{H^{4-2j}}^2 + \int_0^{t_{max}} |u^{(j)}|_{H^{5-2j}}^2 dt\Big) < \infty \tag{6.75}$$

and that the distance from the image of $u(t)$ to ∂W is bounded below away from zero independently of $t \in [0, t_{max})$. The latter follows from the former, however, because (6.75) yields a sup-norm bound for $u(t)$, which therefore takes values in the intersection of $\mathcal{S}$ with a closed ball, and this intersection is compact in W. Thus all three alternatives in Corollary 6.11 will be precluded, thereby showing that t_{max} cannot be strictly less than T.

Let C denote a generic positive constant depending on t_{max}, $C_{sys}(t_{max})$ and $|\varsigma_0|_{H^4} + |\varsigma_1|_{H^2} + |\varsigma_2|_{L^2}$. Define

$$\tilde{g}(t) = g_1(\cdot, t, u(t)) + g_2^i(\cdot, t, u(t))u_{x_i}(t),$$

noting that $\tilde{g}$ satisfies the conclusions of Lemma 6.8(c) with $m = 2$. Theorem 6.10 therefore applies, and its invocation of (5.25) and (5.27) in Theorem 5.5 together with the assumption that the $a^{\eta,\omega}$ are independent of t and u shows that, for $j = 0, 1, 2$ and $t \in (0, t_{max})$,

$$\langle u^{(j)}(t), v\rangle_{L^2} - \langle \varsigma_j, v\rangle_{L^2} + \int_0^t \int_\Omega \left[- \left(a^{\eta,\omega} u^{(j)}_{x_\eta}\right) \cdot v_{x_\omega} + \tilde{g}^{(j)} \cdot v\right] \tag{6.76}$$

for all $v \in \mathcal{H}_1$, where the $g^{(j)}$ are as in Lemma 6.8(c), and that

$$\tfrac{1}{2}|u^{(j)}(\cdot, t)|^2_{L^2} + \int_0^t \int_\Omega \left(a^{\eta,\omega} u^{(j)}_{x_\eta}\right) \cdot u^{(j)}_{x_\omega} = \tfrac{1}{2}|\varsigma_j|^2_{L^2} + \int_0^t \int_\Omega \tilde{g}^{(j)} \cdot u^{(j)}. \tag{6.77}$$

Also, Theorem 3.7 applies to (6.76) by virtue of the symmetry hypothesis in (a); thus for $t \in [0, t_{max}]$ and $j = 0, 1$,

$$\sup_{t \in [0, \bar{t}]} |u^{(j)}(t)|^2_{H^1} + \int_0^t |u^{(j+1)}|^2_{L^2} \leq C\left(|\varsigma_j|^2_{H^1} + \int_0^t |\tilde{g}^{(j)}|^2_{L^2}\right) \tag{6.78}$$

by (3.19). Finally, we have that for the same t,

$$|u^{(j)}(t)|_{H^{4-i}} \leq C\left(|u^{(j)}(t)|_{L^2} + |u^{(j+1)}(t)|_{H^{2-i}} + |\tilde{g}^{(j)}(t)|_{H^{2-i}}\right) \tag{6.79}$$

for $j = 0$ and $i = 0, 1, 2$ and for $j = 1$ and $i = 1, 2$, by the elliptic regularity result (4.16) of Theorem 4.2 applied to the strong form (5.26) of (6.76).

We begin by applying (6.77) with $j = 0$. Our hypotheses on g_1 and g_2 show that $|\tilde{g}(t)| \leq C(\gamma(t) + |u(t)| + |\nabla_x u(t)|)$ pointwise in Ω, where here and throughout the proof γ will denote any partial derivative of g_1 or g_2 with respect to $(x, t) \in \Omega \times [0, t_{max})$ of order less than or equal to three evaluated at $(x, t, 0)$. Any such γ is therefore bounded by C in $L^\infty([0, t_{max}); L^2)$ by hypothesis (b). Absorbing the term involving $|\nabla_x u|$ into the left side of (6.77) and applying Gronwall's inequality, we find that

$$\sup_{0 \leq t < t_{max}} |u(t)|^2_{L^2} + \int_0^{t_{max}} |u(t)|^2_{H^1} \leq C. \tag{6.80}$$

This shows in particular that $\displaystyle\int_0^{t_{max}} |\tilde{g}|^2_{L^2} \leq C$ and therefore by (6.78),

$$\sup_{0 \leq t < t_{max}} |u(t)|^2_{H^1} + \int_0^{t_{max}} |u^{(1)}(t)|^2_{L^2} \leq C. \tag{6.81}$$

In order to proceed we will need expressions for the derivatives of $\tilde{g}^{(1)}$ and $\tilde{g}^{(2)}$ with respect to x in terms of derivatives of $u^{(1)}$ and $u^{(2)}$. The computations are straightforward but somewhat involved, but can be facilitated by the following notational device. First note that if h is a component of either g_1 or g_2, then for (constant) $w \in \mathcal{S}$,

$$h_t(x, t, w) = h_t(x, t, 0) + \nabla_u h_t(x, t, sw) \cdot w$$

for some $s \in (0, 1)$. The first term on the right is a γ-term and $sw \in \mathcal{S}$. Thus if we compute $\tilde{g}^{(1)}$ from the definition of $\tilde{g}$ by the chain rule, for example, we find that its components are linear combinations of the components of γ, u, u_{x_i}, u_t, u_{x_it} and products of components of u_{x_i} and u_t, in which the coefficients are partial derivatives of g_1 or g_2 of order less than or equal to two evaluated at (x, t, z) for some $z \in \mathcal{S}$; these coefficients are therefore bounded pointwise in absolute value by C, by hypothesis (c). This rather complicated statement will be abbreviated

$$\tilde{g}^{(1)} \sim \gamma, u, u_x, u_t,\ u_{xt},\ u_xu_t$$

in which the symbol u_{xt} represents the ensemble of x-derivatives of $u^{(1)}$, for example. Higher order derivatives will be notated similarly.

Now proceeding to the application of (6.77) for $j = 1$ we see immediately from (6.80) and (6.81) that the contributions to the double integral on the right from γ, u, u_x, and u_t are bounded by C. We will henceforth ignore lower-order terms such as these and write, for example, $\tilde{g}^{(1)} \sim u_{xt},\ u_xu_t \dots$. The double integral on the right in (6.77) can then be bounded (symbolically), modulo an additive constant C, as follows:

$$\begin{aligned}\int_0^{t_{max}}\int_\Omega |\tilde{g}^{(1)}||u^{(1)}| \leq& C\iint \big(|u_{xt}||u_t| + |u_x||u_t|^2\big)dxdt\\ \leq& C\int \big(|u_{xt}|_{L^2}|u_t|_{L^2} + |u_x|_{L^2}|u_t|_{L^2}^{1/2}|u_t|_{H^1}^{3/2}\big)dt,\end{aligned}$$

by (6.74). We then apply the bounds in (6.80) and (6.81) to absorb the integrals involving u_{xt} into the left side of (6.77) in the obvious way (notice that the double integral on the left in (6.77) dominates the time integral of $|u^{(1)}|_{H^1}^2$ modulo the time integral of $|u^{(1)}|_{L^2}^2$, the latter having already been bounded in (6.81)). Applying Gronwall's inequality we thus obtain

$$\sup_{0\leq t<t_{max}} |u^{(1)}(t)|_{L^2}^2 + \int_0^{t_{max}} |u^{(1)}|_{H^1}^2 \leq C.$$

Next we apply this bound and the fact that $\tilde{g} \sim \gamma,\ u, u_x$ in (6.79) with $j = 0$ and $i = 2$ to obtain a bound for $|u(t)|_{H^2}$, then combine with the bound above to obtain

$$\sup_{0\leq t<t_{max}} |u^{(1)}|_{L^2}^2 + \sup_{0\leq t<t_{max}} |u|_{H^2}^2 + \int_0^{t_{max}} |u^{(1)}|_{H^1}^2 \leq C. \tag{6.82}$$

Recalling that $\tilde{g}^{(1)} \sim u_{xt},\ u_xu_t, \dots$, we then see that $\int_0^{t_{max}} |\tilde{g}^{(1)}|_{L^2}^2 \leq C$, so that by (6.78) with $j = 1$,

$$\sup_{0\leq t<t_{max}} |u^{(1)}(t)|_{H^1}^2 + \int_0^{t_{max}} |u^{(2)}|_{L^2}^2 \leq C. \tag{6.83}$$

Next we apply (6.77) with $j = 2$. A somewhat lengthy calculation shows that

$$\tilde{g}^{(2)} \sim u_xu_t^2,\ u_{xt}u_t,\ u_xu_{tt},\ u_{xtt}, \dots.$$

The contribution to the right side of (6.77) from the integral of $\tilde{g}^{(2)} \cdot u^{(2)}$, modulo lower order terms and the absorption of the u_{xtt} term into the left side, is therefore bounded by

$$\int_0^{t_{max}} \int_\Omega \big(|u_x||u_t|^2 + |u_{xt}||u_t| + |u_x||u_{tt}|\big)|u_{tt}|.$$

The first term here is bounded by

$$\begin{aligned} C\int_0^{t_{max}} |u_x|_{L^4}|u_t|_{L^4}^2|u_{tt}|_{L^4} &\le C\int_0^{t_{max}} |u|_{H^2}|u_t|_{H^1}^2|u_{tt}|_{H^1} \\ &\le C\Big(\int_0^{t_{max}} |u_{tt}|_{H^1}\Big)^{1/2} \end{aligned}$$

by (6.82) and (6.83), which can be absorbed modulo an additive constant C. The second term above is bounded by

$$\begin{aligned} \int_0^{t_{max}} |u_{xt}|_{L^2}|u_t|_{L^4}|u_{tt}|_{L^4} &\le C\int_0^{t_{max}} |u_t|_{H^1}^2|u_{tt}|_{L^2}^{1/4}|u_{tt}|_{H^1}^{3/4} \\ &\le C\Big(\int_0^{t_{max}} |u_{tt}|_{H^1}\Big)^{3/4} \end{aligned} \tag{6.84}$$

again by (6.82) and (6.83), with the same result. Finally, since $|u_x(t)|_{L^2} \le C$, the third term is bounded by

$$C\int_0^t |u_{tt}|_{L^4}^2 \le C\int_0^t |u_{tt}|_{L^2}^{1/2}|u_{tt}|_{H^1}^{3/2} \le C\Big(\int_0^t |u_{tt}|_{H^1}^2\Big)^{3/4}$$

by (6.83), again with the same result. Gronwall's inequality is then applied to the terms remaining on the right in (6.77). We conclude that

$$\sup_{0\le t<t_{max}} |u^{(2)}(t)|_{L^2}^2 + \int_0^{t_{max}} |u^{(2)}|_{H^1}^2 \le C, \tag{6.85}$$

which provides the bound required for the $j = 2$ terms in (6.75).

To bound the $j = 1$ terms in (6.75) we apply (6.79) with $j = 1$ and σ either zero or one:

$$|u^{(1)}(t)|_{H^{2+\sigma}} \le C\big(|u^{(1)}(t)|_{L^2} + |u^{(2)}(t)|_{H^\sigma} + |\tilde{g}^{(1)}(t)|_{H^\sigma}\big). \tag{6.86}$$

As noted above preceding (6.83), $\tilde{g}^{(1)} \sim u_x u_t, u_{xt}, \ldots$, which is bounded in L^2 by C independently of t by (6.82) and (6.83). Thus the $\tilde{g}^{(1)}$ term above with $\sigma = 0$ is bounded by C and this together with (6.82) and (6.85) shows that

$$\sup_{0\le t<t_{max}} |u^{(1)}(t)|_{H^2}^2 \le C. \tag{6.87}$$

Before treating the other $j = 1$ term in (6.75) we note that at each t,

$$|u|_{H^3} \le C\big(|u|_{L^2} + |u^{(1)}|_{H^1} + |\tilde{g}|_{H^1}\big)$$

by (6.79). It is easy to check that $\nabla_x \tilde{g} \sim u_x^2, u_{xx}, \ldots$, which by (6.82) is bounded in L^2, so that $\tilde{g}$ is bounded in H^1 by C, independently of t. Since the H^3 norm of u dominates the sup norm of u_x, we therefore have that

$$\sup_{0 \le t < t_{max}} |u(t)|_{H^3}, \quad \sup_{0 \le t < t_{max}} |\nabla_x u(t)|_{L^\infty} \le C. \tag{6.88}$$

In order to apply (6.88) with $\sigma = 1$ we compute that

$$\nabla_x \tilde{g}^{(1)} \sim u_x^2 u_t,\ u_x u_{xt},\ u_{xx} u_t,\ u_{xxt}, \ldots.$$

The first two terms here are lower order by the bound immediately above, and (6.86) and (6.87) show that the others are bounded in L^2 by C independently of t. Applying this together with (6.85) and (6.87) we conclude that

$$\sup_{0 \le t < t_{max}} |u^{(1)}(t)|_{H^2}^2 + \int_0^{t_{max}} |u^{(1)}|_{H^3}^2 \le C, \tag{6.89}$$

which provides the bound required for the $j = 1$ terms in (6.75).

Bounds for the $j = 0$ terms in (6.75) are derived in a similar way. First we write (6.79) with $j = 0$ and σ either zero or one:

$$|u(t)|_{H^{4+\sigma}} \le C\big(|u(t)|_{L^2} + |u^{(1)}(t)|_{H^{2+\sigma}} + |\tilde{g}(t)|_{H^{2+\sigma}}\big). \tag{6.90}$$

We then check that $D_x^2 \tilde{g} \sim u_x^3,\ u_x u_{xx},\ u_{xxx} \ldots$, each term here being bounded by C in L^2 independently of t, so that $|\tilde{g}(t)|_{H^2} \le C$. It then follows from this and (6.88) and (6.89) with $\sigma = 0$ that u is bounded in H^4 by C independently of t. Finally we compute that $D_x^3 \tilde{g} \sim u_x^4,\ u_x^2 u_{xx},\ u_x u_{xxx},\ u_{xx}^2, u_{xxxx} \ldots$, all of which are easily seen to be bounded in L^2 by C for almost all t, hence so is $|\tilde{g}|_{H^3}$. Applying this and (6.88) in (6.89) we conclude that

$$\sup_{0 \le t < t_{max}} |u(t)|_{H^4}^2 + \int_0^{t_{max}} |u|_{H^5}^2 \le C,$$

which provides the required bound for the $j = 0$ terms in (6.75), whose proof is now complete. Corollary 6.11 therefore applies to show that $t_{max} = T$.

□

Exercises

6.1 Suppose that hypotheses (a)–(e) and (h) of Theorem 6.2 hold. Show that solutions of (6.1) in the sense of that theorem depend continuously on their initial data in the norm implied by the left side of (6.3).

6.2 Find an example of a problem of the form (6.18)–(6.19) for which the hypotheses of Theorem 6.6 hold with $T = \infty$ and for which there is a $t_0 \in (0, \infty)$ such that for any $\bar{t} \in (t_0, \infty)$ there is a solution whose maximal time of existence is $\bar{t}$ (always with the same ς_0).

6.3 Justify the statement (6.17) in the proof of Theorem 6.5.

6.4 Prove that hypothesis (e) of Theorem 6.10 is always satisfied if $\Omega = \mathbf{R}^n$ and $\mathcal{H}_j = [H^j(\mathbf{R}^n)]^N$. That is, given $(\varsigma_0, \ldots, \varsigma_m) \in \mathcal{H}_{2m} \times \cdots \mathcal{H}_0$, $m \geq 1$, there is an m-regular mapping u on $[0, \infty)$ with respect to these spaces such that $u^{(j)}(0) = \varsigma_j$, $j = 1, \ldots, m$. Suggestion: observe that for all $v \in H^1$,

$$\langle \varsigma_{j+1}, v \rangle_{L^2} = -\langle \varsigma_j, v \rangle_{H^1} + \int_{\mathbf{R}^n} g_j \cdot v$$

where $g_j = \varsigma_{j+1} + \varsigma_j - \Delta \varsigma_j$ (Δ is the Laplace operator). Apply induction to extend $(g_0, \ldots, g_{m-1})$ to an $(m-1)$-regular mapping g with data $(g_0, \ldots, g_{m-1})$ then apply Theorem 5.5.

6.5 Fix an ensemble (5.4) as in Definition 5.1 in which Ω is a proper open subset of $\mathbf{R}^n$ and let $(\varsigma_0, \ldots, \varsigma_m) \in \mathcal{H}_{2m} \times \cdots \mathcal{H}_0$. Show that there is an m-regular mapping on $[0, \infty)$ with respect to these spaces and with these data provided that there is a time-independent bilinear form A on $\mathcal{H}_1 \times \mathcal{H}_1$ satisfying the hypotheses of Lemma 5.3 and therefore of Theorem 5.5 such

$$\begin{cases} L^i \varsigma_j \in \mathcal{H}_{2m-2(j+i)}, & i + j \leq m \\ L_\partial L^i \varsigma_j \perp \mathcal{T}(\mathcal{H}_1) \text{ in } L^2(\partial\Omega), & i + j \leq m - 1, \end{cases}$$

where

$$A(z, w) = \int_{\partial\Omega} L_\partial z \cdot w \, d\sigma - \int_\Omega Lz \cdot w \, dx$$

as in section 3.1 and $\mathcal{T}$ is the trace operator. Suggestion: follow the outline in Problem 6.4. Observe that since $L^i \varsigma_j$ is automatically in $[H^{2m-2(j+i)}]^N$, the assumptions here are that it satisfies both essential and natural boundary conditions.

6.6 Prove that hypothesis (e) of Theorem 6.10 is satisfied for the problem

$$\langle u(t), v \rangle_{L^2} = \langle \varsigma_0, v \rangle_{L^2} + \int_0^t \int_\Omega \big[- (a^{\eta,\omega}(x,t) u_{x_\eta}) \cdot v_{x_\omega} + g(x, t, u) \cdot v \big] \, dx \, dt$$

with $n = 2$ or 3 and $m = 2$ if the other hypotheses of that theorem are satisfied and when $\mathcal{H}_j$ is either $[H^j(\Omega)]^N$ or $[H_0^1(\Omega) \cap H^j(\Omega)]^N$.

Suggestion: first solve a homogeneous linear problem to construct $f \in C([0,t_0];L^2)$ with equivalence class in $L^2([0,t_0];H^1)$ and $f(0) = g(\cdot,0,\varsigma_0)$. Then check that the solution z of the problem

$$\langle z(t), v\rangle_{L^2} = \langle \varsigma_0, v\rangle_{L^2} + \int_0^t \int_\Omega \big[-(a^{\eta,\omega}(x,t) z_{x_\eta}) \cdot v_{x_\omega} \\ + f(t)\cdot v\big]\, dx\, dt$$

is 1-regular with respect to $\mathcal{H}_0, \ldots, \mathcal{H}_3$ with data $(\varsigma_0, \varsigma_1)$. Finally, check that the solution u of the problem

$$\langle u(t), v\rangle_{L^2} = \langle \varsigma_0, v\rangle_{L^2} + \int_0^t \int_\Omega \big[-(a^{\eta,\omega}(x,t) u_{x_\eta}) \cdot v_{x_\omega} \\ + g(\cdot,t,z(t))\cdot v\big]\, dx\, dt$$

is 2-regular with data $(\varsigma_0, \varsigma_1, \varsigma_2)$. The difficult part is the proof that $D_t g(\cdot,t,z(t))^* \in wAC^2([0,t_0];\mathcal{H}_1^*)$; approximation of elements of $\mathcal{H}_1$ by smooth functions is useful here.

6.7 The bound (6.41) in Theorem 6.10 is derived from the bound (2.28) in Theorem 2.7 for abstract linear problems (via the application in Theorem 5.5). The upper bound in the latter includes the time integrals $\displaystyle\int_0^{\bar{t}} |F^{(j)}|^2_{\mathcal{V}_{2m-2j-1}}\, dt$ for the inhomogeneous terms, but the former does not. Explain this apparent discrepancy.

6.8 Construct an example in which the hypotheses of Theorem 6.12 hold with $T = \infty$ but with $\gamma = 0$ in (d) and such that the conclusion of the theorem fails; that is, for every $\kappa > 0$ there is an initial datum ς_0 satisfying the conditions in (e) and for which the maximal existence time t_{max} of the corresponding m-regular solution is finite.

6.9 Consider the weak form (6.27) of the problem

$$\begin{cases} u_t = \Delta u + g(u), \ (x,t) \in \Omega \times (0,T), \\ \nabla_x u(x,t) \cdot \nu(x) = 0, \ (x,t) \in \partial\Omega \times [0,T), \\ u(x,0) = \varsigma_0(x), \ x \in \Omega, \end{cases}$$

where Ω is a bounded open set in $\mathbf{R}^n$, Δ is the Laplace operator, ν is the normal vector field on $\partial\Omega$, and $g(u) = \frac{5}{4}u - \frac{3}{4}u^3$ (thus u is scalar and $N = 1$). Show that, under appropriate regularity conditions on Ω, two of the three roots of g are attracting rest points of the differential equation and boundary condition in the sense of the first paragraph in section 6.5 and that the third is not.

Next let $h(s) = \sqrt{(1-s^2)(1-s^2/4)}$, $H(u) = \displaystyle\int_0^u \frac{ds}{h(s)}$ for $|u| \le 1$, and $L = H(1)$. Show that H has a smooth inverse function $H^{-1} : [-L, L] \to [-1,1]$ and that $u(x) \equiv H^{-1}(x)$ is a steady state solution of the above problem with $n = 1$ and $\Omega = (-L, L)$ (do not be concerned that we have assumed $n \ge 2$ throughout the text). Consider whether or not u

is a stable steady state. Suggestion: show that if $E : \mathcal{H}_1 \to \mathbf{R}$ is the functional $E[v] = \int_{-L}^{L} \left(\frac{1}{2}v_x^2 - G(v)\right)dx$, where G is an antiderivative of g, and if u is a solution, then $E[u(t)]$ is nonincreasing in time.

6.10 Prove the statement in Theorem 6.13 that, under hypothesis (bi), if $\varsigma_0(x) \in \Sigma$ for all $x \in \Omega$, then $u_{\mathfrak{c}}(x,t) \in \Sigma$ for all $(x,t) \in \overline{\Omega} \times [0, t_{max})$.

6.11 Assume that the hypotheses and notations of Theorem 6.13 are in force. Let $w \in \partial\Sigma$ and show that the matrix $\Phi''(w)$ must be nonnegative. Show also that if strict inequality holds in the last item in hypothesis (a) for all nonzero M, then for every $w \in \partial\Sigma$ there is a ball B centered at w such that $B \cap \overline{\Sigma}$ is convex.

6.12 Consider the following system for unknown scalar functions v and w of x and t:

$$\begin{cases} v_t = \mu_1 \Delta v - p(v) - aw + f, \\ w_t = \mu_2 \Delta w + bv - cw + g \end{cases}$$

posed on a bounded open set in $\mathbf{R}^n$ with zero boundary conditions. Here μ_1, μ_2, a, b, c, f and g are positive constants and p is a polynomial of odd degree greater than one whose leading order term has positive coefficient. (If spatial effects are neglected the Laplacean terms are omitted and the resulting system of ordinary differential equations is a prototype which includes the FitzHugh–Nagumo system, an important system arising in mathematical biology; see [**14**].) Show that there is a family of closed rectangles of arbitrary large size centered at the origin of v–u space which are invariant for the above system.

APPENDIX

Selected Topics in Analysis

Here we discuss various topics which are applied in the text but which may be unfamiliar to some readers. Complete proofs are given, with a few exceptions in section A.1 for which standard references are suggested.

§A.1 Density and Imbedding Theorems

In this section we state several standard results concerning approximation and imbeddings in Lebesgue and Sobolev spaces. The first of these, which concerns approximation of elements of $L^p(X, \mu)$ by continuous functions, where X is a topological space and μ a Borel measure on X, was applied in section 3.2 for the case that $(X, \mu) = (\partial\Omega, \sigma)$. Before stating the theorem we recall that a *Radon measure* on a general topological space X is a nonnegative Borel measure which is finite on compact sets and which is outer regular on Borel sets (that is, the measure of a Borel set is the inf of the measures of larger open sets) and inner regular on open sets (the measure of an open set is the sup of the measures of smaller compact sets).

The following standard result is proved in [**13**] Proposition 7.9:

Theorem A.1. *Let X be a locally compact Hausdorff topological space (that is, every point of X has a neighborhood whose closure is compact and open sets in X separate points of X) and let μ be a Radon measure on X. Then the set $C_c(X)$ of continuous real-valued functions on X having compact support is dense in $L^p(X)$ for all $p \in [1, \infty)$.*

Simple examples, which the reader can easily construct, show that the theorem does not hold in general for $p = \infty$.

Next we consider approximation of Sobolev functions by smooth functions. The following result is proved in [**1**], Theorem 3.22:

Theorem A.2. *Let Ω be an open set in $\mathbf{R}^n$, $n \geq 2$, satisfying the segment condition* Definition 1.6(a) *and let k be a nonnegative integer. Then for $p \in [1, \infty)$ the set of restrictions to Ω of elements of $C_c^\infty(\mathbf{R}^n)$ is dense in $W^{k,p}(\Omega)$.*

The following result concerning continuous imbeddings of Sobolev spaces into Lebesgue spaces is applied throughout the text. Part (a) is a subset of the general Sobolev imbedding theorem proved in [**1**] Theorem 4.12, parts (b) and (c) are straightforward applications of (a) involving simple index arithmetic, and the proof of (d) is given following the statement.

Theorem A.3. *Let Ω be an open set in $\mathbf{R}^n$, $n \geq 2$, satisfying the cone condition* Definition 1.6(b) *and let k be a nonnegative integer.*

(a) *The space $H^k(\Omega)$ is continuously included in $L^s(\Omega)$ for all s in the set $\mathcal{I}'_k$ defined by*

$$\mathcal{I}'_k = \begin{cases} [2, 2n/(n-2k)]\,, \ 0 \leq k < n/2, \\ [2, \infty)\,, \ k = n/2, \\ [2, \infty]\,, \ k > n/2. \end{cases}$$

The norm of the inclusion is determined by n, k, the cone constants r and δ in Definition 1.6(b) *and in the case that $k = n/2$ by s.*

(b) *If $u \in H^k(\Omega)$ and $v \in L^2(\Omega)$ then $uv \in L^{s'}(\Omega)$ for all s' whose Hölder conjugate is in the set $\mathcal{I}_k$ defined by*

$$\mathcal{I}_k = \begin{cases} [n/k, \infty], \ 0 \leq k < n/2, \\ (2, \infty], \ k = n/2, \\ [2, \infty], \ k > n/2, \end{cases} \tag{A.1}$$

and $|uv|_{L^{s'}(\Omega)} \leq C|u|_{H^k(\Omega)}|v|_{L^2(\Omega)}$ where C depends on n, k, the cone constants r and δ in Definition 1.6(b) *and in the case that $k = n/2$ on s'.*

(c) *If $u \in H^k(\Omega)$ and $v \in H^1(\Omega)$ then $uv \in L^{s'}(\Omega)$ for all s' whose Hölder conjugate is in the set $\mathcal{K}_k$ defined by*

$$\mathcal{K}_k = \begin{cases} (2, \infty], \ n = 2, \ k = 0, \\ [n/(k+1), \infty], \quad n > 2, \ 0 \leq k < n/2, \\ (2n/(n+2), \infty], \ k = n/2, \\ [2n/(n+2), \infty], \ k > n/2, \end{cases} \tag{A.2}$$

and $|uv|_{L^{s'}(\Omega)} \leq C|u|_{H^k(\Omega)}|v|_{H^1(\Omega)}$ where C depends on the same quantities as the norm of the inclusion in (b).

(d) *Let nonnegative integers m, k and j be given with $m > n/2$ and $j + k \geq 2m - 2$. Then there is a constant C depending on these integers and on the norms of the inclusions $H^m(\Omega) \subset L^\infty(\Omega)$ and $H^{m-1}(\Omega) \subset L^4(\Omega)$ such that if $u \in H^j(\Omega)$ and $v \in H^k(\Omega)$, then $uv \in H^p(\Omega)$ where $p = \min\{j, k, j+k-2m+2\}$, and $|uv|_{H^p} \leq C|u|_{H^j}|v|_{H^k}$.*

(e) *The statements in* (a)–(d) *hold for arbitrary open sets $\Omega \subseteq \mathbf{R}^n$ with constants and norms independent of Ω provided that $H^k(\Omega)$ and $H^1(\Omega)$ are replaced by $H^k_0(\Omega)$ and $H^1_0(\Omega)$ in every instance.*

PROOF of (d). We have to bound $D^\beta u \cdot D^\gamma v$ where $\beta + \gamma = \alpha$ and $|\alpha| \leq p$. If $j - |\beta| \geq m$ then $D^\beta u \in H^{j-\beta} \subset H^m \subset L^\infty$, and the required bound is obvious. The same holds if $k - |\gamma| \geq m$. If $j - |\beta| = k - |\gamma| = m - 1$, then both factors are in L^4 and their product is therefore in L^2, again with the required bound. The remaining cases that $k-|\beta| \le m-1$ and $j-|\gamma| \le m-2$, or the reverse, do not occur because $k+j-|\alpha| \ge 2m-2$.

□

The following bounds are useful for the special case that $n=2$:

Theorem A.4. *Let Ω be an open set in $\mathbf{R}^2$ and $p \in [2,\infty)$.*

(a) *There is a constant C depending only on p such that if $w \in H_0^1(\Omega)$ then*

$$\int_\Omega |w|^p \le C(p)|w|^2_{L^2(\Omega)}|w|^{p-2}_{H^1(\Omega)}.$$

(b) *If Ω satisfies the cone condition then given $\varepsilon \in (0,2)$ there is a constant C depending on Ω, p and ε such that for all $w \in H^1(\Omega)$,*

$$\int_\Omega |w|^p \le C(\Omega,p,\varepsilon)|w|^{2-\varepsilon}_{L^2(\Omega)}|w|^{p-2+\varepsilon}_{H^1(\Omega)}.$$

PROOF. For (a) we begin with the general inequality

$$\int_{\mathbf{R}^2} |\varphi|^2 \le \left(\int_{\mathbf{R}^2} |\nabla\varphi|\right)^2$$

for $\varphi \in C_c^1(\mathbf{R}^2)$. This is proved by bounding $\varphi(x_1,x_2)^2$ by the product of the integrals over $\mathbf{R}$ of $|\varphi_{x_1}|$ and $|\varphi_{x_2}|$ and integrating the result over $\mathbf{R}^2$. We then apply this with φ replaced by φ^r where r is an even integer to obtain

$$\int_{\mathbf{R}^2} \varphi^{2r} \le C(r)\int_{\mathbf{R}^2} \varphi^{2r-2}\int_{\mathbf{R}^2} |\nabla\varphi|^2,$$

then interpolate the L^{2r-2} norm between L^2 and L^{2r} and simplify to obtain the bound in (a) for $p=2r$. The same result then follows for other p by interpolation between L^2 and L^{2r}, and then extends to the H^1 closure of $C_c^1(\Omega)$, which is $H_0^1(\Omega)$.

Part (b) is proved by interpolating the L^p norm between L^2 and L^q for $q>p$, applying Theorem A.3(a) to bound the L^q norm in terms of the H^1 norm, then choosing q large depending on ε.

□

The proof of (a) above does not apply in general with $w \in H^1(\Omega)$ without some further condition on Ω. It suffices for example that Ω be an *extension domain* in the sense that elements of $H^1(\Omega)$ are the restrictions to Ω of elements of $H^1(\mathbf{R}^2)$ with a suitable norm inequality. See for example the discussion in [**1**], pp. 146-155.

The following is a special case of the Rellich-Kondrachov Theorem given in [**1**], Theorem 6.3:

Theorem A.5. *Let Ω be a bounded open set in $\mathbf{R}^n$, $n \ge 2$, satisfying the cone condition* Definition 1.6(b). *Then the inclusion $H^1(\Omega) \subset L^2(\Omega)$ is* *compact; that is, bounded sets in $H^1(\Omega)$ are mapped to compact sets in $L^2(\Omega)$.*

Again, simple examples, which the reader can easily construct, show that the theorem does not hold in general when Ω is unbounded.

The final result of this section concerns imbeddings of Sobolev spaces into spaces of continuously differentiable functions. The following theorem is included in [**1**], Theorem 4.12:

Theorem A.6. *Let $n \geq 2$ and let k_0 be the unique positive integer for which $k_0 - 1 \leq n/2 < k_0$. Define $\lambda = k_0 - n/2$ if $k_0 - 1 < n/2$ and let λ be an arbitrary point in $(0, 1)$ if $k_0 - 1 = n/2$, and finally let $k = k_0 + i$ where i is a nonnegative integer.*

(a) *If Ω is an open set in $\mathbf{R}^n$ then $H_0^k(\Omega)$ is continuously included in $C_{bdd}^{i,\lambda}(\Omega)$.*

(b) *If Ω is a $C_{bdd}^{0,1}$ domain in $\mathbf{R}^n$, not necessarily bounded, then $H^k(\Omega)$ is continuously included in $C_{bdd}^{i,\lambda}(\Omega)$.*

The norms of the inclusions depend on n and i, on λ if $k_0 - n/2 = 1$, and in part (b) *on $M_\Omega^{0,1}$* (Definition 1.7).

Of course, the elements of H^k are Lebesgue equivalence classes, so the statement that H^k is continuously included in $C_{bdd}^{i,\lambda}$ means that each element of H^k contains a representative in $C_{bdd}^{i,\lambda}$ and that the associated mapping is a bounded linear transformation.

§A.2 Weak Solutions of Ordinary Differential Equations

Two results are proved in this section, both of which are well-known for continuously differentiable solutions of systems of ordinary differential equations but which are applied in the text for solutions which are no more than absolutely continuous. These give local existence for nonlinear systems and global existence for linear systems.

Theorem A.7. *Let V be an open set in $\mathbf{R}^N$, $N \geq 1$, let $T \in (0, \infty]$ and assume that $F : V \times [0, T) \to \mathbf{R}^N$ satisfies the following:*

- *F is measurable on $V \times [0, T)$.*
- *If $a : [0, T) \to V$ is measurable on $[0, T)$ then so is the mapping $t \to F(t, a(t))$.*
- *There is a set $E \subseteq [0, T)$ of measure zero such that if $t \notin E$ then $F(t, a)$ is a continuous function of $a \in V$.*

- *V is the union of an increasing sequence of open sets V_j which, if $V \neq \mathbf{R}^n$, are each at a positive distance from ∂V, and $|F(t,\cdot)|_{L^\infty(V_j)}$ is locally integrable on $[0,T)$ (see the introductory discussion to §A.5 for a comment on the meaningfulness of this hypothesis).*

Then given $a_0 \in V$ there is a number $\tau \in (0,T)$ and an absolutely continuous function $a : [0,\tau] \to V$ such that

$$a(t) = a_0 + \int_0^t F(s, a(s))ds \tag{A.3}$$

for all $t \in [0,\tau]$.

PROOF. There is a j and an $r > 0$ such that the closure of the ball B of radius r centered at a_0 is contained in V_j. We can then choose $\tau \in (0,T)$ so that $\int_0^\tau |F(\cdot,t)|_{L^\infty(V_j)}dt < r$ and a continuous function $\omega : [0,\tau] \to [0,\infty)$ such that $\omega(0) = 0$ and $\int_{t_1}^{t_2} |F(\cdot,t)|_{L^\infty(V_j)}dt < \omega(|t_2 - t_1|)$ for $0 \leq t_1 \leq t_2 \leq \tau$. We will obtain a solution of (A.3) as a fixed point of a map on the set $\mathcal{S}$ of functions $a : [0,\tau] \to \bar{B}$ satisfying $a(0) = a_0$ and having modulus of continuity ω, regarded as a subset of the Banach space X of continuous $\mathbf{R}^N$-valued functions on $[0,\tau]$ with the usual maximum norm. Observe that the constant function a_0 is in $\mathcal{S}$, which is therefore nonempty, $\mathcal{S}$ is clearly convex and $\mathcal{S}$ is a compact subset of X by the Ascoli-Arzela theorem. It is easy to check that if $\mathcal{F}$ is the map

$$\mathcal{F}(a)(t) = a_0 + \int_0^t F(s, a(s))ds$$

then $\mathcal{F}$ maps $\mathcal{S}$ to $\mathcal{S}$. To show that $\mathcal{F}$ is continuous in the norm of X we let a_i be a sequence in $\mathcal{S}$ which converges in X to a. Then $a \in \mathcal{S}$ and for $t \in [0,\tau]$,

$$|\mathcal{F}(a_i)(t) - \mathcal{F}(a)(t)| \leq \int_0^\tau |F(s, a_i(s)) - F(s, a(s))|ds.$$

The integrand on the right here goes to zero for almost all s and is bounded by $2|F(\cdot,s)|_{L^\infty(V_j)}$, which is integrable and which is independent of both i and t. This proves that $\mathcal{F}(a_i) \to \mathcal{F}(a)$ uniformly on $[0,\tau]$ and therefore that $\mathcal{F}$ is continuous in the norm of X. Schauder's theorem ([**12**] pg. 538) then applies to show that $\mathcal{F}$ has a fixed point $a \in \mathcal{S}$, which is therefore a solution of (A.3) and, being the antiderivative of an integrable function, is absolutely continuous.

□

Solutions in the generality of the theorem need not be unique, of course. For example, if $V = (-1,1)$, $F(a) = a^{2/3}$, and $a_0 = 0$, then both $a_1(t) \equiv 0$ and $a_2(t) = t^3/27$ are weak solutions of (A.3).

It is often important to know whether a solution can be extended past the time τ given in the theorem and if not, to understand the nature of the obstruction. We will say that $\tilde{a}$ is an *extension* of a relative to $[0, T) \times V$ if $\tilde{a}$ is a solution of (A.3) on an interval $[0, \tilde{\tau}]$ where $\tilde{\tau} \in [\tau, T)$, $\tilde{a}$ takes values in V and $\tilde{a}$ agrees with with a on $[0, \tau]$ (thus a is an extension of itself). The *maximal time of existence* t_{max} relative to $[0, T) \times V$ of the solution a is then the supremum of times $\tilde{\tau} \in [\tau, T)$ for which such an extension exists on $[0, \tilde{\tau}]$. Observe that if $t \in [0, t_{max})$ then the image $a([0, t])$ is a compact subset of V and so is contained in V_j for some minimal j denoted by $J_a(t)$, which is nondecreasing in t.

The following result is often useful in obtaining a lower bound for t_{max}:

Corollary A.8. *Let* T*,* V*,* F *and* a_0 *be as in* Theorem A.7 *and let* a *be a solution of* (A.3) *on* $[0, \tau]$ *in the sense of that theorem with maximal time of existence* t_{max} *relative to* $[0, T) \times V$*. Then either* $t_{max} = T$ *or* $\lim_{t \to t_{max}^-} J_a(t) = \infty$*.*

Proof. If not, then $t_{max} < T$ and the limit in the hypothesis is an integer j_1. It follows that the image $a([0, t_{max}))$ is contained in V_{j_1}, and since $|F(t, \cdot)|_{L^\infty(V_{j_1})}$ is integrable on $[0, t_{max}]$, a is uniformly continuous on $[0, t_{max})$, by (A.3). Consequently a has a continuous extension a_1 at $t = t_{max}$ satisfying

$$a_1 = a_0 + \int_0^{t_{max}} F(s, a(s)) ds.$$

Since $a_1 \in \overline{V}_{j_1} \subset V$, Theorem A.7 applies to show that there is a solution a' to the problem

$$a'(t) = a_1 + \int_{t_{max}}^{t} F(s, a'(s)) ds$$

for $t \in [t_{max}, \tilde{\tau}]$ for some $\tilde{\tau} \in (t_{max}, T)$ and taking values in V. Concatenating the solutions a and a' we then obtain an extension $\tilde{a}$ of a to the interval $[0, \tilde{\tau}]$, contradicting the definition of t_{max}.

□

The Corollary must not be confused with its converse: it is possible, for example, that a solution approaches a point $\overline{a} \in \partial V$ as t approaches a time $t_{max} < T$ but that $\overline{a}$ is interior to a larger set $V' \supset V$ on which the hypotheses of Theorem A.7 hold. In this case the maximal time of existence relative to $V' \times [0, T)$ is strictly larger than that relative to $V \times [0, T)$. The usefulness of the Corollary therefore depends on judicious choices for V and T. See the application in the proof of Theorem 6.2 for example.

When the function F in (A.3) is *linear* in a, solutions are global in time:

Theorem A.9. *Let* $N \geq 1$*,* $T \in (0, \infty]$*,* M *a locally integrable mapping from* $[0, T)$ *into the set of* $N \times N$ *matrices and* $F \in L^1_{loc}([0, T); \mathbf{R}^N)$*. Then given*

$a_0 \in \mathbf{R}^N$ there is a unique absolutely continuous function $a : [0,T) \to \mathbf{R}^N$ satisfying

$$a(t) = a_0 + \int_0^t [M(s)a(s) + F(s)]ds \tag{A.4}$$

for all $t \in [0,T)$. If M and F are absolutely continuous on $[0,T)$ then so is the t-derivative $a'(t)$.

PROOF. The solution is constructed by a standard iteration procedure: let $a^0 = a_0$ and

$$a^{k+1}(t) = a_0 + \int_0^t [M(s)a^k(s) + F(s)]ds$$

for integers $k \geq 0$. If by induction a^k is continuous on $[0,T)$ then $Ma^k + F$ is locally integrable and therefore a^{k+1} is continuous on $[0,T)$ as well. To prove convergence let $E^k(t) = |a^{k+1}(t) - a^k(t)|$ and $h(t) = \int_0^t |M(s)|ds$. Fix $\bar{t} \in [0,T)$ and $C(\bar{t})$ such that $E^0(t) \leq C(\bar{t})$ for $t \in [0,\bar{t}]$. If by induction $E^k(t) \leq C(\bar{t})h(t)^k/k!$ for these t, then

$$E^{k+1}(t) \leq \int_0^t |M(s)|E^k(s)ds \leq C(\bar{t}) \int_0^t |M(s)|\frac{h(s)^k}{k!}ds = C(\bar{t})\frac{h(t)^{k+1}}{(k+1)!}.$$

Since the terms on the right are summable in k, the approximations a^k converge uniformly on $[0,\bar{t}]$ for all $\bar{t} \in [0,T)$ to a continuous mapping $a : [0,T) \to \mathbf{R}^N$. It follows immediately that a satisfies (A.4) and therefore that a is absolutely continuous. To prove uniqueness we let $a = a_2 - a_1$ be the difference between two such solutions and note that $|a|^2$ is absolutely continuous and that $D_t|a(t)|^2 \leq |M(t)|a(t)|^2$ for almost all t. It follows that $a \equiv 0$. The last statement in the theorem follows from the fact that $u' = Mu + F$ for all t if M and F are continuous.

□

§A.3 Applications of Spectral Measures

In this section we prove three results involving the spectral measure associated to the inclusion map $\mathcal{H}_1 \subseteq \mathcal{H}_0$, where $\mathcal{H}_0$ and $\mathcal{H}_1$ are separable Hilbert spaces with $\mathcal{H}_1$ dense and continuously included in $\mathcal{H}_0$, as in Chapter 2. The first characterizes $\mathcal{H}_1$ and its norm in terms of the spectral measure, the second gives the construction of the special orthonormal basis for $\mathcal{H}_0$ applied in the proof of Theorem 2.1, and the third gives a product rule for time dependent operators defined by the associated spectral integrals.

Let $S \in \mathcal{L}(\mathcal{H}_0, \mathcal{H}_1)$ be the operator satisfying

$$\langle f, v\rangle_{\mathcal{H}_0} = \langle Sf, v\rangle_{\mathcal{H}_1} \tag{A.5}$$

for $f \in \mathcal{H}_0$ and $v \in \mathcal{H}_1$, as in (2.59). Since S is symmetric as an element of $\mathcal{L}(\mathcal{H}_0)$ there is a spectral measure E mapping the Borel sets in $\mathbf{R}$ into the set of orthogonal projections on $\mathcal{H}_0$ and (2.60)–(2.62) hold with associated nonnegative measures $E_{(z,w)}$. The following theorem contains the first two results of this section (but see Exercise 2.2 for a more accessible proof for the case that $\mathcal{H}_1$ is compactly contained in $\mathcal{H}_0$):

Theorem A.10. *Let $\mathcal{H}_0$, $\mathcal{H}_1$, S and E be as above and let $L = \|S\|_{\mathcal{L}(\mathcal{H}_0)}$.*

(a) *If $v \in \mathcal{H}_0$, then $v \in \mathcal{H}_1$ if and only if $\int_{(0,L]} \mu^{-1} dE_{(v,v)} < \infty$, and if so then this integral equals $|v|^2_{\mathcal{H}_1}$.*

(b) *Given $\varepsilon > 0$ there is an $\mathcal{H}_0$-orthonormal basis $\{\varphi_k\}$ for $\mathcal{H}_0$ whose elements are in $\mathcal{H}_1$ and with the following property: if $v \in \mathcal{H}_1$ then the $\mathcal{H}_0$-projections $v^K = \sum_1^K \langle v, \varphi_k\rangle_{\mathcal{H}_0}\varphi_k$ converge to v strongly in $\mathcal{H}_1$ and $|v^K|_{\mathcal{H}_1} \leq (1+\varepsilon)|v|_{\mathcal{H}_1}$ for all K.*

Proof. Throughout the proof we abbreviate operators $\int_{(0,L]} \psi(\mu) dE(\mu)$ by the simpler notation $\mathcal{I}(\psi)$ or $\mathcal{I}(\psi(\mu))$ and we freely apply (2.60)–(2.62) as required.

We need two preliminary results for the proof of (a). First we claim that if $\delta \in (0, L)$ and χ_δ is the characteristic function of $(\delta, L]$, and if $v \in \mathcal{H}_0$ and $\chi_\delta\psi$ is bounded and measurable on $(0, L)$, then $\mathcal{I}(\chi_\delta\psi)v \in \mathcal{H}_1$ and

$$|\mathcal{I}(\chi_\delta\psi)v|^2_{\mathcal{H}_1} = \int_{(0,L]} \mu^{-1}\chi_\delta(\mu)\psi(\mu)^2 dE_{(v,v)}(\mu). \tag{A.6}$$

To prove this we compute $\mathcal{I}(\chi_\delta\psi)v = \mathcal{I}(\mu)\mathcal{I}(\mu^{-1}\chi_\delta\psi)v = S\mathcal{I}(\mu^{-1}\chi_\delta\psi)v$, which is in the range of S and therefore is in $\mathcal{H}_1$. Then by (A.5)

$$|\mathcal{I}(\chi_\delta\psi)v|^2_{\mathcal{H}_1} = \langle S\mathcal{I}(\mu^{-1}\chi_\delta\psi)v, \mathcal{I}(\chi_\delta\psi)v\rangle_{\mathcal{H}_1} = \langle \mathcal{I}(\mu^{-1}\chi_\delta\psi)v, \mathcal{I}(\chi_\delta\psi)v\rangle_{\mathcal{H}_0},$$

which equals the term on the right side of (A.6).

Next we note that if $v \in \mathcal{H}_1$ then in the decomposition

$$v = \mathcal{I}(\chi_\delta)v + \mathcal{I}(1-\chi_\delta)v \tag{A.7}$$

two of the three terms, hence all three, are in $\mathcal{H}_1$; and the two terms on the right are orthogonal in $\mathcal{H}_1$ because

$$\begin{aligned}
\langle \mathcal{I}(\chi_\delta)v, \mathcal{I}(1-\chi_\delta)v\rangle_{\mathcal{H}_1} &= \langle S\mathcal{I}(\mu^{-1}\chi_\delta)v, \mathcal{I}(1-\chi_\delta)v\rangle_{\mathcal{H}_1} \\
&= \langle \mathcal{I}(\mu^{-1}\chi_\delta)v, \mathcal{I}(1-\chi_\delta)v\rangle_{\mathcal{H}_0} \\
&= \langle \mathcal{I}(\mu^{-1}\chi_\delta(1-\chi_\delta))v, v\rangle_{\mathcal{H}_0} = 0.
\end{aligned}$$

In particular, the Pythagorean theorem holds in the $\mathcal{H}_1$-norm in (A.7).

We can now prove (a). If $v \in \mathcal{H}_1$ then by the two facts proved above,

$$\int \mu^{-1}\chi_\delta(\mu) dE_{(v,v)} = |\mathcal{I}(\chi_\delta)|^2_{\mathcal{H}_1} \leq |v|^2_{\mathcal{H}_1}$$

so that

$$\int \mu^{-1} dE_{(v,v)} \leq |v|^2_{\mathcal{H}_1}. \tag{A.8}$$

Conversely suppose that $v \in \mathcal{H}_0$ and that the integral in (a) is finite. Let $v_l = \mathcal{I}(\chi_{\delta_l})v$ where δ_l is a sequence decreasing to zero. Then $v_l \to v$ in $\mathcal{H}_0$ and for $0 < \delta_l < \delta_{l'}$,

$$\begin{aligned}|v_{\delta_l'} - v_{\delta_l}|^2_{\mathcal{H}_1} =& \langle S\mathcal{I}(\mu^{-1}(\chi_{\delta_l} - \chi_{\delta_{l'}}))v, \mathcal{I}(\chi_{\delta_l} - \chi_{\delta_{l'}})v\rangle_{\mathcal{H}_1} \\ =& \langle \mathcal{I}(\mu^{-1}(\chi_{\delta_l} - \chi_{\delta_{l'}}))v, \mathcal{I}(\chi_{\delta_l} - \chi_{\delta_{l'}})v\rangle_{\mathcal{H}_0} = \int_{(\delta_l, \delta_{l'}]} \mu^{-1} dE_{(v,v)}\end{aligned}$$

which goes to zero as $\delta_l, \delta_{l'} \to 0$. This proves that $\{v_l\}$ is Cauchy in $\mathcal{H}_1$ and therefore that its limit v is in $\mathcal{H}_1$. In particular $|v|^2_{\mathcal{H}_1} = \lim |v_\delta|^2_{\mathcal{H}_1}$, which by (A.6) is bounded above by $\displaystyle\int \mu^{-1} dE_{(v,v)}$. This together with (A.8) completes the proof of (a).

To prove (b) we fix a parameter $r \in (0,1)$ and for $m \geq 1$ define sets $V_m = (Lr^m, Lr^{m-1}]$ and projections $P_m = E(V_m)$. We make three preliminary observations: first for $m \neq m'$ the ranges of P_m and $P_{m'}$ are orthogonal in $\mathcal{H}_1$; second, if v is in the range of P_m then

$$|v|^2_{\mathcal{H}_1} = |P_m v|^2_{\mathcal{H}_1} = \int_{V_m} \mu^{-1} dE_{(v,v)} \leq (Lr^m)^{-1} |v|^2_{\mathcal{H}_0}; \tag{A.9}$$

and third, for $v \in \mathcal{H}_1$, $\displaystyle\sum_1^M P_m v \to v$ strongly in $\mathcal{H}_1$ as $M \to \infty$. The first two of these are obvious, and for the third we let $M < M'$ and apply (A.9):

$$\Big|\sum_{m=M+1}^{M'} P_m v\Big|^2_{\mathcal{H}_1} = \sum_{m=M+1}^{M'} \int_{V_m} \mu^{-1} dE_{(v,v)} = \int_{(Lr^{M'}, Lr^M]} \mu^{-1} dE_{(v,v)}$$

which goes to zero as $M \to \infty$ by (a).

Now let $\{\psi_j^m\}_j$ be an $\mathcal{H}_0$-orthonormal basis for the range of P_m and let Q_J^m be the $\mathcal{H}_0$-orthogonal projection onto the span of $\{\psi_1^m, \ldots, \psi_J^m\}$. Define the set $\{\varphi_k\}_k = \{\psi_j^m\}_{m,j}$, enumerating by the standard diagonal process. Then given K there is a unique minimal M and numbers J_m for $m \leq M$ such that $\{\varphi_k : k \leq K\} = \{\psi_j^m : m \leq M \text{ and } j \leq J_m\}$. We can now prove the bound in (b) by applying the above three observations. Let $v \in \mathcal{H}_1$ and let v^K be its $\mathcal{H}_0$-orthogonal projection onto the span of $\{\psi_1, \ldots, \psi_K\}$. Then since $v^K = \sum_1^M Q_{J_m}^m v$ is an orthogonal sum in $\mathcal{H}_1$ and $Q_{J_m}^m = P_m Q_{J_m}^m P_m$,

$$
\begin{aligned}
|v^K|_{\mathcal{H}_1}^2 &= \sum_1^M |P_m Q_{J_m}^m P_m v|_{\mathcal{H}_1}^2 \le \sum_1^M (Lr^m)^{-1} |Q_{J_m}^m P_m v|_{\mathcal{H}_0}^2 \\
&\le \sum_1^M (Lr^m)^{-1} |P_m v|_{\mathcal{H}_0}^2 = \sum_1^M (Lr^m)^{-1} \int_{V_m} dE_{(v,v)} \qquad \text{(A.10)} \\
&\le r^{-1} \sum_1^M \int_{V_m} \mu^{-1} dE_{(v,v)} \le r^{-1} |v|_{\mathcal{H}_1}^2.
\end{aligned}
$$

The bound in (b) for v^K therefore follows with $r^{-1} = 1 + \varepsilon$.

To prove that $v^K \to v$ in $\mathcal{H}_1$ we again let M and J_m be as above for a given K. Then given M_0 and J arbitrarily large, we can stipulate that if K_0 is sufficiently large and $K \ge K_0$ then $M > M_0$ and $J_m \ge J$ for $m \le M_0$. Then for such K,

$$
\text{(A.11)} \quad v - v^K = \Big(v - \sum_1^{M_0} P_m v\Big) + \sum_1^{M_0} P_m (P_m v - Q_{J_m}^m v) - \sum_{M_0+1}^{M} P_m Q_{J_m}^m v.
$$

The square of the $\mathcal{H}_1$-norm of the middle term on the right is bounded by

$$
\sum_1^{M_0} (Lr^m)^{-1} |P_m v - Q_{J_m}^m v|_{\mathcal{H}_0}^2 \le (Lr^{M_0})^{-1} \sum_1^{M_0} |P_m v - Q_J^m v|_{\mathcal{H}_0}^2
$$

by (A.9), and the square of the $\mathcal{H}_1$-norm of the third term on the right is bounded by $r^{-1} \int_{(Lr^M, Lr^{M_0-1}]} \mu^{-1} dE_{(v,v)}$ by the same argument as in (A.10) above. Thus $|v - v^K|_{\mathcal{H}_1}$ can be made arbitrarily small by first choosing M_0 sufficiently large that the first and third terms on the right side of (A.11) are small in $\mathcal{H}_1$, then choosing J sufficiently large depending on M_0 so that the second term is small. This proves that $v^K \to v$ strongly in $\mathcal{H}_1$ and together with the fact that $\mathcal{H}_1$ is dense in $\mathcal{H}_0$ shows that the set $\{\varphi_k\}$ is complete in $\mathcal{H}_0$.

□

The following abstract product rule is applied in the proof of Theorem 2.13:

Theorem A.11. *Let $\mathcal{H}_0$, $\mathcal{H}_1$, S, E and L be as in* Theorem A.10 *and let I be a nontrivial interval in* $\mathbf{R}$. *Assume that $z \in wAC(I; \mathcal{H}_0)$ with derivative z' and that $\psi = \psi(\mu, t)$ and its t-derivative ψ_t are continuous and bounded on $(0, L] \times I$ (the derivative is one-sided at endpoints in I, if any). Let*

$M(t) = \int_{(0,L]} \psi(\mu, t) dE(\mu)$ *and* $M'(t) = \int_{(0,L]} \psi_t(\mu, t) dE(\mu)$ *for* $t \in I$. *Then* $M(t)z(t) \in wAC(I; \mathcal{H}_0)$ *with derivative* $M'z + Mz'$.

PROOF. First if z is independent of t then for $v \in \mathcal{H}_0$ and $t_1, t_2 \in I$,

$$\begin{aligned}\langle M(\cdot)z, v\rangle_{\mathcal{H}_0}\Big|_{t_1}^{t_2} &= \int \psi(\mu, \cdot)\Big|_{t_1}^{t_2} dE_{(z,v)} = \int\int_{t_1}^{t_2} \psi_t(\mu, t) dt\, dE_{(z,v)} \\ &= \int_{t_1}^{t_2}\int \psi_t(\mu, t) dE_{(z,v)} dt = \int_{t_1}^{t_2} \langle M'(t)z, v\rangle_{\mathcal{H}_0} dt.\end{aligned}$$

(The application of Fubini's theorem in the third step is justified by the continuity of ψ_t on the product space.) This proves that $Mz \in wAC(I; \mathcal{H}_0)$ with derivative $M'z$. Then for nonconstant z and for $v \in \mathcal{H}_0$ we have

$$\begin{aligned}\langle M(\cdot)z(\cdot), v\rangle_{\mathcal{H}_0}\Big|_{t_1}^{t_2} &= \langle z(\cdot), M(\cdot)v\rangle_{\mathcal{H}_0}\Big|_{t_1}^{t_2} \\ &= \int_{t_1}^{t_2} \big(\langle z', Mv\rangle_{\mathcal{H}_0} + \langle z, M'v\rangle_{\mathcal{H}_0}\big) dt = \int_{t_1}^{t_2} \langle Mz' + M'z, v\rangle_{\mathcal{H}_0} dt\end{aligned}$$

by Exercise 2.2. □

§A.4 Boundary Measures: Proofs of Theorems 3.1–3.3

In this section we construct surface measures and outer normal vector fields on the boundaries of Lipschitz domains and we prove the trace and divergence theorems stated in Theorems 3.2 and 3.3. Important use will be made of Rademacher's theorem, which states that a Lipschitz mapping of an open set in $\mathbf{R}^n$ into $\mathbf{R}^m$ is differentiable in the sense of calculus almost everywhere and that the derivative is essentially bounded in terms of the Lipschitz constant. A self-contained proof of Rademacher's theorem is given in Ziemer [**46**], Theorem 2.2.1.

We begin with the following technical fact concerning compatibility on overlapping coordinate domains:

Lemma A.12 *Let* Ω *be a Lipschitz domain in* $\mathbf{R}^n$ *as described in* Definition 1.7(a), *let* $\mathcal{B}_i$ *be as in that definition and* $T_i : \mathcal{B}_i \to \partial\Omega$ *as in* (1.6). *Then* T_i *is bi-Lipschitz from* $\mathcal{B}_i$ *onto* $\partial\Omega \cap \mathcal{U}_i$. *In addition, if* $\partial\Omega \cap \mathcal{U}_i \cap \mathcal{U}_k \neq \phi$ *and* $y(z) \equiv (T_i^{-1} \circ T_k)(z)$ *for* $z \in T_k^{-1}(\partial\Omega \cap \mathcal{U}_i \cap \mathcal{U}_k)$, *then* ψ_k *is differentiable at some such* z *if and only if* ψ_i *is differentiable at* $y(z)$, *and in this case* y *is differentiable at* z *and*

$$\Big|\det \frac{dy}{dz}(z)\Big| \sqrt{1 + |\nabla\psi_i(y(z))|^2} = \sqrt{1 + |\nabla\psi_k(z)|^2}. \tag{A.12}$$

PROOF. The first assertion follows from the fact that if $w = T_i(y)$ then $|w|^2 = |y|^2 + |\psi_i(y)|^2$. To prove the differentiability statement we solve the relation $T_i(y(z)) = T_k(z)$ to express $\psi_i(y)$ and the components of y as polynomials in $\psi_k(z)$ and the components of z. This shows that if ψ_k is differentiable at a particular z_0, then so is y. Since the mapping $z \to y(z)$ is bi-Lipschitz, this derivative is invertible, and then ordinary calculus shows that the inverse mapping $y \to z(y)$ is differentiable at $y(z_0)$. Therefore $\psi_i(y(z))$, which was expressed as a polynomial in z and $\psi_k(z)$, is differentiable at z_0 by the chain rule.

To prove (A.12) we again write $T_i(y) = T_k(z)$ and differentiate with respect to z to obtain

$$\mathcal{Q}_i \begin{bmatrix} I \\ \nabla\psi_i^{tr}(y(z)) \end{bmatrix} y'(z) = \mathcal{Q}_k \begin{bmatrix} I \\ \nabla\psi_k^{tr}(z) \end{bmatrix} \tag{A.13}$$

where $\mathcal{Q}_i$ and $\mathcal{Q}_k$ are as in the first item in Definition 1.7. The relation (A.12) will follow by equating the lengths of the cross products of the $n-1$ columns of the matrices on either side of (A.13). Recall that the cross product of vectors $x_1, \ldots, x_{n-1}$ in $\mathbf{R}^n$ is the unique vector v with the property that, for all $x \in \mathbf{R}^n$, $v \cdot x$ is the determinant of the matrix whose columns are $x_1, \ldots, x_{n-1}, x$. Given such x let $w = \mathcal{Q}_i^{tr} x$ and $w^{tr} = [\tilde{w}^{tr}\ w_n.]$ The matrix on the left in (A.13) augmented by x then has determinant

$$\begin{aligned} \det \begin{bmatrix} y' & \tilde{w} \\ \nabla\psi_i^{tr} y' & w_n \end{bmatrix} &= \det \left(\begin{bmatrix} I & 0 \\ \nabla\psi_i^{tr} & 1 \end{bmatrix} \begin{bmatrix} y' & \tilde{w} \\ 0 & w_n - \nabla\psi_i \cdot \tilde{w} \end{bmatrix} \right) \\ &= (\det y') \begin{bmatrix} -\nabla\psi_i \\ 1 \end{bmatrix} \cdot w = \left((\det y') \mathcal{Q}_i \begin{bmatrix} -\nabla\psi_i \\ 1 \end{bmatrix} \right) \cdot x. \end{aligned}$$

The cross product in question is therefore the vector in the parentheses on the right, whose length is $|\det y'(z)| \sqrt{1 + |\nabla\psi_i(y)|^2}$. A similar computation shows that the length of the cross product of the columns on the right side of (A.13) is $\sqrt{1 + |\nabla\psi_k(z)|^2}$.

$\square$

Boundary Measures and Normal Vectors: Proof of Theorem 3.1.

First, the sum defining $\mathcal{I}(f)$ for $f \in C_c(\partial\Omega)$ in the statement in (a) is finite because the collection $\{\mathcal{W}_i\}_i$ is locally finite. Thus $\mathcal{I} : C_c(\partial\Omega) \to \mathbf{R}$, $\mathcal{I}$ is linear in f and $\mathcal{I}(f) \geq 0$ if f is nonnegative. A standard version of the Riesz representation theorem therefore applies to show that there is a Radon measure σ as described in (a) (see Folland [**13**] Theorem 7.2, for example). The separability assertion (b) is then a straightforward application of Theorem A.1.

We prove (c) in a sequence of steps. First assume that $f \in C_c(\partial\Omega)$ and that E is compact in $\partial\Omega \cap \mathcal{U}_i$. There is then a sequence χ_l of functions in $C_c(\partial\Omega)$ taking values in $[0, 1]$, having support in $\partial\Omega \cap \mathcal{U}_i$ and converging pointwise to the characteristic function of E. Then if $f_l = f\chi_l$,

$$\int_E f\,d\sigma = \lim_l \int_{\partial\Omega} f_l\,d\sigma = \lim_l \mathcal{I}(f_l)$$
$$= \lim_l \sum_k \int_{T_k^{-1}(\partial\Omega\cap\mathcal{W}_k)} (f_l \circ T_k)(z)\sqrt{1+|\nabla\psi_k(z)|^2}dz.$$

Since f_l has support in $\partial\Omega \cap \mathcal{U}_i$, the integral here may be taken over $T_k^{-1}(\partial\Omega \cap \mathcal{W}_k \cap \mathcal{U}_i)$. Then making the change of variables $y = (T_i^{-1} \circ T_k)(z)$ and applying Lemma A.12 we find that

$$\int_E f\,d\sigma = \lim_l \sum_k \int_{T_i^{-1}(\partial\Omega\cap\mathcal{W}_k\cap\mathcal{U}_i)} (f_l \circ T_i)(y)\sqrt{1+|\nabla\psi_i(y)|^2}dy$$
$$= \lim_l \int_{T_i^{-1}(\partial\Omega\cap\mathcal{U}_i)} (f_l \circ T_i)(y)\sqrt{1+|\nabla\psi_i(y)|^2}dy$$
$$= \int_{T_i^{-1}(E)} (f \circ T_i)(y)\sqrt{1+|\nabla\psi_i(y)|^2}dy,$$

as required. The same result then holds for $f \in C_c(\Omega)$ and Borel measurable $E \subseteq \partial\Omega \cap \mathcal{U}_i$ because σ, being a Radon measure, is inner regular. Next suppose that $f \in (L^1 \cap L^\infty)(\partial\Omega)$. Then f is the limit, both in $L^1(\partial\Omega)$ and pointwise a.e., of elements of $C_c(\partial\Omega)$, which may be taken to be uniformly bounded (see Folland [**13**] Proposition 7.9 for example). The relation (3.7) applies to the approximants and is easily seen to persist in the limit. Finally, assume that f is measurable and nonnegative. If its integral over E is $+\infty$, then given a number M there is a simple function $\varphi \leq f$ whose integral is greater than M. Applying the previous step to φ, we then find that the integral on the right side of (3.7) is greater than M as well; this integral is therefore $+\infty$. If the integral of f over E is finite, then we apply the previous step to $f^R \equiv \min\{f, R\} \in L^1 \cap L^\infty$ and then take the limit as $R \to \infty$ using the dominated convergence theorem. This completes the proof of (c), and (d) follows easily from (c).

To prove (e) we again recall notations from Definition 1.7: $T_i(y) = R_i(y, \psi_i(y))$ for $y \in \mathcal{B}_i \subset \mathbf{R}^{n-1}$, where R_i is the rigid motion on $\mathbf{R}^n$ given by $R_i(w) = a_i + \mathcal{Q}_i w$ where a_i is constant and $\mathcal{Q}_i$ is orthogonal. Since each T_i is bi-Lipschitz from $\mathcal{B}_i$ to $\partial\Omega \cap \mathcal{U}_i$, T_i and its inverse take sets of measure zero to sets of measure zero. Thus for σ-a.a. $x = T_i(y) \in \partial\Omega \cap \mathcal{U}_i$ we can define

$$\nu(x) = (1 + |\nabla\psi_i(y)|^2)^{-1/2}\mathcal{Q}_i \begin{bmatrix} -\nabla\psi_i(y) \\ 1 \end{bmatrix}. \tag{A.14}$$

The assertions in (e) are easily checked, but we must also prove that the above definition is independent of i when x is in more than one of the covering sets. To do this fix $x = T_i(y) = T_k(z)$ where ψ_i and ψ_k are differentiable at y and z respectively (recall the differentiability statements in Lemma A.12) and let ν_i and ν_k be the respective normal vectors defined as above. For given $l \in \{1, \ldots, n-1\}$ define a curve $y(s) = (T_i^{-1} \circ T_k)(z + se_l)$ for s in a neighborhood of zero and e_l a standard basis vector. Then

$$R_i(y(s), \psi_i(y(s))) = R_k(z + se_l, \psi_k(z + se_l))$$

and y is differentiable at $s = 0$. Therefore

$$\mathcal{Q}_i \begin{bmatrix} \dot{y} \\ \nabla\psi_i \cdot \dot{y} \end{bmatrix} = \mathcal{Q}_k \begin{bmatrix} e_l \\ (\psi_k)_{z_l} \end{bmatrix}.$$

By inspection, ν_i is perpendicular to the vector on the left, hence to the vectors on the right for every l, as is ν_k. Since these $n-1$ vectors are linearly independent, ν_i and ν_k are parallel unit vectors; and since both point outward at x, they must agree.

□

Trace Theorem: Proof of Theorem 3.2.

Let ε_0 and L be as in Definition 1.7(b) and let χ be a smooth function on $\mathbf{R}$ taking values in $[0,1]$ with $\chi(s) = 0$ for $s \leq \varepsilon_0/2$ and $\chi(s) = 1$ for $s \geq \varepsilon_0$. Then for $f \in C_c^1(\mathbf{R}^n)$,

$$\begin{aligned}
\int_{\partial\Omega} |f|\, d\sigma &= \sum_i \int_{T_i^{-1}(\partial\Omega\cap\mathcal{W}_i)} (|f| \circ R_i)(\tilde{y}, \psi_i(\tilde{y}))\sqrt{1 + |\nabla\psi_i(\tilde{y})|^2}\, d\tilde{y} \\
&\leq C(L) \sum_i \int_{T_i^{-1}(\partial\Omega\cap\mathcal{W}_i)} \left| \int_0^{\psi_i(\tilde{y})} \frac{d}{ds}\big[\chi(s)(f \circ R_i)(\tilde{y}, s)\big]\, ds \right| d\tilde{y} \\
&\leq C(L, \varepsilon_0) \sum_i \int_{R_i^{-1}(\Omega\cap\mathcal{W}_i)} (|f \circ R_i| + |\nabla_x f \circ R_i|)\, dy \\
&\leq C(L, \varepsilon_0) \int_\Omega (|f| + |\nabla_x f|)\, dx.
\end{aligned}$$

A similar result holds with f replaced by $|f|^p$ (recall that for weak derivatives, $|f|_{x_i} = f_{x_i}\mathrm{sgn}(f)$) and therefore $|\mathcal{T}(f)|_{L^p(\partial\Omega)} \leq C|f|_{W^{1,p}(\Omega)}$ for $f \in C_c^\infty(\mathbf{R}^n)$. Since the set of restrictions to Ω of such f is dense in $W^{1,p}(\Omega)$ (Theorem A.2) and $L^p(\partial\Omega)$ is complete, $\mathcal{T}$ extends to a map in $\mathcal{L}(W^{1,p}(\Omega), L^p(\partial\Omega))$. The fact that $\mathcal{T}(f) = f\big|_{\partial\Omega}$, which holds for $f \in C_c^\infty(\mathbf{R}^n)$, then follows for $f \in C_c(\mathbf{R}^n)$ by a standard approximation argument.

□

Divergence Theorem: Proof of Theorem 3.3.

It suffices to prove the theorem for $f \in C_c^\infty(\mathbf{R}^n)$, the latter being dense in $W^{1,1}(\Omega)$ by Theorem A.2. First, if the support of f is contained in Ω, then both sides in (3.8) are zero. Suppose next that, in the notation of Definition 1.7, the support of $f\big|_{\overline{\Omega}}$ is contained in $\overline{\Omega} \cap \mathcal{U}_i$ for some i. Let $v(z) = f(R_i(z))$ for $z \in \mathcal{B}_i \times (0, s_i)$. Then $(\nabla_x f)(R_i(z)) = \mathcal{Q}_i \nabla_z v(z)$ so that

(A.15) $$\int_\Omega f_{x_j}\, dx = \int_{R_i^{-1}(\Omega\cap\mathcal{U}_i)} \mathcal{Q}_i^{jk} v_{z_k}(z)\, dz = \mathcal{Q}_i^{jk} \int_{\mathcal{B}_i} \int_0^{\psi_i(y)} v_{z_k}(y, s)\, ds\, dy$$

where y now denotes a point in $\mathcal{B}_i \subset \mathbf{R}^{n-1}$. For $k < n$ the double integral on the right is

$$\int_{\mathcal{B}_i} \frac{\partial}{\partial y_k}\Big[\int_0^{\psi_i(y)} v(y,s)ds\Big]dy - \int_{\mathcal{B}_i} v(y,\psi_i(y))\frac{\partial \psi_i}{\partial y_k}(y)dy.$$

The integral on the left is zero because the support of the term in brackets is compact in $\mathcal{B}_i$. Thus from (3.7) and (A.14),

$$\int_\Omega f_{x_j}dx = \int_{\mathcal{B}_i}\Big[-\sum_{k=1}^{n-1}\mathcal{Q}_i^{jk}v(y,\psi_i(y))\frac{\partial \psi_i}{\partial y_k}(y) + \mathcal{Q}_i^{jn}v(y,\psi_i(y))\Big]dy$$

$$= \int_{\mathcal{B}_i} v(y,\psi_i(y))\nu^j(R_i(y))\sqrt{1+|\nabla_y\psi_i(y)|^2}dy = \int_{\partial\Omega} f\nu^j d\sigma.$$

For general $f \in C_c^\infty(\mathbf{R}^n)$ we append to $\{\mathcal{U}_i\}_1^\infty$ an open set $\mathcal{U}_0$ to obtain an open cover $\{\mathcal{U}_i\}_0^\infty$ of $\overline{\Omega}$, then construct a partition of unity $\{\varphi_l\}$ of $\overline{\Omega}$ subordinate to this cover and apply the above results to each $f\varphi_l$.

□

§A.5 The Spaces $L^{p,q}$

In this section we discuss the spaces $L^{p,q}(X\times I)$, specific examples of which were introduced in section 3.2. Three main results are proved: the first shows that, if X is a fairly general measure space and I is a nontrivial interval in $\mathbf{R}$ then there is a one-to-one correspondence between elements $u \in L^p(I;L^q(X))$ and elements $u_{\mathfrak{m}} \in L^{p,q}(X \times I)$, the latter being equivalence classes of functions which are measurable with respect to the product measure and for which a certain norm is finite; this correspondence is given in Corollary A.16 below. The second and third results give characterizations of elements of $L^p(I;W^{k,q}(\Omega))$ and $wAC(I;W^{k,q}(\Omega))$ in terms of the weak $\Omega\times I^\circ$–derivatives of $u_{\mathfrak{m}}$; these are proved in Theorems A.23 and A.24.

The following assumptions will include the case that X is an open set in $\mathbf{R}^n$ as well as the case that X is the boundary of a Lipschitz domain in $\mathbf{R}^n$:

(A.16)

- I is a nontrivial interval in $\mathbf{R}$
- (X,μ) is a σ–finite measure space
- $L^r(X,\mu)$ is separable for $r \in [1,\infty)$
- λ is the product measure on $X \times I$
- $p \in [1,\infty]$ and $q \in (1,\infty]$.

Recall that in Definition 1.3 the space $L^p(I; L^q(X))$ was defined to be the set of equivalence classes f on I of weakly measurable functions $\varphi : I \to L^q(X)$ such that the mapping $t \to |\varphi(t)|_{L^q(X)}$ is in $L^p(I)$. For such f the integral

$$\int_I \Big[\int_X f(t)v\, d\mu\Big]\psi(t)dt \tag{A.17}$$

is defined (that is, is independent of representative φ and is finite) for all $v \in L^{q'}(X)$ and $\psi \in L^{p'}(I)$, where prime denotes Hölder conjugate. We now consider a similar but different space in which this same integral is well-defined:

Definition A.13. *Let I, (X, μ), λ, p and q be as in* (A.16). *Then $L^{p,q}(X \times I)$ is the normed linear space of λ-equivalence classes F of λ-measurable functions $\varphi : X \times I \to \mathbf{R}$ for which the L^p norm on I of $|\varphi(\cdot, t)|_{L^q(X)}$ is finite. The norm is defined by $|F|_{L^{p,q}(X\times I)} \equiv \big||\varphi(\cdot,t)|_{L^q(X)}\big|_{L^p(I)}$ for any such representative φ. The vector space $L^{p,q}_{(t)loc}(X \times I)$ is the set of equivalence classes of functions which are measurable on $\Omega \times I$ and whose restrictions to $X \times I'$ are in $L^{p,q}(\Omega \times I')$ for every compact interval $I' \subseteq I$.*

Observe that the definition depends on Tonelli's theorem: for $q < \infty$ and φ as in the definition, the mapping $(x, t) \to |\varphi(x, t)|^q$ is λ-measurable, and therefore the mapping $t \to |\varphi(\cdot, t)|_{L^q(X)}$ is measurable on I. For $q = \infty$, the same reasoning shows that $t \to |\varphi(t)|_{L^r(E)}$ is measurable on I for $r < \infty$ and $E \subseteq X$ of finite measure. Letting $r \to \infty$ and then applying the fact that (X, μ) is σ-finite we again conclude that $t \to |\varphi(\cdot, t)|_{L^\infty(X)}$ is measurable on I. Similar considerations based on the Fubini and Tonelli theorems show that for $F \in L^{p,q}$ the integral

$$\int_I \Big[\int_X F(\cdot,t)v\, d\mu\Big]\psi(t)dt \tag{A.18}$$

is well-defined for the same v and ψ as in (A.17). We will show that there is a natural correspondence $f \leftrightarrow F$ between elements of $L^p(I; L^q(X))$ and $L^{p,q}(X \times I)$ for which the integrals in (A.17) and (A.18) agree. We begin with the following uniqueness result:

Lemma A.14. *Let I, (X, μ), λ, p and q be as in* (A.16) *and let $f \in L^p(I; L^q(X))$. If the integral in* (A.17) *is zero for all $v \in L^{q'}(X)$ and $\psi \in L^{p'}(I)$, where prime denotes Hölder conjugate, then f is zero as an element of $L^p(I; L^q(X))$. Similarly, if $F \in L^{p,q}(X \times I)$ and if the integral in* (A.18) *is zero for all such v and ψ, then F is zero as an element of $L^{p,q}(X \times I)$.*

PROOF. Let $\{v_j\}$ be countable and dense in $L^{q'}(X)$ and let $\varphi : I \to L^q(X)$ be a specific representative of f. There is then a set $E \subset I$ of measure zero such that if $t_0 \in I - E$ then $|\varphi(t_0)|_{L^q(X)}$ is finite and t_0 is a Lebesque point of each of the maps $t \to \int_X \varphi(t) v_j d\mu$. Let t_0 be such a point and $\psi_\delta \in L^{p'}(I)$ be δ^{-1} times the characteristic function of a small interval of length δ contained in I and containing t_0. Applying the hypothesis that the integral in (A.17) is zero for the pair v_j, ψ_δ and letting $\delta \to 0$, we find that $\int_X \varphi(t_0) v_j d\mu$ is zero for all j, so that $\varphi(t_0)$ is zero as an element of $L^q(X)$. That is, $\varphi = 0$ almost everywhere on I and therefore f is zero as an element of $L^p(I; L^q(X))$.

Now let φ denote a specific representative of $F \in L^{p,q}(X \times I)$. Thus $\varphi : X \times I \to \mathbf{R}$ is a λ-measurable function and

$$S \equiv \{(x,t) \in X \times I \text{ such that } \varphi(x,t) \neq 0\}$$

is a λ-measurable set. Since the mappings $t \to \int_X \varphi(\cdot, t) v_j d\mu$ and $t \to |\varphi(\cdot, t)|_{L^q(X)}$ are in $L^p(I)$, there is again a null set $E \subset I$ such that if $t_0 \in I - E$ then t_0 is a Lebesque point of each of these maps and $|\varphi(\cdot, t_0)|_{L^q(X)}$ is finite. Letting ψ_δ be as above and applying the hypothesis that the integral in (A.18) is zero for the pair v_j, ψ_δ, we again find that $\int_X \varphi(\cdot, t_0) v_j d\mu$ is zero for all j. It follows that for almost all t_0 the set

$$S_{t_0} \equiv \{x \in X \text{ such that } \varphi(x, t_0) \neq 0\}$$

has μ-measure zero. Then since S is λ-measurable,

$$\lambda(S) = \int_I \Big(\int_{S_t} d\mu \Big) dt = 0.$$

This proves that $\varphi = 0$ λ-almost everywhere on $X \times I$ and therefore that F is zero as an element of $L^{p,q}(X \times I)$.

□

In the following theorem we show that the spaces $L^p(I; L^q(X))$ and $L^{p,q}(X \times I)$ are linearly isomorphic in a natural way:

Theorem A.15. *Let I, (X, μ), λ, p and q be as in* (A.16).

(a) *Given $F \in L^{p,q}(X \times I)$ there is a unique $f \in L^p(I; L^q(X))$ such that*

$$\text{(A.19)} \qquad \int_I \Big[\int_X f(t) v \, d\mu \Big] \psi(t) dt = \int_I \Big[\int_X F(\cdot, t) v \, d\mu \Big] \psi(t) dt = \int_{X \times I} F v \psi \, d\lambda$$

for all $v \in L^{q'}(X)$ and $\psi \in L^{p'}(I)$, where prime denotes Hölder conjugate. The mapping from $F \to f$ is a linear isometry.

(b) *Assume in addition to* (A.16) *that X is a locally compact metric space in which compact sets have finite measure and that for every $\varepsilon > 0$ there is a countable partition $\{V_k\}_k$ of X by sets of finite μ-measure and* *diameter less than ε. Then given $f \in L^p(I; L^q(X))$ there is a unique $F \in L^{p,q}(X \times I)$ such that* (A.19) *holds for all $v \in L^{q'}(X)$ and $\psi \in L^{p'}(I)$.*

PROOF. To prove (a) we let $\{v_j\}$ be countable and dense in $L^{q'}(X)$ and fix an element $F \in L^{p,q}$ and a representative $\varphi : X \times I \to \mathbf{R}$ of F. Then as in the proof of Lemma A.14 there is a set $E \subset I$ of measure zero such that every $t_0 \in I - E$ is a Lebesque point of the maps $t \to \int_X \varphi(\cdot, t) v_j d\mu$ and $t \to |\varphi(\cdot, t)|_{L^q(X)}$, and $|\varphi(\cdot, t_0)|_{L^q(X)}$ is finite. Then for such t_0, $\left| \int_X \varphi(\cdot, t_0) v_j d\mu \right| \leq |\varphi(\cdot, t_0)|_{L^q} |v_j|_{L^{q'}}$ for all j, so that the X-equivalence class $\Phi(t_0)$ of $\varphi(\cdot, t_0)$ is in $L^q(X)$. It is then straightforward to check that $\Phi : I \to L^q(X)$ is weakly measurable and that its I-equivalence class f is an element of $L^p(I; L^q(X))$, that (A.19) holds, and that $|f|_{L^p(I;L^q(X))} = |F|_{L^{p,q}(X \times I)}$.

We prove (b) first for $p \in (1, \infty]$. Let f be an element of $L^p(I; L^q(X))$, $\varphi : I \to L^q(X)$ a representative of f, $\varepsilon > 0$ and $\{V_j^\varepsilon\}$ the corresponding partition of X. Define

$$\varphi_\varepsilon(x, t) = \sum_j \left[(\mu_j^\varepsilon)^{-1} \int_X \varphi(t) \chi_{V_j^\varepsilon} d\mu \right] \chi_{V_j^\varepsilon}(x)$$

where $\mu_j^\varepsilon = \mu(V_j^\varepsilon)$, $\chi_{V_j^\varepsilon}$ is the characteristic function of V_j^ε and the sum is over sets V_j^ε of nonzero measure. The summand on the right is the product of functions measurable in t and in x respectively so that φ_ε is λ-measurable on $X \times I$ and $|\varphi_\varepsilon(\cdot, t)|_{L^q(X)} \leq |\varphi(t)|_{L^q(X)}$ for every $t \in I$.

Now choose $r \in (1, p \wedge q) \subseteq (1, \infty)$ and sets $S \subseteq X$ and $I' \subseteq I$ of finite measure. Then by Hölder's inequality there is a constant C depending on S and I' such that $|\varphi_\varepsilon|_{L^r(S \times I')} \leq C |f|_{L^p(I;L^q(X))}$ independently of ε. Since $L^r(X \times I)$ and $L^{r'}(X \times I)$ are separable spaces (which we will check below) there is a sequence $\varepsilon_i \to 0$ such that φ_{ε_i} converges weakly in $L^r(S \times I')$. A diagonal process then shows that there is a subsequence, still denoted $\varepsilon_i \to 0$, such that φ_{ε_i} converges weakly in $L^r(S \times I')$ for all such sets $S \times I'$, say to $F \in L^r_{loc}(X \times I)$. F is thus an equivalence class of λ-measurable functions on $X \times I$.

We will show that $F \in L^{p,q}(X \times I)$ and that the unique element of $L^p(I; L^q(X))$ associated to F by part (a) is the f with which we began. First let $v \in C_c(X)$ (so that the support of v has finite measure) and let $\psi : I \to \mathbf{R}$ be bounded and measurable with compact support. Then applying λ-measurability and Fubini's theorem where required, we find that

(A.20)

$$\begin{aligned}
&\int_{X\times I} Fv\psi\,d\lambda = \lim_{\varepsilon_i\to 0}\int_{X\times I}\varphi_{\varepsilon_i}v\psi\,d\lambda\\
&= \lim_{\varepsilon_i\to 0}\int_I\Big[\int_X\Big(\sum_j(\mu_j^{\varepsilon_i})^{-1}\Big[\int_{V_j^{\varepsilon_i}}\varphi(t)d\mu\Big]\chi_{V_j^{\varepsilon_i}}(x)\Big)v(x)d\mu(x)\Big]\psi(t)dt\\
&= \lim_{\varepsilon_i\to 0}\int_I\Big[\int_X\varphi(t)\Big(\sum_j(\mu_j^{\varepsilon_i})^{-1}\Big[\int_{V_j^{\varepsilon}}v\,d\mu\Big]\chi_{V_j^{\varepsilon_i}}(x)\Big)d\mu(x)\Big]\psi(t)dt\\
&= \int_I\Big[\int_X\varphi\,v\,d\mu\Big]\psi(t)dt = \int_I\Big[\int_X f\,v\,d\mu\Big]\psi(t)dt
\end{aligned}$$

(the continuity of v was applied in the second last step). This proves that (A.19) holds for $v\in C_c(X)$ and ψ bounded and measurable with compact support. Next, since $C_c(X)$ is dense in the separable space $L^{q'}(X)$ (Theorem A.1), there is a countable dense subset $\{v_j\}\subset C_c(X)$. Then if Φ is a λ-measurable representative of F, the maps $t\to\int_X\Phi(\cdot,t)v_j d\mu$ and $t\to\int_X\varphi(t)v_j d\mu$ are in $L^p(I)$, so there is a null set $E\subset I$ whose complement consists of Lebesgue points of these maps and in which these integrals are finite. We fix $t_0\in I-E$ and let ψ_δ be as in the proof of Lemma A.14. Then applying (A.20) with $v=v_j$ and $\psi=\psi_\delta$ and letting $\delta\to 0$ we find that $\int_X\Phi(x,t_0)v_j(x)d\mu=\int_X\varphi(t_0)v_j d\mu$. Since this holds for all j, $\varphi(t_0)$ and $\Phi(\cdot,t_0)$ represent the same element of $L^q(X)$, and this in turn implies first that the λ-equivalence class of Φ, which is F, is in $L^{p,q}(X\times I)$, and second that (A.20) holds for all $v\in L^{q'}(X)$ and $\psi\in L^{p'}(I)$. Thus by Lemma A.14, f is the unique element of $L^{p,q}(X\times I)$ associated to F as in (a). This completes the proof of (b) for $p\in(1,\infty]$.

For the case $p=1$ we let $f\in L^1(I;L^q(X))$ and choose a representative $\varphi: I\to L^q(X)$ of f. Then the equivalence class over I of the mapping $t\to\varphi(t)/\big(1+|\varphi(t)|_{L^q(X)}\big)$ is in $L^\infty(I;L^q(X))$ and so by the $p=\infty$ case proved above there is a $G\in L^{\infty,q}$ such that

$$\int_I\Big[\int_X G(\cdot,t)v\,d\mu\Big]\tilde{\psi}(t)dt = \int_I\Big[\int_X\varphi(t)v\,d\mu\Big]\frac{\tilde{\psi}(t)}{1+|\varphi(t)|_{L^q}}dt \tag{A.21}$$

for all $v\in L^{q'}(X)$ and $\tilde{\psi}\in L^1(I)$. It follows by an argument similar to that given above following (A.20) that $|G(\cdot,t)|_{L^q(X)}=|\varphi(t)|_{L^q(X)}/\big(1+|\varphi(t)|_{L^q(X)}\big)$ for almost all t (the null set depending on the representative) and therefore that $F\equiv(1+|\varphi|_{L^q(X)})G$ is in $L^{1,q}$. Then given $v\in L^{q'}(X)$ and $\psi\in L^\infty(I)$ with compact support, we can take $\tilde{\psi}(t)=(1+|\varphi(t)|_{L^q(X)})\psi(t)\in L^1(I)$ and apply (A.21) to obtain

$$\int_I \Big[\int_X F(\cdot,t)v\,d\mu\Big]\psi(t)dt = \int_I \Big[\int_X \varphi(t)v\,d\mu\Big]\psi(t)dt.$$

The assumption that ψ has compact support is easily removed by truncating and applying the dominated convergence theorem.

There remains to check that the spaces $L^r(X \times I))$ are separable for $r \in (1,\infty)$. First, since $X \times I$ is locally compact, $C_c(X \times I)$ is dense in $L^r(X \times I)$ (Theorem A.1), so it suffices to show that $C_c(X \times I)$ is a separable subset of $L^r(X \times I)$. Let S be countable and dense in $L^r(X)$, define points $t_j^J = j/J$ for positive integers J and integers or half-integers j, and for each such J define a partition of unity $\{\chi_j^J\}_{j\in\mathbf{Z}}$ on $\mathbf{R}$ consisting of piecewise linear functions χ_j^J supported on $[t^J_{j-\frac{1}{2}}, t^J_{j+\frac{1}{2}}]$. We then leave to the reader to check that an element of C_c with support in $X \times [-K,K]$ can be approximated to arbitrary accuracy in L^∞, hence in L^r, by a function $\sum_{|j\leq KJ} v_j(x)\chi_j^J(t)$ for suitably chosen $v_j \in S$, and furthermore that the set of such approximants is countable.

□

It will be useful to adopt a more suggestive notation for the unique element $F \in L^{p,q}$ in Theorem A.15(b) in terms of the given $f \in L^p(I;L^q))$. Specifically, we will replace the notations f, F with u, u_{m} and then combine Lemma A.14 and Theorem A.15 to conclude the following:

Corollary A.16. *Let I, (X,μ), λ, p and q be as in* (A.16) *and assume that X satisfies the hypotheses in* Theorem A.15(b). *Then there is a one-to-one isometry $u \leftrightarrow u_{\mathrm{m}}$ between elements $u \in L^p(I;L^q(X))$ and elements $u_{\mathrm{m}} \in L^{p,q}(X \times I)$ in which*

$$\text{(A.22)} \qquad \begin{aligned} \int_I \Big[\int_X u(t)v\,d\mu\Big]\psi(t)dt &= \int_I \Big[\int_X u_{\mathrm{m}}(\cdot,t)v\,d\mu\Big]\psi(t)dt \\ &= \int_{X\times I} u_{\mathrm{m}} v\psi\,d\lambda \end{aligned}$$

for all $v \in L^{q'}$ and $\psi \in L^{p'}$.

Note that if $u \in C(I;L^q(X))$ then its equivalence class $[u]_I$ is in $L^\infty_{loc}(I;L^q(X))$; no ambiguity will result from our writing u_{m} in place of $[u]_{\mathrm{m}}$. Observe also that the notation u_{m} does not explicitly reference the particular indices p and q. A simple argument based on the following extension of Theorem A.15 shows that no ambiguity arises when u belongs to more than one space $L^p(I;L^q(X))$. This extension will also be applied in the proof of Theorem A.19.

Corollary A.17. *Let I, (X,μ), λ, p and q be as in* (A.16) *and assume that X satisfies the hypotheses in* Theorem A.15(b). *Let $u \in L^p(I;L^q(X))$ and*

$v \in C_c(X \times I)$. *Then*

$$\int_I \Big[\int_X u(t)v(\cdot,t)\,d\mu \Big] dt = \int_I \Big[\int_X u_{\mathrm{m}}(x,t)v(x,t)d\mu \Big] dt.$$

PROOF. Choose $\varepsilon > 0$ and for $j \in \mathbf{Z}$ let χ_j^ε be the characteristic function of the set $I \cap [j\varepsilon, (j+1)\varepsilon]$. Then by (A.19)

$$\sum_j \int_I \Big[\int_X u(t)v(\cdot, j\varepsilon)d\mu \Big] \chi_j^\varepsilon(t)dt = \sum_j \int_I \Big[\int_X u_{\mathrm{m}}(x,t)v(x,j\varepsilon)d\mu \Big] \chi_j^\varepsilon(t)dt,$$

and the result follows in the limit as $\varepsilon \to 0$.

□

Bochner Space Differentiability and Absolute Continuity

Next we let Ω be an open set in $\mathbf{R}^n$ and examine a concept of spatial differentiability of an element $u \in L^p(I; L^q(\Omega))$ in terms of the existence of weak derivatives on $\Omega \times I^\circ$ of its measurable counterpart u_{m}. We begin with the following formulation:

Definition A.18. *Let* $u \in L^p(I; L^q(\Omega))$ *and* $w \in L^r(I; L^s(\Omega))$ *where* Ω *is an open set in* $\mathbf{R}^n$, I *is a nontrivial interval in* $\mathbf{R}$, $p, r \in [1, \infty]$ *and* $q, s \in (1, \infty]$, *and let* α *be a multi-index. Then* $w = D_x^\alpha u$ *in the Bochner space sense if for every pair of representatives* $\varphi : I \to L^q(\Omega)$ *and* $\psi : I \to L^s(\Omega)$ *of* u *and* w *there is a null set* $E \subset I$ *such that for* $t \in I - E$, $\psi(t) = D_x^\alpha \varphi(t)$ *in the sense of weak derivatives on* Ω.

The following equivalent formulation in terms of measurable counterparts is applied throughout the text:

Theorem A.19. *Let* $u \in L^p(I; L^q(\Omega))$ *and* $w \in L^r(I; L^s(\Omega))$ *where* Ω *is an open set in* $\mathbf{R}^n$, I *is a nontrivial interval in* $\mathbf{R}$, $p, r \in [1, \infty]$ *and* $q, s \in (1, \infty]$, *and let* α *be a multi-index. Then* $w = D_x^\alpha u$ *in the Bochner space sense if and only if* $w_{\mathrm{m}} = D_x^\alpha u_{\mathrm{m}}$ *in the sense of weak derivatives in* $\Omega \times I^o$. *If so, then* $(D_x^\alpha u)_{\mathrm{m}} = D_x^\alpha (u_{\mathrm{m}})$.

PROOF. First assume that $w = D_x^\alpha u$ in the Bochner space sense and let $v \in C_c^{|\alpha|}(\Omega \times I^o)$. Then from Corollary A.17 and Fubini's theorem, each applied twice, we have

$$\begin{aligned}
\int_{\Omega \times I} w_{\mathrm{m}} v &= \int_I \Big[\int_\Omega w_{\mathrm{m}}(x,t)v(x,t)dx \Big] dt \\
&= \int_I \Big[\int_\Omega w(t)v(\cdot,t)dx \Big] dt = (-1)^{|\alpha|} \int_I \Big[\int_\Omega u(t) D_x^\alpha v(\cdot,t)dx \Big] dt \\
&= (-1)^{|\alpha|} \int_I \Big[\int_\Omega u_{\mathrm{m}}(x,t) D_x^\alpha v(x,t)dx \Big] dt = (-1)^{|\alpha|} \int_{\Omega \times I} u_{\mathrm{m}} D_x^\alpha v,
\end{aligned}$$

which proves that $w_{\mathrm{m}} = D_x^\alpha u_{\mathrm{m}}$ in the weak sense on $\Omega \times I^o$.

Now assume that $w_{\mathrm{m}} = D_x^\alpha u_{\mathrm{m}}$ in the weak sense on $\Omega \times I^o$, let $\varphi : I \to L^q$ and $\gamma : I \to L^s$ be representatives of u and w respectively and let $\{v_j\}_j \subset C_c^{|\alpha|}(\Omega)$ be a countable set with the property that if $v \in C_c^{|\alpha|}(\Omega)$ then there is a sequence $v_{j_i} \to v$ in $L^{s'}$ such that $D_x^\alpha v_{j_i} \to D_x^\alpha v$ in $L^{q'}$ (the v_j may be taken to be elements of a set in $C_c^\infty(\mathbf{R}^n)$ which is dense in $W^{|\alpha|,q'\vee s'}(\mathbf{R}^n)$ multiplied by suitable cutoff functions, for example; see Theorem A.2). There is then a null set $E \subset I$ such that each $t_0 \in I - E$ is a Lebesgue point of the maps $t \to \int_\Omega \varphi(t) D_x^\alpha v_j$ and $t \to \int_\Omega \gamma(t) v_j$. Fix $t_0 \in I - E$ and let $\psi^\varepsilon : I \to \mathbf{R}$ be a nonnegative function supported on an interval of length ε contained in I and containing t_0 and whose integral over I equals one. Then applying Corollary A.17 and Fubini's theorem, each twice, we obtain

$$\begin{aligned}
\int_\Omega \varphi(t_0) D_x^\alpha v_j dx &= \lim_{\varepsilon\to 0} \int_I \Big[\int_\Omega \varphi(t) D_x^\alpha v_j dx \Big] \psi^\varepsilon(t) dt \\
&= \lim_{\varepsilon\to 0} \int_I \Big[\int_\Omega u_{\mathrm{m}}(\cdot, t) D_x^\alpha v_j dx \Big] \psi^\varepsilon(t) dt \\
&= \lim_{\varepsilon\to 0} \int_{\Omega\times I} u_{\mathrm{m}} D_x^\alpha (v_j \psi^\varepsilon) = (-1)^{|\alpha|} \lim_{\varepsilon\to 0} \int_{\Omega\times I} w_{\mathrm{m}} v_j \psi^\varepsilon \\
&= \ldots = (-1)^{|\alpha|} \int_\Omega \gamma(t_0) v_j dx.
\end{aligned}$$

This proves that, for almost all $t_0 \in I$, $D_x^\alpha \varphi(t_0) = \gamma(t_0)$ in the sense of weak derivatives on Ω and therefore that $D_x^\alpha u = w$ in the Bochner space sense of Definition A.18.

□

We now consider the relationship between the derivative of an absolutely continuous mapping $u \in wAC(I; L^q(\Omega))$ in the sense of Definition 2.2 and the weak time-derivative of its measurable counterpart u_{m}:

Theorem A.20. *Let Ω be an open set in $\mathbf{R}^n$, I a nontrivial interval in $\mathbf{R}$ and $q \in (1, \infty]$.*

(a) *If $u \in L^1_{loc}(I; L^q(\Omega))$ then its equivalence class on I contains an element $u_{AC} \in wAC(I; L^q(\Omega))$ with derivative $(u_{AC})'$ if and only if the t-derivative $D_t u_{\mathrm{m}}$ exists in the weak sense on $\Omega\times I^\circ$ and is in $L^{1,q}_{(t)loc}(\Omega\times I)$. If so, then $((u_{AC})')_{\mathrm{m}} = D_t u_{\mathrm{m}}$.*

(b) *Let $\mathcal{H}_1$ be a closed subspace of $H^1(\Omega)$ which is dense in $L^2(\Omega)$ and let $u_{\mathrm{m}} \in L^{\infty,2}_{(t)loc}(\Omega\times I)$. Assume that there is an $F \in L^2_{loc}(I; \mathcal{H}_1^*)$ such that for every $v \in \mathcal{H}_1$ and $\psi \in C_c^1(I)$ (having one-sided derivatives at endpoints of I if any)*

$$\int_{\Omega\times I} u_{\mathrm{m}}(x,t) v(x)\, \psi_t(t) = \int_I (F(t)\cdot v) \psi(t)\, dt.$$

Then u_{m} is the measurable counterpart of an element $u \in C(I; L^2_{weak}(\Omega))$ for which $u^* \in wAC^2(I; \mathcal{H}_1^*)$ with derivative F (recall that for $z \in L^2$, $z^* \in \mathcal{H}_1^*$ is defined by $z^* \cdot v = \langle z, v\rangle_{L^2}$.)

PROOF. To prove (a) we first assume that u_{AC} exists as described in the statement and let $\psi \in C_c^1(\Omega \times I^\circ)$. Then by Fubini's theorem and Corollary A.17,

$$\begin{aligned}
\int_{\Omega\times I^\circ} u_{\mathrm{m}}\psi_t &= \int_I \Big(\int_\Omega u\,\psi_t\,dx\Big)dt = \int_I \Big(\int_\Omega u_{AC}\psi_t dx\Big)dt \\
&= \lim_{h\to 0} h^{-1} \int_I \Big(\int_\Omega u_{AC}(x,t)[\psi(x,t+h) - \psi(x,t)]dx\Big)dt \\
&= -\lim_{h\to 0} h^{-1} \int_I \Big(\int_\Omega [u_{AC}(x,t) - u_{AC}(x,t-h)]\psi(x,t)dx\Big)dt \\
&= -\lim_{h\to 0} h^{-1} \int_I \int_{t-h}^t \Big(\int_\Omega (u_{AC})'(s)\psi(\cdot,t)\Big)ds\,dt \\
&= -\lim_{h\to 0} h^{-1} \int_I \int_s^{s+h} \Big(\int_\Omega (u_{AC})'(s)\psi(\cdot,t)\Big)dt\,ds \\
&= -\int_I \Big(\int_\Omega ((u_{AC})')_{\mathrm{m}}(s)\psi(\cdot,s)\Big)\,ds = -\int_{\Omega\times I^\circ} ((u_{AC})')_{\mathrm{m}}\psi.
\end{aligned}$$

This proves that $((u_{AC})')_{\mathrm{m}}$ is the t-derivative of u_{m} in the weak sense on $\Omega \times I^\circ$.

Now suppose that the weak derivative $D_t u_{\mathrm{m}}$ exists in the weak sense as described in the statement of the theorem and let $\{\varphi_j\}_j \subset C_c^1(\Omega)$ be countable and dense in $L^{q'}(\Omega)$ where $q' \in [1, \infty)$ is the Hölder conjugate of q. Abusing notation slightly, we let u denote a specific representative of its equivalence class. There is then a null set $E \subset I$ such that if $t_0 \in I - E$ then $|u(t_0)|_{L^q(\Omega)} < \infty$ and t_0 is a Lebesgue point of each of the maps $t \to \int_\Omega u(t)\varphi_j dx$. Fix $t_0 \in I^\circ - E$ and observe that, for any $t \in I$,

$$\Big|\int_\Omega u(t_0)\varphi_j dx + \int_{t_0}^t \int_\Omega D_t u_{\mathrm{m}}\, \varphi_j dx\, ds\Big| \le C(t)|\varphi_j|_{L^{q'}(\Omega)}.$$

It follows that there is an element of $L^q(\Omega)$, which we denote by $u_{AC}(t)$, such that

$$\int_\Omega u_{AC}(t)v\,dx = \int_\Omega u(t_0)v\,dx + \int_{t_0}^t \int_\Omega D_t u_{\mathrm{m}}\, v\,dx\,ds \tag{A.23}$$

for all $v \in L^{q'}(\Omega)$. Apparently $u_{AC} \in wAC(I; L^q(\Omega))$ with derivative satisfying $((u_{AC})')_{\mathrm{m}} = D_t u_{\mathrm{m}}$, which justifies the notation.

To complete the proof we need to show that $u_{AC}(t) = u(t)$ for almost all t. Let $t_1 \in I^\circ - E$ and assume without less of generality that $t_1 > t_0$. For $h > 0$ define $\psi^h \in C_c^1(I^\circ)$ by taking $\psi = 0$ in the complement of $[t_0, t_1]$,

$\psi = -1$ on $[t_0 + h, t_1 - h]$, and ψ monotone in the remaining two sets. Then by Corollary A.17, for all j,

$$\int_\Omega [u(t_1) - u(t_0)]\varphi_j dx = \lim_{h\to 0} \int_I \Big(\int_\Omega u(t)\varphi_j dx \Big) \psi_t^h(t) dt$$
$$= \lim_{h\to 0} \int_{\Omega\times I^\circ} u_{\mathrm{m}}\, \varphi_j \psi_t^h = - \lim_{h\to 0} \int_{\Omega\times I^\circ} D_t u_{\mathrm{m}}\, \varphi_j \psi^h$$
$$= \int_{t_0}^{t_1} \Big(\int_\Omega D_t u_{\mathrm{m}}\, \varphi_j dx \Big) dt = \int_\Omega [u_{AC}(t_1) - u(t_0)]\varphi_j dx$$

by (A.23). It follows that $u_{AC}(t_1) = u(t_1)$ for all $t_1 \in I^\circ - E$, so that u_{AC} and u agree almost everywhere on I as elements of $L^q(\Omega)$.

The proof of (b) is nearly identical. We fix a specific representative u_{m}, a set $\{\varphi_j\}$ which is dense in $\mathcal{H}_1$ and a corresponding null set E as in the proof of (a), and repeat the argument above to construct an element $G \in wAC^2(I; \mathcal{H}_1^*)$ whose derivative is F and which agrees with $u_{\mathrm{m}}(\cdot, t)^*$ for $t \in I - E$. Then for $v \in \mathcal{H}_1$, I' compact in I and $t \in I' - E$,

$$|G(t) \cdot v| = |u_{\mathrm{m}}^*(\cdot, t) \cdot v| \le |u_{\mathrm{m}}(\cdot, t)|_{L^{\infty,2}(\Omega\times I')} |v|_{L^2}.$$

The bound on the right extends to all $t \in I'$ by the absolute continuity of G, and this together with the fact that $\mathcal{H}_1$ is dense L^2 shows that for each $t \in I$ there is an element $u(t) \in L^2(\Omega)$ such that $G(t) = u(t)^*$. Also, $u \in C(I; L^2_{weak}(\Omega))$ because $G(t) \cdot v$ is continuous in t for $v \in \mathcal{H}_1$ and $\mathcal{H}_1$ is dense in L^2; and the proof that u_{m} is the measurable counterpart of u is essentially identical to that of the final step in the proof of (a).

□

A similar result for mappings into $L^q(\partial\Omega)$ can be derived by applying Theorem A.20 in the coordinate domains associated to $\partial\Omega$. To do this we recall the notation for Lipschitz domains in Definition 1.7(a) and the definition (1.6) of the operators T_i and define the composition operator τ_i mapping functions $f : \partial\Omega \cap \mathcal{U}_i \to \mathbf{R}$ into functions $\tau_i f : \mathcal{B}_i \to \mathbf{R}$ by the relation $(\tau_i f)(\tilde{y}) = f(T_i(\tilde{y}))$. The operator τ_i then maps σ-equivalence classes (σ is defined in Theorem 3.1) on $\partial\Omega$ into Lebesgue equivalence classes on $\mathcal{B}_i$ and extends as an isometry from $L^q(\partial\Omega \cap \mathcal{U}_i)$ onto $L^q(\mathcal{B}_i)$ (provided the measure in the latter space is $\sqrt{1 + |\nabla\psi_i(\tilde{y})|^2 d\tilde{y}}$). For mappings f of an interval $I \subseteq \mathbf{R}$ into $L^q(\partial\Omega \cap \mathcal{U}_i)$ we define $(\tau_i f)(t) \equiv \tau_i(f(t))$. Again τ_i extends as an isometry of $L^p(I; L^q(\partial\Omega\cap\mathcal{U}_i))$ onto $L^p(I; L^q(\mathcal{B}_i))$ and preserves weak absolute continuity. All the above statements hold as well for the inverse mapping τ_i^{-1}, *mutatis mutandi.*

The following theorem gives a useful characterization of weakly absolutely continuous mappings into $L^q(\partial\Omega)$:

Theorem A.21. *Let* $u \in L^1_{loc}(I; L^q(\partial\Omega))$ *where* I *is a nontrivial interval in* $\mathbf{R}$, Ω *is a Lipschitz domain in* $\mathbf{R}^n$ *and* $q \in (1, \infty]$; *and let* $\mathcal{U}_i, T_i, \mathcal{B}_i$ *and* ψ_i *be as in* Definition 1.7 *and* (1.6). *Let* τ_i *be as above and define* $v_i = (\tau_i u)_{\mathrm{m}}$, *which is an element of* $L^{1,q}(\mathcal{B}_i \times I')$ *for every compact interval* $I' \subseteq I$. *Then the Lebesgue equivalence class* u *contains an element* $u_{AC} \in wAC(I; L^q(\partial\Omega))$ *if and only if for each* i *the weak derivative* $D_t v_i$ *with respect to* $t \in I$ *exists in the weak sense on* $\mathcal{B}_i \times I^\circ$ *and*

$$\text{(A.24)} \qquad \Big[\sum_i |D_t v_i|^q_{L^q(\mathcal{B}_i)}\Big]^{1/q} \in L^1_{loc}(I)$$

(the measure on $\mathcal{B}_i$ *is* $\sqrt{1 + |\nabla\psi_i(\tilde{y})|^2 d\tilde{y}}$*) or*

$$\text{(A.25)} \qquad \sup_i |D_t v_i|_{L^\infty(\mathcal{B}_i)} \in L^1_{loc}(I)$$

according as $q < \infty$ *or* $q = \infty$. *If so, then for each* i,

$$[(\tau_i u_{AC})']_{\mathrm{m}} = D_t[\tau_i(u_{\mathrm{m}})]$$

as elements of $L^{1,q}_{(t)loc}(\mathcal{B}_i \times I)$. *The proof of* (A.24) *for the "only if" statement requires the additional hypothesis that there is a nonnegative integer* i' *such that no* $\mathcal{U}_i$ *intersects more than* i' *of the other covering sets* $\mathcal{U}_i$.

Proof. First note that the statements in (A.24) and (A.25) are meaningful by Fubini's theorem because v_i is measurable on the product domain $\mathcal{B}_i \times I$. If the weak derivatives $D_t v_i$ exist as in the statement then by Theorem A.20 there are functions v_{iAC} in the equivalence class $\tau_i u$ which are in $wAC(I; L^q(\mathcal{B}_i))$ with derivatives $v_i' \in L^1_{loc}(I; L^q(\mathcal{B}_i))$. We then define

$$u_{AC} = \sum_i \chi_i \cdot \tau_i^{-1} v_{iAC} \text{ and } w = \sum_i \chi_i \cdot \tau_i^{-1} v_i'$$

where χ_i is the characteristic function of $\mathcal{W}_i$ (which is defined in Theorem 3.1(a)), and then check that u_{AC} is in the equivalence class u and that $u_{AC} \in wAC(I; L^q(\partial\Omega))$ with derivative w. The details are straightforward and so are left to the reader.

To prove the converse we suppose that there is a function $u_{AC} \in wAC(I; L^q(\partial\Omega))$ which is in the Lebesgue equivalence class of u on I. Then $v_{iAC} \equiv \tau_i u_{AC}$ is in $wAC(I; L^q(\mathcal{B}_i))$ and v_{iAC} is in the equivalence class on I of $\tau_i u$, whose measurable counterpart $(\tau_i u)_{\mathrm{m}}$ is v_i. Theorem A.20 therefore applies to show that the weak derivatives $D_t v_i$ exist in the weak sense on $\mathcal{B}_i \times I^\circ$. The statements in (A.24) and (A.25) then follow easily. Note, however, that since the covering sets $\{\mathcal{U}_i\}$ are not disjoint, the proof of (A.24) requires an application of the following general fact: if $\{E_i\}$ is a countable collection of sets in an arbitrary σ-algebra on which there is a nonnegative measure μ, and if no E_i intersects more than i' of the other E_i, then

$$\sum_i \mu(E_i) \leq (1 + i')\mu(\cup_i E_i).$$

□

The spaces $L^p(I; W^{k,q}(\Omega))$ and $wAC(I; W^{k,q}(\Omega))$

We conclude this section by defining the above two spaces and by characterizing their elements in terms of the weak derivatives of their measurable counterparts; this will give succinct and useful versions of Theorems A.19 and A.20. Notice that Definitions 1.3 and 2.2 do not apply here because we cannot readily identify $W^{k,q}(\Omega)$ with the dual of a separable Banach space X. It can, however, be identified with a closed subspace of such a space, as follows. Let Ω be an open set in $\mathbf{R}^n$, $p \in [1, \infty]$, $q \in (1, \infty]$, q' the Hölder conjugate of q and k a positive integer. Let K be the number of multi-indices of length less than or equal to k and define the norm of an element $v = \{v_\alpha : |\alpha| \le k\} \in [L^s(\Omega)]^K$ by $\Big[\sum_{|\alpha| \le k} |v_\alpha|^s_{L^s(\Omega)}\Big]^{1/s}$ or by $\max_{|\alpha|\le k} |v_\alpha|_{L^\infty(\Omega)}$ according as $s \in [1, \infty)$ or $s = \infty$. Then since $q' < \infty$, $X \equiv [L^{q'}(\Omega)]^K$ is a separable space and its dual X^* can be identified with $[L^q(\Omega)]^K$ in the obvious way. (An easy exercise based on Hölder's inequality shows that the norm of an element in $[L^q(\Omega)]^K$ is the same as the norm in X^* of the corresponding linear functional.) The spaces $L^p(I; [L^q(\Omega)]^K)$ and $wAC(I; [L^q(\Omega)]^K)$ are therefore included in Definitions 1.3 and 2.2, and the spaces of interest can be defined as closed subspaces:

Definition A.22. *Let Ω, p, q, q', k and K be as above and let I be a nontrivial interval in $\mathbf{R}$. Then $L^p(I; W^{k,q}(\Omega))$ is the set of Lebesgue equivalence classes over I of functions $\varphi : I \to W^{k,q}(\Omega)$ which are weakly measurable in the sense that the mappings $t \to \sum_{|\alpha|\le k} \int_\Omega D^\alpha_x \varphi(\cdot, t)\, v_\alpha dx$ are measurable on I for all $v = \{v_\alpha\} \in [L^{q'}(\Omega)]^K$ and such that the equivalence class of mappings $t \to |\varphi(t)|_{W^{k,q}(\Omega)}$ is in $L^p(I)$. The space $L^p_{loc}(I; W^{k,q}(\Omega))$ is the set of elements $u \in L^p_{loc}(I; L^q(\Omega))$ such that $u \in L^p(I'; W^{k,q}(\Omega))$ for every compact interval $I' \subseteq I$.*

The space $wAC(I; W^{k,q}(\Omega))$ is the the set of elements in $wAC(I; [L^q(\Omega)]^K)$ which take values in $W^{k,q}(\Omega)$ for all $t \in I$.

The reader will see that the measurability condition in this definition is equivalent to the requirement that $t \to \int_\Omega D^\alpha_x \varphi(\cdot, t)\, v\, dx$ is measurable on I for all $v \in L^{q'}(\Omega)$ and $|\alpha| \le k$. Observe also that if $u \in wAC(I; W^{k,q}(\Omega))$, then $u' \in L^1_{loc}(I; [L^q(\Omega)]^K)$, but there is no *a priori* stipulation that u' must be an element of $L^1_{loc}(I; W^{k,q}(\Omega))$. This is always true, however, as the reader can easily verify.

Theorem A.19 can now be applied to give useful alternative characterizations of the spaces $L^p(I; W^{k,q}(\Omega))$. The proof consists in straightforward checking of details and is therefore omitted.

Theorem A.23. *Let Ω be an open set in $\mathbf{R}^n$, I a nontrivial interval in $\mathbf{R}$, $u \in L^p(I; L^q(\Omega))$ where $p \in [1,\infty]$ and $q \in (1,\infty]$, and k a positive integer. The following three statements are then equivalent:*

(a) *$u \in L^p(I; W^{k,q}(\Omega))$.*

(b) *For every multi-index α of length less than or equal to k the derivative $D^\alpha u$ exists in the Bochner space sense of* Definition A.17 *and is an element of $L^p(I; L^q(\Omega))$.*

(c) *For every multi-index α of length less than or equal to k the derivative $D_x^\alpha u_{\mathrm{m}}$ with respect to $x \in \Omega$ exists in the weak sense on $\Omega \times I^\circ$ and is an element of $L^{p,q}(\Omega \times I)$.*

In the same way we can apply Theorem A.20 to give a useful characterization of absolutely continuous mappings into $W^{k,q}(\Omega)$ in terms of their measurable counterparts. Again, the proof is elementary and so is omitted.

Theorem A.24. *Let $u \in L^1_{loc}(I; L^q(\Omega))$ where Ω is an open set in $\mathbf{R}^n$, I is a nontrivial interval in $\mathbf{R}$ and $q \in (1,\infty]$. Then the Lebesgue equivalence class u contains an element $u_{AC} \in wAC(I; W^{k,q}(\Omega))$ if and only if for every α with $|\alpha| \le k$ the derivatives $D_x^\alpha u_{\mathrm{m}}$ and $D_t D_x^\alpha u_{\mathrm{m}}$ with respect to $x \in \Omega$ and $t \in I$ exist in the weak sense in $\Omega \times I^\circ$ and are elements of $L^{1,q}(\Omega \times I')$ for every compact interval $I' \subseteq I$.*

The following specific application of Theorems A.23 and A.24 is particularly useful for transferring information about solutions of the problems considered in Chapters 5 and 6 between their Bochner–Sobolev space descriptions, which have an intuitive dynamical systems appeal, and the measurable counterpart descriptions, for which the calculus of weak derivatives is especially convenient:

Corollary A.25. *Let $w \in L^\infty_{loc}(I; L^2(\Omega))$ where Ω is an open set in $\mathbf{R}^n$ and I is a nontrivial interval in $\mathbf{R}$, let w_{m} be its measurable counterpart and let $m \ge 1$. The following two statements are then equivalent:*

(a) *For $j = 0, \ldots, m$ the weak derivatives $D_t^j D_x^\alpha w_{\mathrm{m}}$ on $\Omega \times I^\circ$ are in $L^{\infty,2}_{(t)loc}(\Omega \times I)$ for $2j + |\alpha| \le 2m$ and in $L^{2,2}_{(t)loc}(\Omega \times I)$ for $2j + |\alpha| \le 2m+1$.*

(b) *There are maps $w^{(0)} = w, w^{(1)}, \ldots, w^{(m)}$ such that the equivalence class over I of $w^{(j)}$ is in $L^\infty_{loc}(I; H^{2m-2j}) \cap L^2_{loc}(I; H^{2m-2j+1})$ for $j \le m$ (we abbreviate $H^k(\Omega) = H^k$); and for $j \le m-1$,*

$$w^{(j)} \in Lip_{loc}(I; H^{2m-2j-2}) \cap wAC^2(I; H^{2m-2j-1}) \cap C(I; H^{2m-2j}_{weak})$$

and the derivative of $w^{(j)}$ is the the equivalence class of $w^{(j+1)}$.

PROOF: The proof is a straightforward iteration of the results of Theorems A.23 and A.24 with the the exception of the weak continuity of $w^{(j)}$

into H^{2m-2j} in (b). To prove this we first note that $w^{(j)} \in H^{2m-2j}$ for almost all t with the bound independent of t, hence for all t, because weak H^{2m-2j}–subsequential limits of sequences $w^{(j)}(t_i)$ are unique by the absolute continuity assumed in (b). The same two facts then show that if $t_i \to t_0$ in I then the weak H^{2m-2j}–limit of $w^{(j)}(t_i)$ is $w^{(j)}(t_0)$, as required.

□

§A.6 A Representation Theorem for Boundary Integrals

The following result is useful in a variety of contexts and in particular was applied in chapters 4 and 5 to express integrals over the boundary of a spatial domain as a sum of integrals over its interior. Parts (a)–(d) give Bochner–Sobolev space formulations of the relationship between the respective kernels and part (e) is a reformulation in terms of weak derivatives.

Theorem A.26.

(a) *Let Ω be a $C^{0,1}_{bdd}$ domain in $\mathbf{R}^n$, $n \geq 2$, not necessarily bounded, $q \in (1, \infty]$ and q' its Hölder conjugate. Then there exist operators $S^1 \in \mathcal{L}(L^q(\partial\Omega), L^q(\Omega))$ and $S^2 \in \mathcal{L}(L^q(\partial\Omega), [L^q(\Omega)]^n)$ such that for $f \in L^q(\partial\Omega)$ and $v \in W^{1,q'}(\Omega)$,*

$$\int_{\partial\Omega} fv\, d\sigma = \int_{\Omega} \big[(S^1 f)\, v + (S^2 f) \cdot \nabla_x v\big]\, dx. \tag{A.26}$$

The norms of S^1 and S^2 depend on n, q and the constant $M^{0,1}_{\Omega}$ in Definition 1.7.

Now assume either that Ω is a $C^{0,1}_{bdd}$ domain in $\mathbf{R}^n$, $n \geq 2$, in which case we set $k = 1$, or that Ω is a C^k_{bdd} domain with $k \geq 2$. Let $l \in \{0, \ldots, k-1\}$, $q \in (1, \infty]$ with Hölder conjugate q', $p \in [1, \infty]$ and I a nontrivial interval in $\mathbf{R}$.

(b) *If $f \in L^q(\partial\Omega)$ with $\langle f \rangle^{\alpha}_{L^q,\partial\Omega} < \infty$ for every $|\alpha| \leq l$ (see* (5.6)*) then $S^1 f$ and the components of $S^2 f$ are in $W^{l,q}(\Omega)$ with norms bounded by $C \sum_{|\alpha| \leq l} \langle f \rangle^{\alpha}_{L^q,\partial\Omega}$ where the constant C depends on n, q, l and either $M^{0,1}_{\Omega}$ or M^k_{Ω}, according as $k = 1$ or $k > 1$.*

(c) *If $f \in L^p(I; L^q(\partial\Omega))$ and if $\langle f \rangle^{(\alpha,0)}_{L^q,\partial\Omega} \in L^p(I)$ for every $|\alpha| \leq l$ (see* (5.7)*), then $S^1 f$ and the components of $S^2 f$ are in $L^p(I; W^{l,q}(\Omega))$ with norms bounded by $C \sum_{|\alpha| \leq l} \langle f \rangle^{(\alpha,0)}_{L^q,\partial\Omega}$ where C is as in* (b).

(d) *If $f \in wAC^p(I; L^q(\partial\Omega))$ with derivative f' and if for every $|\alpha| \leq l$ $\sup_I \langle f \rangle^{\alpha}_{L^q,\partial\Omega} < \infty$ and $\langle f' \rangle^{(\alpha,0)}_{L^q,\partial\Omega} \in L^p_{loc}(I)$, then $S^1 f$ and the components* *of $S^2 f$ are in $wAC^p(I; W^{l,q}(\Omega))$ with derivatives $S^1 f'$ and the components of $S^2 f'$ respectively.*

(e) *If $f \in L^1(I; L^q(\partial\Omega))$ and if for some $i \geq 0$ and all $|\alpha| \leq l$, $\langle f\rangle^{(\alpha,i)}_{L^q,\partial\Omega} \in L^p(I)$, then $D^i_t D^\alpha_x (S^1 f)_{\mathfrak{m}}$ and the components of $D^i_t D^\alpha_x (S^2 f)_{\mathfrak{m}}$ are in $L^{p,q}(\Omega \times I)$ for $|\alpha| \leq l$ with norms bounded by $C\Big|\sum_{|\alpha|\leq l} \langle f\rangle^{(\alpha,i)}_{L^q,\partial\Omega}\Big|_{L^p(I)}$ where C is as in* (b) *and D_t and D_x denote weak derivatives on $\Omega \times I^\circ$.*

Observe that by the trace theorem, Theorem 3.2, the left side of (A.26) defines the action of an element of the dual of $W^{1,q'}(\Omega)$ on v, so that for the case $p = p' = 2$, say, the representation (A.26) is already known: the self-duality of H^1 implies that there is an element F in $H^1(\Omega)$ such that

$$\text{(A.27)} \qquad \int_{\partial\Omega} f v \, d\sigma = \langle F, v\rangle_{H^1} = \int_\Omega (Fv + \nabla F \cdot \nabla_x v) dx$$

for all $v \in H^1(\Omega)$. The important point of (b) is therefore the regularity of F, or in the general case, of $S^1 f$ and $S^2 f$. For example, since the right side of (A.27) defines an element of $\mathcal{B}(\mathcal{H}_1 \times \mathcal{H}_1)$, exactly of the the type considered in Chapter 4, the application of Theorem 4.2, were its proof independent of Theorem A.26, would show that $F \in H^2(\Omega)$ so that (A.27) would hold with $S^1 f = F$ and $S^2 f = \nabla_x F$. Observe also that there may be multiple choices for $S^1 f$ and $S^2 f$: again in the case $W^{l,q} = H^1$ there will be solutions $\gamma \in H^2 \cap H^1_0$ of the weak form of the eigenvalue equation for the Laplace operator, $\Delta\gamma = \lambda\gamma$. Thus if the pair $(F, \nabla F)$ satisfies (A.27) then so too does the pair $(F + \lambda\gamma, \nabla F + \nabla\gamma)$. The other important point of the theorem is therefore that specific, explicit choices $S^1 f$ and $S^2 f$ are made so that $S^1 f$ and $S^2 f$ are in the appropriate Bochner spaces in the time–dependent case.

PROOF of Theorem A.26. We will omit the proof of (a)–(b) for the $C^{(0,1)}_{bdd}$ case, this proof being essentially the same as that of (b) for C^1_{bdd} domains but with the ψ_i in Definition 1.7 having L^∞ rather than bounded continuous derivatives. This distinction plays no role in the proof. The notations of Definition 1.7 will be applied freely throughout, C will denote a generic constant depending on n, q, k and l, and Ω will be a C^k_{bdd} domain with $k \geq 1$.

To prove (a) and (b) we let v be the restriction to Ω of an element of $C^\infty_c(\mathbf{R}^n)$, the latter being dense in $W^{1,q'}$, and a nondecreasing C^∞ function $\chi = \chi(s)$ on $\mathbf{R}$ which is zero for $s \leq \varepsilon_0/2$ and identically one for $s \geq \varepsilon_0$ (ε_0 is defined in Definition 1.7(b)). Let $\{\varphi_i\}$ be the partition of unity constructed in Theorem 1.8 with $X = \partial\Omega$ subordinate to the cover $\{\mathcal{U}_i\}$. Then by

Theorem 3.1,

$$\begin{aligned}\int_{\partial\Omega} fv\,d\sigma &= \sum_i \int_{\partial\Omega\cap\mathcal{U}_i} f\,v\,\varphi_i\,d\sigma \\ &= \sum_i \int_{\mathcal{B}_i} \Big[(f\circ T_i)(\tilde{y})\sqrt{1+|\nabla\psi_i(\tilde{y})|^2} \\ &\qquad \times \int_0^{\psi_i(\tilde{y})} D_{y_n}\Big(\chi(y_n)\big((v\varphi_i)\circ R_i\big)(\tilde{y},y_n)\Big)\,dy_n\Big]\,d\tilde{y}.\end{aligned}$$

Carrying out the D_{y_n}-differentiation, rearranging and writing the double integral as an integral over $\Omega\cap\mathcal{U}_i$, we find that

$$\int_{\partial\Omega} fv\,d\sigma = \sum_i \int_{\Omega\cap\mathcal{U}_i} (F_i^1 v + F_i^2\cdot\nabla v)\,dx, \tag{A.28}$$

where explicit expressions for F_i^1 and F_i^2 in terms of f, ψ_i, T_i, R_i and χ, which the reader can easily compute, show by inspection that F_i^1 and the components of F_i^2 depend linearly on f and extend to elements of $W^{l,q}(\Omega)$ with supports contained in the support of φ_i and with

$$|F_i^1|_{W^{l,q}(\Omega)},\ |F_i^2|_{[W^{l,q}(\Omega)]^n} \le C|f\circ T_i|_{W^{l,q}(\mathcal{B}_i)}. \tag{A.29}$$

Now if $q\in(1,\infty)$ and $|\alpha|\le l$, then for specific representatives F_i^1 and for almost all $x\in\Omega$,

$$\Big|\sum_{i_1}^{i_2} D^\alpha F_i^1(x)\Big|^q \le \Big(\sum_{i_1}^{i_2}|D^\alpha F_i^1(x)|\Big)^q \le C(i_0,q)\sum_{i_1}^{i_2}|D^\alpha F_i^1(x)|^q$$

by the last hypothesis in Definition 1.7 and the fact that the support of φ_i, hence of F_i^1, is contained in $\mathcal{U}_i$. Integrating, summing over α and applying (A.29) we thus find that

$$\Big|\sum_{i_1}^{i_2} F_i^1\Big|^q_{W^{l,q}(\Omega)} \le C\sum_{i_1}^{i_2}|f\circ T_i|^q_{W^{l,q}(\mathcal{B}_i)},$$

which goes to zero as $i_1, i_2\to\infty$ by the assumption in (b). This proves that $\sum_i F_i^1$ converges in $W^{l,q}(\Omega)$ to an element $F^1\equiv S^1 f$. A similar argument applies to $\sum_i F_i^2\equiv F^2\equiv S^2 f$, and the representation (A.26) then follows immediately from (A.28). For the case that $q=\infty$ we apply the fact that each $x\in\Omega$ has a neighborhood in which the sums $\sum F_i^1$ and $\sum F_i^2$ have no more than i_0 nonzero summands; the representations for F^1 and F^2 then follow easily. This proves (a) and (b).

To prove (c) we write $S_i^1 f$ in place of F_i^1 and re-examine the computation leading to (A.28) to find that, for $v \in C_c^\infty(\mathbf{R}^n)$,

$$\int_\Omega (S_i^1 f)\, v\, dx = \int_{\partial\Omega} f\, L_i^1 v\, d\sigma \tag{A.30}$$

where L_i^1 is a linear map from $C_c(\mathbf{R}^n)$ to $C_c(\partial\Omega \cap \mathcal{U}_i)$. Thus if f temporarily denotes a specific representative of its equivalence class, then since the right side above is measurable in t, so is the left, and hence so is the sum defining $\int_\Omega (S^1 f)\, v\, dx$. This proves that $S^1 f$ is weak-$*$ measurable as a map into $L^q(\Omega)$, hence so are its Bochner-space derivatives $D_x^\alpha(S^1 f)$ described in (b). It then follows from the L^p bound assumed in (c) that $S^1 f \in L^p(I; W^{l,q}(\Omega))$ with the required bound, and a similar argument applies to $S^2 f$ to complete the proof of (c).

To prove (d) we let f and f' be as in the hypotheses so that by (b) and (c) $S^1 f$ is a bounded map from I into $W^{l,q}(\Omega)$ and $S^1 f' \in L^p_{loc}(I; W^{l,q}(\Omega))$. Let $v \in C_c^\infty(\mathbf{R}^n)$, $|\alpha| \leq l$ and $t_1, t_2 \in I$. Then applying (A.30) twice, we obtain that

$$\begin{aligned}\int_\Omega (S^1 f)(D_x^\alpha v)\, dx\Big|_{t_1}^{t_2} &= \sum_i \int_\Omega (S_i^1 f)(D_x^\alpha v) dx\Big|_{t_1}^{t_2} = \sum_i \int_{\partial\Omega} f\,(L_i^1 D_x^\alpha v)\, d\sigma\Big|_{t_1}^{t_2} \\ &= \sum_i \int_{t_1}^{t_2}\int_{\partial\Omega} f'(L_i^1 D_x^\alpha v)\, d\sigma dt = \sum_i \int_{t_1}^{t_2}\int_\Omega (S_i^1 f')(D_x^\alpha v) dx\, dt \\ &= \int_{t_1}^{t_2}\int_\Omega (S^1 f')(D_x^\alpha v) dx\, dt.\end{aligned}$$

This proves that the Bochner space derivative $D_x^\alpha(S^1 f)$ is a weakly absolutely continuous map into $L^q(\Omega)$ with derivative $D_x^\alpha(S^1 f')$. A similar argument applies to $S^2 f$, completing the proof of (d).

We give the proof of (e) for the case $q < \infty$, the case $q = \infty$ being similar. Reexamining again the computation that led to (A.28) and (A.29), we find that we can write

$$F_i^1 = g_i \cdot (f \circ T_i \circ \Pi \circ R_i^{-1})$$

where if $k \geq 2$ (resp. $k = 1$) then $g_i \in C_{bdd}^{k-1}(\mathcal{U}_i)$ (resp. $L^\infty(\mathcal{U}_i)$) with compact support in $\mathcal{U}_i$ and with derivatives up to order $k-1$ bounded independently of i; and Π is the projection $\Pi(\tilde{y}, y_n) = (\tilde{y}, 0)$. Now fix α with $|\alpha| \leq l$ and define

$$H_i = D_x^\alpha\Big(g_i \cdot \big[D_t^i(f \circ T_i)_{\mathfrak{m}} \circ \Pi \circ R_i^{-1}\big]\Big)$$

Choosing particular representatives, we then have that for almost all $t \in I$,

$$|H_i|_{L^q(\Omega)} \leq C\Big(\sum_{|\beta|<l} |D_{\tilde{y}}^\beta D_t^i[(f \circ T_i)_{\mathfrak{m}}]|^q_{L^q(\mathcal{B}_i)}\Big)^{1/q}$$

for all i. The sum on the right is finite by hypothesis so that, as in the proof of (b) above, $\sum_{1}^{j_0} H_i$ converges strongly in $L^q(\Omega)$ as $j_0 \to \infty$ for all such t and the limit H is in $L^p(I; L^q(\Omega)$ with norm bounded by $C \sum_{|\alpha| \le l} \langle f \rangle_{L^q, \partial\Omega}^{(\alpha,i)}$. The same bound therefore applies to $|H_{\mathrm{m}}|_{L^{p,q}(\Omega \times I)}$.

To complete the proof we have to show that $H_{\mathrm{m}} = D_t D_x^\alpha (S^1 f)_{\mathrm{m}}$. To do this let $\varphi \in C_c^\infty(\Omega \times I^\circ)$ and compute

$$\begin{aligned}
\int_{\Omega \times I} (S^1 f)_{\mathrm{m}} \, D_t^i D_x^\alpha \varphi &= \sum_i \int_{(\Omega \cap \mathcal{U}_i) \times I} (F_i^1)_{\mathrm{m}} \, D_t^i D_x^\alpha \varphi \\
&= (-1)^{i+|\alpha|} \sum_i \int_{(\Omega \cap \mathcal{U}_i) \times I} D_x^\alpha \Big(g_i \cdot \big[D_t^i (f \circ T_i)_{\mathrm{m}} \circ \Pi \circ R_i^{-1} \big] \Big) \varphi \\
&= (-1)^{i+|\alpha|} \sum_i \int_{(\Omega \cap \mathcal{U}_i) \times I} H_i \varphi = (-1)^{i+|\alpha|} \int_{\Omega \times I} H \varphi,
\end{aligned}$$

so that $D_t^i D_x^\alpha (S^1 f)_{\mathrm{m}} = H_{\mathrm{m}} \in L^{p,q}(\Omega \times I)$ with the required bound. A similar argument applies to $S^2 f$ and completes the proof of (e).

□

Notes and References

The material of this text depends to a considerable extent on the foundation laid by the remarkable development of mathematical analysis in the first half of the twentieth century. The Lebesgue measure, weak differentiation, Hilbert and Banach spaces, Sobolev spaces and spectral theory have all proved to be essential concepts and structures for modern applied analysis. An excellent exposition of the history of this development is given in Naumann [**31**]. Our primary source for the details of the theory of Sobolev spaces has been Adams [**1**], more compact treatments being given in Ziemer [**46**], Gilbarg and Trudinger [**19**] and Evans [**12**].

The brief account in Chapter 4 of elliptic regularity in Sobolev spaces follows a standard approach included in many texts, most of which amplify the material in important ways. Those of Ladyzhenskaya and Uralceva [**27**], Bers, John and Schechter [**5**], Nečas [**33**], Grisvard [**20**], Gilbarg and Trudinger [**19**] and Evans [**12**] have emerged as standards. The definitive work on $C^{k,\alpha}$ regularity of solutions of elliptic systems of arbitrary order is given in Douglas and Nirenberg [**10**] and Agmon, Douglis and Nirenberg [**2**] and [**3**].

These elementary expositions of elliptic systems typically reflect an elegant and efficient application of fundamental concepts in Hilbert space theory to the construction of solutions via the basic theory of Sobolev spaces. The exposition in Chapters 2, 3 and 5 of the present text seeks to duplicate this approach for time–dependent problems. In particular, the abstract Hilbert space formulations in chapter 2 allow for application to problems well beyond the second order systems considered in chapters 3 and 5 (Exercises 3.2 and 3.3 for example). More standard approaches to parabolic equations are given in Ladyzhenskaya et. al. [**26**], Friedman [**15**] and [**16**], Lieberman [**30**] and Evans [**12**]. In particular [**26**] includes a treatment of parabolic systems and has considerable overlap with Chapter 5 of the present text.

Sections 3.2 and A.4 develop material first encountered by the author in the original edition of Nečas [**33**]: Rademacher's theorem and the characterization of the dual of $C_0(X)$ provide derivations of the basic facts of calculus on hypersurfaces which are quite straightforward and yet sufficiently general to accommodate boundaries with corners and edges. This material deserves greater visibility in the curriculum.

The spaces $L^{p,q}$ introduced in sections 3.2 and A.5 have received little attention heretofore but seem natural, perhaps even necessary in the present context. Indeed, one may regard the solution of the basic problem (3.1) as a mapping from the time axis into the set of *actions* on the test function space, with the Bochner space representation and its measurable counterpart simply being alternative *representations* of this mapping. Both are useful. In particular, the measurable counterpart representation defined in terms of the spaces $L^{p,q}$ avoids the pitfalls associated with the Sierpinski example and also simplifies the calculus required for the treatment of nonlinear problems, as in section 6.4.

Theorems 6.4 and 6.5 in section 6.2 are standard versions of the first and most basic results in the mathematical theory of the Navier–Stokes equations for incompressible flow, those of Leray [**29**] and Hopf [**23**]. Similar expositions are given in the texts of Temam [**41**], Galdi [**18**] and Tsai [**44**], among many others, each extending the basic theory in a particular direction. The key step in the construction is Theorem 6.4, which provides the required strong local convergence of approximate solutions. This result depends in turn on the inequality (6.13) which first appears perhaps in Friedrichs [**17**]. A very substantial research effort proceeding from this basic theory ensued in the second half of the twentieth century, generating a considerable literature. The survey articles of Temam [**42**] and Lemarié-Rieusset [**28**] review these developments, but a corresponding review for the subsequent decades has yet to appear.

Section 6.5 and Exercise 6.9 give the most basic result and a representative example in the very broad research area concerned with the asymptotic behavior of solutions of quasilinear systems. A basic reference is that of Henry [**22**], in which a theory of analytic semigroups of operators is developed and which includes a number of applications, including to parabolic systems of reaction-diffusion equations. There is a considerable literature addressing the stability question of Section 6.5 when the underlying known solution is a one-dimensional traveling wave. Representative results and examples are given in Sattinger [**37**] and [**38**], Jones [**25**], Texier and Zumbrun [**43**] and Zumbrun [**47**]. A completely different approach seeks to characterize the ensemble of asymptotic states corresponding to the entire set of allowable initial data for a given system. Representative discussions are given in Constantin et al. [**7**] and Eden et al. [**11**].

Section 6.6 follows closely the paper of Chueh, Conway and Smoller [**6**], which extended earlier work of Weinberger [**45**] on invariant regions. [**6**] also gives partial results on the necessity of hypotheses (a)–(c) of Theorem 6.13 and provides further examples. The numerous citations of [**6**] speak to its applicability, although in many cases invariance is more easily established directly by adapting the proof of the theorem rather than by checking its hypotheses.

Bibliography

[1] Robert A. Adams, *Sobolev spaces*, Academic Press, New York-London, 1975. Pure and Applied Mathematics, Vol. 65.

[2] S. Agmon, A. Douglis, and L. Nirenberg, *Estimates near the boundary for solutions of elliptic partial differential equations satisfying general boundary conditions. I*, Comm. Pure Appl. Math. **12** (1959), 623–727.

[3] S. Agmon, A. Douglis, and L. Nirenberg, *Estimates near the boundary for solutions of elliptic partial differential equations satisfying general boundary conditions. II*, Comm. Pure Appl. Math. **17** (1964), 35–92.

[4] G. K. Batchelor, *An introduction to fluid dynamics*, Second paperback edition, Cambridge Mathematical Library, Cambridge University Press, Cambridge, 1999.

[5] Lipman Bers, Fritz John, and Martin Schechter, *Partial differential equations*, American Mathematical Society, Providence, R.I., 1979. With supplements by Lars Gårding and A. N. Milgram, With a preface by A. S. Householder, Reprint of the 1964 original, Lectures in Applied Mathematics, 3A.

[6] K. N. Chueh, C. C. Conley, and J. A. Smoller, *Positively invariant regions for systems of nonlinear diffusion equations*, Indiana Univ. Math. J. **26** (1977), no. 2, 373–392.

[7] P. Constantin, C. Foias, B. Nicolaenko, and R. Temam, *Integral manifolds and inertial manifolds for dissipative partial differential equations*, Applied Mathematical Sciences, vol. 70, Springer-Verlag, New York, 1989.

[8] R. Courant and K.O. Friedrichs, *Supersonic flow and shock waves*, Springer-Verlag, New York-Heidelberg, 1976. Reprinting of the 1948 original, Applied Mathematical Sciences, Vol. 21.

[9] J. Diestel and J. J. Uhl Jr., *Vector measures*, American Mathematical Society, Providence, R.I., 1977. Mathematical Surveys, No. 15.

[10] Avron Douglis and Louis Nirenberg, *Interior estimates for elliptic systems of partial differential equations*, Comm. Pure Appl. Math. **8** (1955), 503–538.

[11] A. Eden, C. Foias, B. Nicolaenko, and R. Temam, *Exponential attractors for dissipative evolution equations*, RAM: Research in Applied Mathematics, vol. 37, Masson, Paris; John Wiley & Sons, Ltd., Chichester, 1994.

[12] Lawrence C. Evans, *Partial differential equations*, 2nd ed., Graduate Studies in Mathematics, vol. 19, American Mathematical Society, Providence, RI, 2010.

[13] Gerald B. Folland, *Real analysis*, 2nd ed., Pure and Applied Mathematics (New York), John Wiley & Sons, Inc., New York, 1999. Modern techniques and their applications; A Wiley-Interscience Publication.

[14] R. FitzHugh, *Mathematical Models of Threhold Phenomena in the Nerve Membrane*, Bull. Math. Biophysics **17** (1955), 257-278.

[15] Avner Friedman, *Partial differential equations*, Holt, Rinehart and Winston, Inc., New York-Montreal, Que.-London, 1969.

[16] ———, *Partial differential equations of parabolic type*, Prentice-Hall, Inc., Englewood Cliffs, N.J., 1964.

[17] Kurt Friedrichs, *Spektraltheorie halbbeschränkter Operatoren und Anwendung auf die Spektralzerlegung von Differentialoperatoren*, Math. Ann. **109** (1934), no. 1, 685–713.

[18] G. P. Galdi, *An introduction to the mathematical theory of the Navier-Stokes equations*, Second Edition, Springer Monographs in Mathematics, Springer, New York, 2011.

[19] David Gilbarg and Neil S. Trudinger, *Elliptic partial differential equations of second order*, 2nd ed., Grundlehren der Mathematischen Wissenschaften, vol. 224, Springer-Verlag, Berlin, 1983.

[20] P. Grisvard, *Elliptic problems in nonsmooth domains*, Monographs and Studies in Mathematics, vol. 24, Pitman (Advanced Publishing Program), Boston, MA, 1985.

[21] Paul R. Halmos, *Introduction to Hilbert space and the theory of spectral multiplicity*, AMS Chelsea Publishing, Providence, RI, 1998. Reprint of the second (1957) edition.

[22] Daniel B. Henry, *Geometric Theory of Semilinear Parabolic Equations*, Lecture Notes in Mathematics, vol. 840, Springer, 1981.

[23] Eberhard Hopf, *Über die Anfangswertaufgabe für die hydrodynamischen Grundgleichungen*, Math. Nachr. **4** (1951), 213–231.

[24] Kelsey Houston-Edwards, *Numbers Game*, Sci. Amer. **Sept.** (2019), 35-40.

[25] Christopher K. R. T. Jones, *Stability of the travelling wave solution of the FitzHugh-Nagumo system*, Trans. Amer. Math. Soc. **286** (1984), no. 2, 431–469.

[26] O. A Ladyženskaja, V. A. Solonnikov, and N. Ural′ceva, *Linear and quasilinear equations of parabolic type*, Translated from the Russian by S. Smith. Translations of Mathematical Monographs, Vol. 23, American Mathematical Society, Providence, R.I., 1968.

[27] Olga A. Ladyzhenskaya and Nina N. Ural'tseva, *Linear and quasilinear elliptic equations*, Translated from the Russian by Scripta Technica, Inc. Translation editor: Leon Ehrenpreis, Academic Press, New York-London, 1968.

[28] P. G. Lemarié-Rieusset, *Recent developments in the Navier-Stokes problem* **431** (2002), xiv+395.

[29] Jean Leray, *Sur le mouvement d'un liquide visqueux emplissant l'espace*, Acta Math. **63** (1934), no. 1, 193–248.

[30] Gary M. Lieberman, *Second order parabolic differential equations*, World Scientific Publishing Co., Inc., River Edge, NJ, 1996.

[31] J. Naumann, *Notes on the prehistory of Sobolev spaces*, Bol. Soc. Port. Mat. **63** (2010), 13–55. English language version available at http://edoc.hu-berlin.de/series/mathematik-preprints/2002-2/PDF/2.pdf.

[32] M. Navier, *Sur les Lois du Mouvement des Fluides*, Mémoires de L'Académie Royale des Sciences, Paris. **Tome VI** (1823), 390-440.

[33] Jindřich Nečas, *Direct methods in the theory of elliptic equations*, Springer Monographs in Mathematics, Springer, Heidelberg, 2012. Translated from the 1967 French original by Gerard Tronel and Alois Kufner.

[34] H. L. Royden, *Real analysis*, 3rd ed., Macmillan Publishing Company, New York, 1988.

[35] Walter Rudin, *Functional analysis*, Second, International Series in Pure and Applied Mathematics, McGraw-Hill, Inc., New York, 1991.

[36] ———, *Real and complex analysis*, 3rd ed., McGraw-Hill Book Co., New York, 1987.

[37] D. H. Sattinger, *On the stability of waves of nonlinear parabolic systems*, Advances in Math. **22** (1976), no. 3, 312–355.

[38] ———, *Weighted norms for the stability of traveling waves*, J. Differential Equations **25** (1977), no. 1, 130–144.

[39] Joel Smoller, *Shock waves and reaction-diffusion equations*, Second Edition, Grundlehren der Mathematischen Wissenschaften [Fundamental Principles of Mathematical Sciences], vol. 258, Springer-Verlag, New York, 1994.

[40] Ivar Stakgold, *Branching of solutions of nonlinear equations*, SIAM Rev. **13** (1971), 289–332.

[41] Roger Temam, *Navier-Stokes equations*, AMS Chelsea Publishing, Providence, RI, 2001.

[42] ———, *Some developments on Navier-Stokes equations in the second half of the 20th century*, Development of mathematics 1950–2000, 2000, pp. 1049–1106.

[43] Benjamin Texier and Kevin Zumbrun, *Relative Poincaré-Hopf bifurcation and galloping instability of traveling waves*, Methods Appl. Anal. **12** (2005), no. 4, 349–380.

[44] Tai-Peng Tsai, *Lectures on Navier-Stokes equations*, Graduate Studies in Mathematics, vol. 192, American Mathematical Society, Providence, RI, 2018.

[45] H. Weinberger, *Invariant sets for weakly coupled parabolic and elliptic systems*, Rend. Mat. Univ. Roma **VI** (1975), no. 8, 295–310.

[46] William P. Ziemer, *Weakly differentiable functions*, Graduate Texts in Mathematics, vol. 120, Springer-Verlag, New York, 1989.

[47] Kevin Zumbrun, *Stability and dynamics of viscous shock waves*, Nonlinear conservation laws and applications, 2011, pp. 123–167.

Index

Selected Published Titles in This Series

251 **David Hoff,** Linear and Quasilinear Parabolic Systems: Sobolev Space Theory, 2020
249 **Nikolai M. Adrianov, Fedor Pakovich, and Alexander K. Zvonkin,** Davenport–Zannier Polynomials and Dessins d'Enfants, 2020
248 **Paul B. Larson and Jindrich Zapletal,** Geometric Set Theory, 2020
247 **István Heckenberger and Hans-Jürgen Schneider,** Hopf Algebras and Root Systems, 2020
246 **Matheus C. Bortolan, Alexandre N. Carvalho, and José A. Langa,** Attractors Under Autonomous and Non-autonomous Perturbations, 2020
245 **Aiping Wang and Anton Zettl,** Ordinary Differential Operators, 2019
244 **Nabile Boussaïd and Andrew Comech,** Nonlinear Dirac Equation, 2019
243 **José M. Isidro,** Jordan Triple Systems in Complex and Functional Analysis, 2019
242 **Bhargav Bhatt, Ana Caraiani, Kiran S. Kedlaya, Peter Scholze, and Jared Weinstein,** Perfectoid Spaces, 2019
241 **Dana P. Williams,** A Tool Kit for Groupoid C^*-Algebras, 2019
240 **Antonio Fernández López,** Jordan Structures in Lie Algebras, 2019
239 **Nicola Arcozzi, Richard Rochberg, Eric T. Sawyer, and Brett D. Wick,** The Dirichlet Space and Related Function Spaces, 2019
238 **Michael Tsfasman, Serge Vlăduţ, and Dmitry Nogin,** Algebraic Geometry Codes: Advanced Chapters, 2019
237 **Dusa McDuff, Mohammad Tehrani, Kenji Fukaya, and Dominic Joyce,** Virtual Fundamental Cycles in Symplectic Topology, 2019
236 **Bernard Host and Bryna Kra,** Nilpotent Structures in Ergodic Theory, 2018
235 **Habib Ammari, Brian Fitzpatrick, Hyeonbae Kang, Matias Ruiz, Sanghyeon Yu, and Hai Zhang,** Mathematical and Computational Methods in Photonics and Phononics, 2018
234 **Vladimir I. Bogachev,** Weak Convergence of Measures, 2018
233 **N. V. Krylov,** Sobolev and Viscosity Solutions for Fully Nonlinear Elliptic and Parabolic Equations, 2018
232 **Dmitry Khavinson and Erik Lundberg,** Linear Holomorphic Partial Differential Equations and Classical Potential Theory, 2018
231 **Eberhard Kaniuth and Anthony To-Ming Lau,** Fourier and Fourier-Stieltjes Algebras on Locally Compact Groups, 2018
230 **Stephen D. Smith,** Applying the Classification of Finite Simple Groups, 2018
229 **Alexander Molev,** Sugawara Operators for Classical Lie Algebras, 2018
228 **Zhenbo Qin,** Hilbert Schemes of Points and Infinite Dimensional Lie Algebras, 2018
227 **Roberto Frigerio,** Bounded Cohomology of Discrete Groups, 2017
226 **Marcelo Aguiar and Swapneel Mahajan,** Topics in Hyperplane Arrangements, 2017
225 **Mario Bonk and Daniel Meyer,** Expanding Thurston Maps, 2017
224 **Ruy Exel,** Partial Dynamical Systems, Fell Bundles and Applications, 2017
223 **Guillaume Aubrun and Stanisław J. Szarek,** Alice and Bob Meet Banach, 2017
222 **Alexandru Buium,** Foundations of Arithmetic Differential Geometry, 2017
221 **Dennis Gaitsgory and Nick Rozenblyum,** A Study in Derived Algebraic Geometry, 2017
220 **A. Shen, V. A. Uspensky, and N. Vereshchagin,** Kolmogorov Complexity and Algorithmic Randomness, 2017
219 **Richard Evan Schwartz,** The Projective Heat Map, 2017

For a complete list of titles in this series, visit the
AMS Bookstore at **www.ams.org/bookstore/survseries/**.